VISION AND ITS INSTRUMENTS

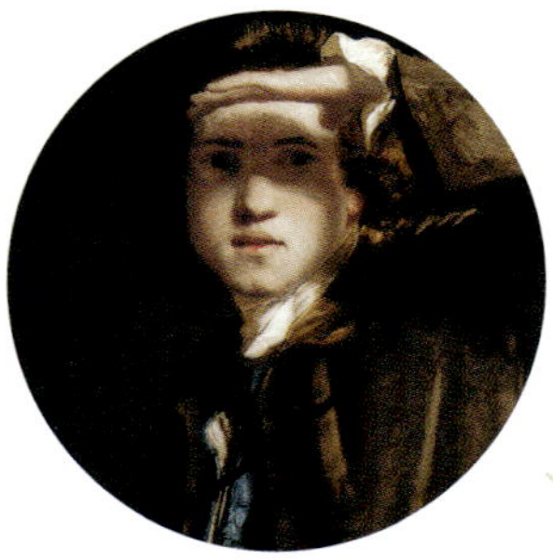

VISION

AND ITS

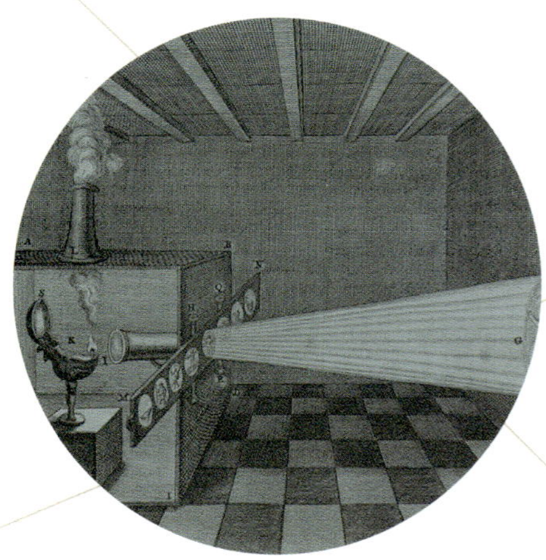

THE PENNSYLVANIA STATE UNIVERSITY PRESS, UNIVERSITY PARK, PENNSYLVANIA

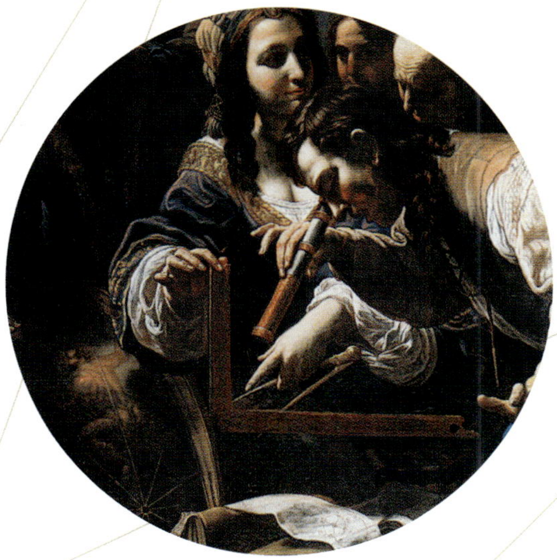

INSTRUMENTS

Art, Science, and Technology in Early Modern Europe

EDITED BY ALINA PAYNE

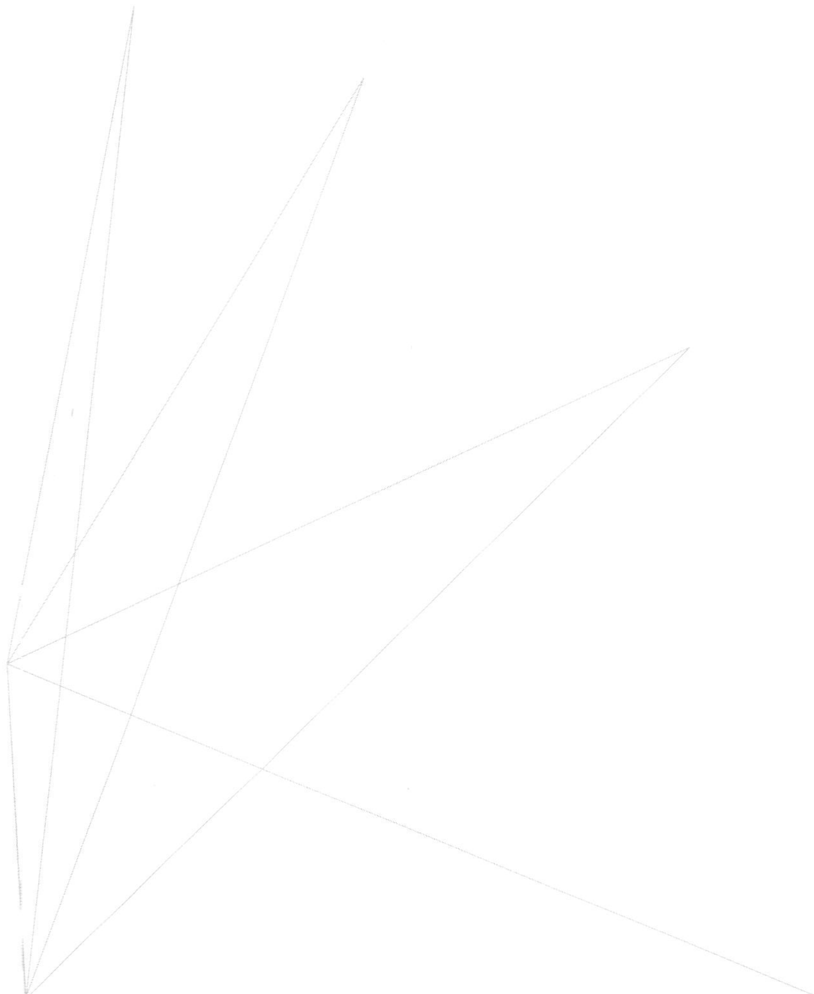

Library of Congress Cataloging-in-Publication Data

Vision and its instruments : art, science, and technology in early modern Europe / edited by Alina Payne.
pages cm
Summary: "A collection of essays investigating the early modern debates on the nature of sight and its epistemic value"—Provided by publisher.
Includes bibliographical references and index.
ISBN 978-0-271-06389-8
(cloth : alk. paper)
1. Visual perception.
2. Art and science—Europe.
3. Art and technology—Europe.
I. Payne, Alina Alexandra, editor.

N7430.5.V485 2015
701'.8—dc23
2014013692

Printed in Korea by Pacom
Published by The Pennsylvania State University Press,
University Park, PA 16802-1003

The Pennsylvania State University Press is a member of the Association of American University Presses.

It is the policy of The Pennsylvania State University Press to use acid-free paper. Publications on uncoated stock satisfy the minimum requirements of American National Standard for Information Sciences—Permanence of Paper for Printed Library Material, ANSI Z39.48–1992.

Designed by Regina Starace

Additional credits: pages ii–iii, details of Raimondi, *Raphael in His Studio* (fig. 8.11), "Laterna magica" (fig. 1.3), Tornioli, *The Astronomers* (fig. I.2); page 10, detail of Fly Engraving (fig. 3.2); page 66, detail of Codex Huygens, "Undezima figura" (fig. 5.9); page 164, detail of Juncker, *Pear and Insects* (fig. 9.6); page 238, detail of advertising photography for *Rear Window* (fig. 12.1).

Contents

Illustrations

Acknowledgments

The project of which this volume is a physical trace started in 2006 when I received the Max Planck and Alexander von Humboldt Prize for the Humanities and conceived the project "The Object as Event." With this generous financial support, I was able to organize two workshops—one in 2007 at the KHI / Max Planck Institute in Florence and one in 2008 at the Radcliffe Institute for Advanced Study at Harvard—with a group of art historians and historians of science, philosophy, and literature to look at vision in arts and sciences across objects/instruments. For a variety of reasons, not least of which was the financial crisis of 2008/9, which slowed down the publication process considerably, the volume of essays arising from these two workshops is only now seeing the light of print. I am grateful to the authors for staying the course and for believing in the value of this collective work. Our debates and exchanges over these years made this volume an even richer and more exciting outcome than I had originally envisaged.

I am particularly grateful to the institutions that supported this work financially: the Max-Planck-Gesellschaft, the Alexander von Humboldt Foundation, the KHI / Max Planck Institute in Florence, and the Radcliffe Institute and Department of History of Art and Architecture at Harvard. The Hertziana Library in Rome and the Villa I Tatti in Florence, just like the KHI, offered me their physical resources and help, as well as abode, at various points in the preparation of this volume. For this a heartfelt thanks!

Many friends, students, and colleagues participated and helped in so many ways—without them this project would never have gotten off the ground. I wish to thank and record the support and help of Aliki Economides, Giovanni Fara, Federica Favino, Elizabeth Kieven, David Kim, Andreea Nicoara, Alessandro Nova, Katharine Park, Lino Pertile, David Pullins, Cara Rachele, Michael Rocke, Sybille Ebert Schifferer, Nicola Suthor, Aaron Wile, and Gerhard Wolf. As ever, family comes last in records of indebtedness, but they are certainly not least—I am grateful to my parents and my daughter, Julia, for love and support and for keeping up my morale when I needed it most.

I. Introduction

ALINA PAYNE

The starting points for this volume were two historical events—one in the sciences, the other in the arts—that mark inaugural moments in the history of vision. One point of departure was a completely forgotten, and indeed unknown, event in the career of Galileo. In the corpus of manuscripts left by the Sienese polymath Teofilo Gallaccini (1564–1641), more specifically in his treatise on astronomy ("La monade celeste"), lies buried a passage in which he described an experiment he had witnessed in the summer of 1633.[1] After the publication of his *Dialogue of Two World Views* (1630) and his 1633 condemnation, Galileo had taken refuge in Siena for several months at the invitation of Archbishop Ascanio Piccolomini. In August 1633 the archbishop organized a soirée on his palace rooftop loggia to allow Galileo to display the workings of his telescope upon the night sky. Naturally, he invited his entourage of learned friends, among them Gallaccini. As a scientist (astronomer, medical doctor, and mathematician) but also as an architecture critic and literary figure of some prominence at the academies and the University of Siena, he was in a position to appreciate and comment upon what Galileo had to show. In this exquisite Renaissance frame, upon this stage, literally, the experiment of looking at the moon took place for six nights in a row. And we know of this event because Gallaccini illustrated six images of the moon as he saw them through the telescope on those consecutive nights (fig. I.1). The effects of seeing the Aristotelian conception of the moon as a pure silver disc annihilated by the observation of its topography clearly vibrates from the page. Indeed, in these drawings the surface of the moon looks even more convulsed than the images Galileo himself had shown in his *Sidereus nuncius* (1610), and they suggest by this very exaggeration the excitement that was felt on that distant rooftop, all those centuries ago.

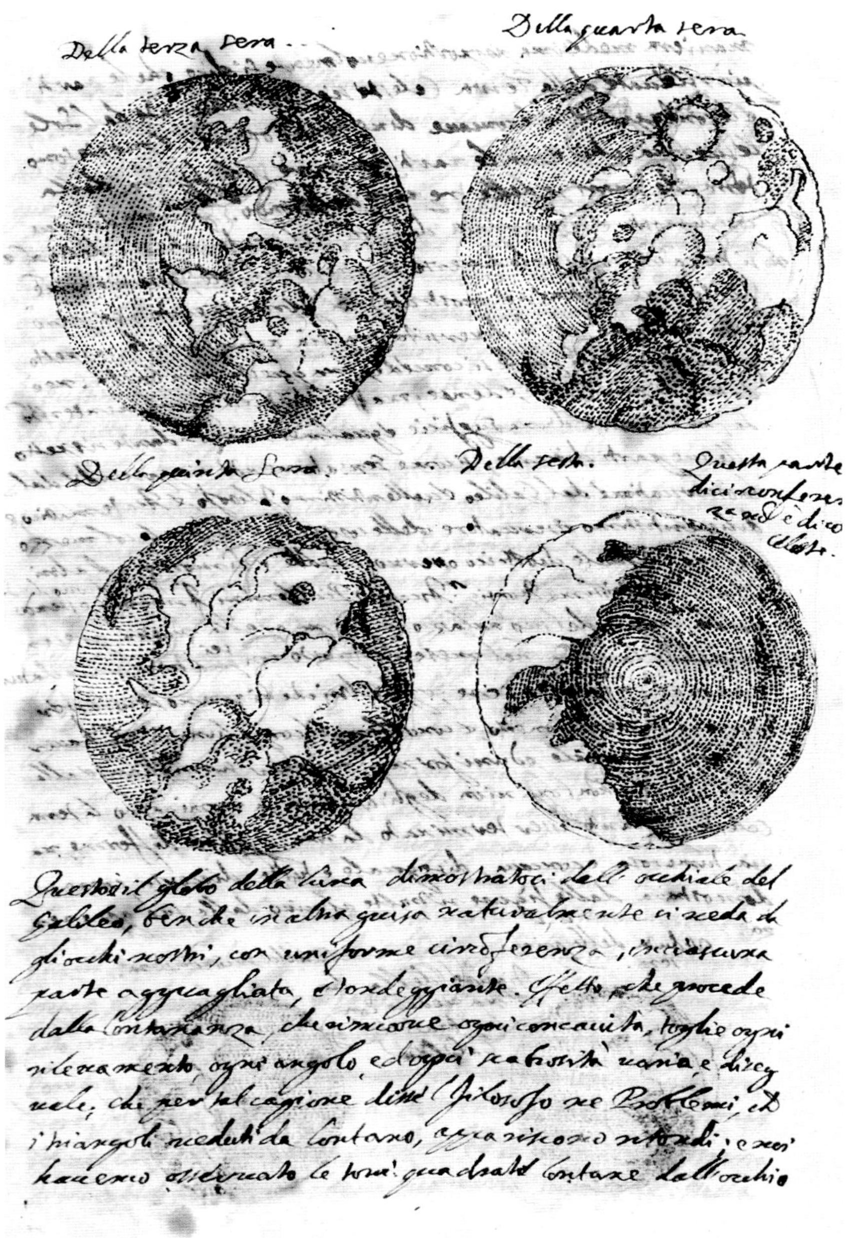

I.1 Teofilo Gallaccini, four views of the moon. In "Dell'unità del cielo [La monade celeste]," Biblioteca comunale degli Intronati, Siena, MS L.vi.31, fol. 108v. © Biblioteca comunale degli Intronati, Siena (29.8.2013).

However, what was on display beyond the moon and beyond Galileo's finely ground lens was *sight itself.* Indeed, the issues that accompanied the debates and confrontations caused by Galileo's discoveries were fundamentally about sight: What can be seen, how far, how well? Can sight be trusted, can its prostheses be trusted, can the eye witness upon itself? Thus, beyond a triumph for science, Galileo's moment was also a moment of crisis, where the testimony of sight over thought was propelled to the center of scientific discourse. For here, sight was objectified, was made visible as an act, and it was made thus by an instrument: the telescope. The tube, its lens, its adjustments showed where sight passed through; it literally contained sight and made the passage of the visual rays a palpable object. In short, the instrument, as it were, made the act of seeing physically apprehensible. Upon this small stage, then, surrounded by a group of excited people not unlike Niccolò Tornioli's astronomers—his painting may well have been modeled on this event—vision became an event; it went on display (fig. I.2).[2]

Galileo's was an inaugural moment of conflict and disjunction, a coming of age of the cognitive problems vision had posed for some time—a moment of recognition that sight was both a strength and a weakness—and it was tied to the sciences and to instruments. By contrast, the second inaugural moment, the one that had set in motion the process that found such a climax in Galileo's life, was one of acclaim and recognition, and was associated with the arts, with paints, brushes, and mirrors. Far better known than the performance on the Piccolomini *altana* and predating it by a good two hundred years, was Filippo Brunelleschi's experiment in front of the Baptistery of Florence, sometime before 1413. This is *the* moment for the Renaissance as we have come to construct it in the arts, the paradigmatic moment of the invention of perspective that is the mathematization of sight. Of course, it is not really the moment of invention—these "thick" moments need to be unpacked, as they are never really just single instances—but it is the moment of its public performance, of the presentation of vision as event. Standing in the main entrance of Florence Cathedral and intent on painting the view of the Baptistery, Brunelleschi—with his panel, his easel, his brushes, his paints, his measuring instruments and drafting equipment, and his *camera oscura*—took up a lot of physical space. One has to imagine friends and colleagues around him looking at his work, looking through the hole in his painting to see how lifelike it all was, how true to nature—a solid group, moving about, bending over, leaning over his shoulder and generally drawing attention to themselves and blocking the entrance to Florence's principal religious establishment. In short, a performance of artistic seeing was being presented to the city and its citizens on the elevated platform of the cathedral stoop. The excitement on that day in Florence must have been similar to that in August 1633.

Whatever else this event inaugurated, it was also the moment when Brunelleschi proposed painting as an instrument—an instrument that was useful in several ways. First, it allowed images to become more convincing because the space of the fictive action was presented *as if* coextensive with the space of the viewer; Second, and more fundamentally, it provided a means of representing reality "correctly," of representing what is seen *as it is*; it provided a means to explore nature by way of representation. Granted, this was a view from a distance—not the close-up view of the scientist who gets more and more myopic and close to his objects, the better to see them, until he is literally inside them, like the early modern microscopists. Nor is this the view of the astronomer whose long-distance sight gets stronger and stronger, as the telescope enables him to see the distant heavens up close. But it was the first step in that quest, too, for Brunelleschi literally turned painting into an epistemic instrument by way of perspective.

Granted, as instruments go, perspective is an invisible one—a mental process and construction that is erased in the image, unlike telescopes and microscopes, which are solid and can be lifted and held. But it is an instrument nevertheless. Painting here promises to represent *reality*, and it affords lengthy and close examination by detaching individual frames from the continuum of nature. Indeed, two centuries later, perspective helped Galileo understand sunspots, that is, helped him develop yet another and decisive demonstration of the corruptibility of the heavens.[3] Thus painting, by way of perspective, steps onto the stage, like Brunelleschi's group of viewers and doubting Thomases in the door of the cathedral. Like

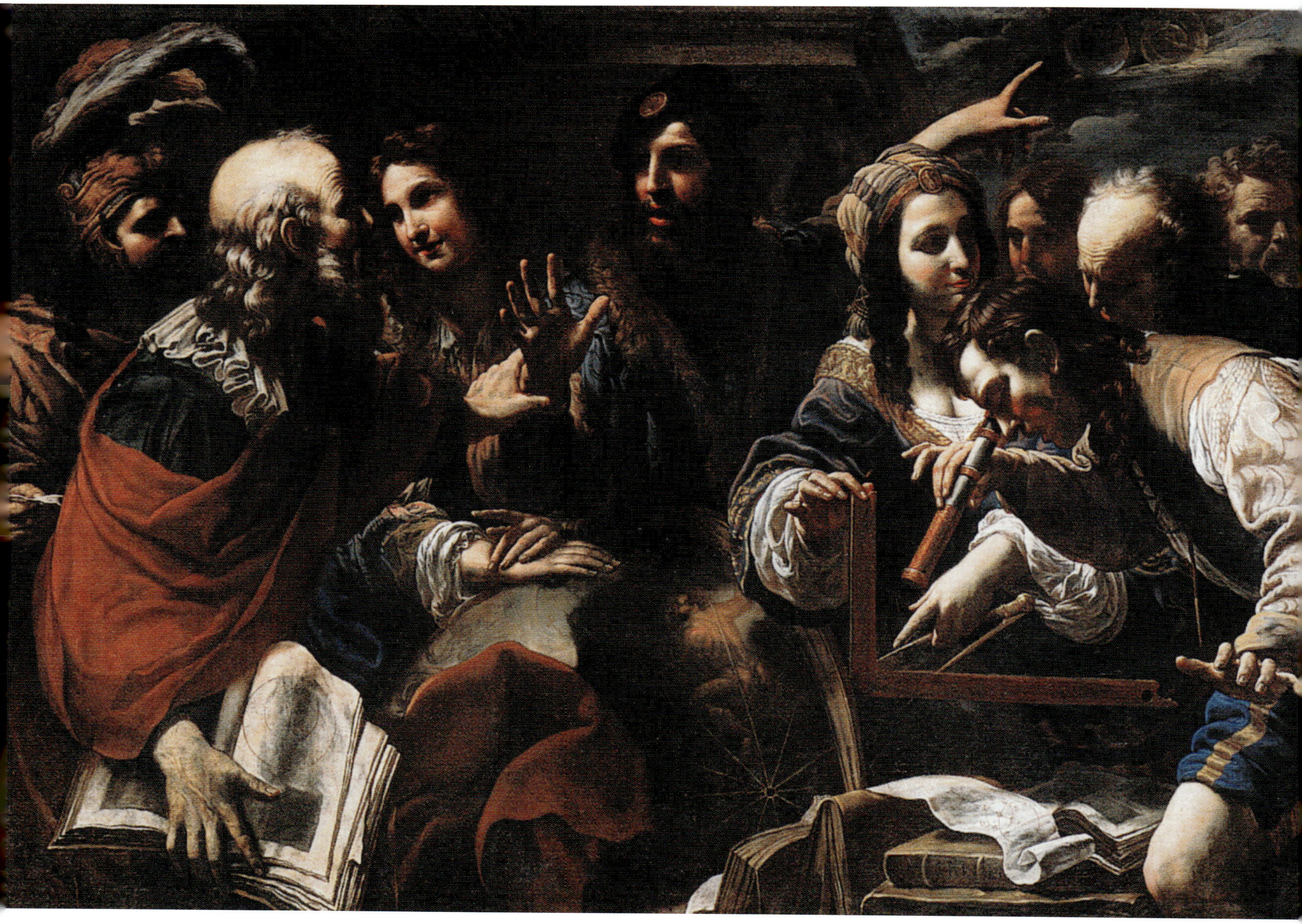

I.2 Niccolò Tornioli, *The Astronomers*, ca. 1645. Oil on canvas, 148 x 218.5 cm. Inv. 269. Galleria Spada, Rome.

them, it obstructs the normal flow of traffic and causes a "disturbance." In this story, painting emerges as a disturbance in the apprehension of nature—a disturbance that must be noticed and acknowledged, like the small group blocking the entrance to Florence Cathedral.

The questions about vision and its instruments around which the essays in this volume cohere, then, start with the kinds of problems these two moments pose. Both represent sight as performance—as it is looked at, as it is turned upon itself, as it is made the object of vision, of spectatorship, and is judged. In one case, sight is physically distilled into an object—the telescope—whereas in the other, it is present by way of an abstract construction that allows physical things (paintings) to come into being. Both pose the question of the eye: What sort of instrument is it? How does it interact with the world of manufactured instruments? What happens when it does? What problems arise from an epistemological perspective?

Of course, these two moments or stories are also about two apparently distinct provinces—art and science. But their place of overlap is the eye, as instrument and as the origin of more instruments, such as painting and the telescope. They are very different instruments, but therein lies the point of this volume. In the early modern economy of the eye, painting (that is, representation) and telescopes are instruments of equal value for knowledge. Both are seen to provide equal access to a truth, both allow investigation, both push the envelope of knowledge further.

By this meandering route, the larger issues that this volume addresses begin to come into focus. To be sure, in recent years the topic of vision and visuality has attracted substantial interest among historians of art as well as historians of science. Traditionally, perspective, which governed representation and its laws, had been the most frequently addressed theme, for being both a paradigmatic invention of the Renaissance and the most obvious convergence point between artists, architects, and mathematicians, and thus inaugurating a relationship between art and science that would become increasingly complex in the modern period. The string of publications, from Erwin Panofsky's seminal "Perspektive als symbolische Form" (1924–25), to Martin Kemp's *Science of Art* (1990), and, more recently, to such collective works as *The Treatise on Perspective: Published and Unpublished* (ed. Lyle Massey, 2003), testifies to the persistence of this line of inquiry. In a parallel move, the categorical shift from a text- to an image-based culture that characterizes society in the era of the computer screen and mass advertising has led the field of art history to subject images to a renewed and intense scrutiny, from Hans Belting's *Bild und Kult: Eine Geschichte des Bildes vor dem Zeitalter der Kunst* (1990) and William Mitchell's *Picture Theory* (1994), to Caroline Jones and Peter Galison's *Picturing Science, Producing Art* (1998), to the projects currently under way at the universities of Basel and Humboldt (Berlin)—"Bildkritik" and "Eikones" under Gottfried Boehm, and "Bildakt" under Horst Bredekamp.[4] A related line of inquiry has emerged from the perspective of the history of science, looking toward the contribution of the arts, in particular in the project on observation under way at the Max Planck Institute in Berlin.[5] Also related is the recent work connecting cognitive science, and in particular the pathology of the brain, with art making and art history.[6] Together, these efforts showcase how interest in the exchanges between the arts and sciences has shifted from perspective to the processes of image making and reception, to the physiology of the body and the mechanics of instruments.

Despite the wide-ranging quality of this work and its interdisciplinary nature, most inquiries have tended to focus on the period following the Scientific Revolution, with particular emphasis on the modern interaction between the arts and sciences, while the early modern investigations have remained focused on perspective.[7] However, the economy of the eye has not had a linear development that climaxed in the modern world. Yet the course of the debates on the nature of sight and seeing, on sight as vehicle as well as limit to knowledge, and on the eye itself as ultimate instrument of nature (often challenged and hence unstable in its privileged position as offering primary access to truth) in this period still provides much uncharted ground and reveals it to have been both complex and complicated. More important, in its cumulative form this discourse has continued to inform and still lies embedded within modern concepts and prejudices, like old sins with long shadows. Not that the pejorative implication is warranted. Different ways of "seeing" analogically may yet contain insights that modernity would do well to revisit. Indeed, the centuries-old tradition of comparing living organisms and their makeup with machines and inanimate nature so as to extract principles of lightness and aerodynamics, for example is experiencing a comeback.

Among this cornucopia of offerings, then, space for a different kind of investigation needs to be cleared, one that teases out the multiple strands of the discourse about sight and investigates particular moments of discovery and crisis, insight and mutual illumination, that the arts and the sciences shared in their perpetual effort to understand nature through the eye in the early modern period. In this context, the tools and prostheses that extended or modified sight, that enhanced and transformed it, and that turned sight into *an event* are particularly relevant to the questions addressed here. It is the ambition of this volume, then, not to limit the definitions of sight and its instruments to modern formulations, but to explore the full range of early modern conceptions of vision in which *mal'occhio* (evil eye), witchcraft, spiritual visions, and phantasms, as well as the artist's brush or the architect's compass, were seen to provide knowledge equal to, and in certain cases better than, newly developed scientific instruments and practices. Put simply, this volume aims to revisit the complexity of the early modern economy of the image, of the eye, and of its instruments, and to provide a frame that acknowledges the porous walls of discourses that nowadays, in our culture if not in our discipline, are seen as fundamentally unrelated. From art and architecture, to science and philosophy, to literature and film, the essays in this volume provide a finely textured series of analyses of vision, and its instruments.

In part I, "Epistemic Images," the authors show how seeing—and the representation of what was seen—was subordinated to concepts, whether epistemic or moral. Lorraine Daston argues that scientific seeing "before objectivity" was not what we might think—not realistic but universalist—and that epistemic images were understood to be more real than the individual exemplar they purported to depict. Thus she argues that eighteenth-century botanists perceived color as a variable to be removed from their representations, a misleading bit of visual information. Botanists, she shows, intervened to correct images produced by artists precisely in order to prevent them from drawing exactly what they saw. Vision was useful but also suspect. By looking at the physician Conrad Gessner (1516–1565), Sachiko Kusukawa raises the question of the "icona absoluta," the perfect picture of the "complete plant": a visual compendium of all conditions of its existence that recorded ultimately not so much the specimen as the total process of seeing it. Once again, vision and representation do not overlap, despite the charge between them. Catherine Wilson looks at the epistemic image from the perspective of philosophy and wonders what appeal it had for classes of viewers and producers of images—not only for scientists but also for artists and art lovers. Why did they like images that were so "micrographical" in nature? Beyond individual taste and curiosity, she identifies a moral stance toward observation, a statement that inquiry into natural forms was an "improving" discipline.

The authors of part II, "Seeing the Unseeable," explore the conditions of seeing: What are its limits? What happens when seeing fails? How can knowledge then be corroborated or nature even understood? Where do art and science meet, and what truth-value do extreme conditions of seeing—such as mystic visions—offer? Looking at Leonardo, Frank Fehrenbach explores the persistent, intertwined trust and doubt associated with the testimony of the eye. By analyzing the point, Leonardo, Fehrenbach argues, explores the smallest element of painting and visible reality (physical boundaries are made of an aggregate of points) and raises the question of where the border between seeing and not seeing, between something and nothing, occurs. Connecting science and art, Leonardo uses them to reflect upon each other.

In my essay, I also address the limits of visibility, as experienced by the Sienese mathematician and architecture critic Teofilo Gallaccini. Unlike Leonardo and the botanists who focused on visible phenomena, Gallaccini attempted to see (and thus understand) that which cannot be seen—the working of forces (in mechanics). His efforts to use visual (and drawn) analogies to this end failed, thus demonstrating the wall a scientist hits when pushing the visual to its limits. Looking at the *Divina Commedia*, Gerhard Wolf reveals an intricate fabric that weaves together poetry, painting, and science and provides Dante with ways to describe the unseeable: through mystical visions. A form of poeticized science allows him to probe the limits of the visible/invisible and representable/unrepresentable. Finally, Carla Mazzio looks at the ultimate unseeable: air. In an argument that connects Dürer (*Melencolia I*), Shakespeare (*Hamlet*), and Donne (*Devotions*), she identifies a discourse about air that is a *mise-en-abîme* of sight itself and that produces a particular affect that literature records.

In part III, "The Painter's Brush and the Mind's Eye," the authors turn to the painter's act of seeing and representing as a confrontation that likewise pushes vision to its limits. Beyond perspective, they argue, painting itself—the application of paint to canvas—can be an instrument of scientific exploration. Looking at the work of Correggio, Arent de Gelder, and Pietro Testa, Nicola Suthor presents painting as an extreme act of mediation that transcends the mind and is physically driven. The hand, the brush, and the brushstroke emerge as vehicles of knowledge that are closer to the act of seeing and ahead of thinking itself. In Suthor's argument, seeing becomes a multisensorial experience that artists alone grasp and allow us to recognize.

Karin Leonhard also explores the boundaries between sight and other senses. By looking at baroque Italian still life paintings she argues that painters use color—the seen—to communicate and enhance unseen qualities and elicit a form of synesthesia. Conversely, she argues that according to scientists, color could be sensed even without its being seen. Bringing together optics, scientific color theory, and painting practice, Leonhard makes a compelling case that the insights of painters corroborated contemporary scientific insights.

Finally, looking at a host of artists, from Schongauer to Michelangelo, Claudia Swan turns to another dimension of painting's engagement with visible reality: its ability to depict the visionary as if real, to make phantasms appear naturalistic. The connection between seeing and believing was a question ever present in early modern thought, and chimeras challenged the truth-value of painting and, by extension, the painter's imagination, in a world in which witchcraft and magic called into question the reliability of visual perception. Could the false be taken for real through a painter's artistry? If so, was the reliability of vision and naturalistic representation not also suspect?

In part IV, "Looking Back: From Photography and Film to Alberti," the final two essays turn toward modern instruments of vision—the photo lens and the movie camera—and examine the layers of deposits left by early modern culture within our own. Michael Cole argues that fixing the moving image by way of the camera projects a sense of instantaneity even to earlier, prephotographic objects. Focusing on the sculpture of Michelangelo and Bernini, he argues that our way of seeing through the lens since the invention of photography subverts the manner of seeing that lay at the foundations of early modern art. To disengage the two instruments—the eye and the camera—means to differentiate sculpture from its photographic double and relinquish the notion of narration in favor of sculptural permanence, of action that did not entail motion. In short, it means unthinking—or unseeing—modern instruments.

In the volume's final essay, Victor Stoichita presents film as the end product of a long tradition of seeing and gazing that starts with Alberti and his famous simile of the painting as an open window. Offering a close reading of Hitchcock's *Rear Window*, he not only posits the Albertian device of the window as a semiotic space, and looking as an eroticized activity, but reflects on the voyeur as the ultimate gazer in all visual media, both modern and premodern. In his telling, the window emerges as yet another (overlooked) instrument that modifies vision and thus brings the argument full circle to its early modern origins.

Taken together, the essays in this volume explore the exchanges between the domains of natural history, philosophy and religion, alchemy, astronomy, and instrument making (from telescope lenses and convex/concave mirrors to microscopes and cameras), and the artists who subjected the eye (both as instrument and as mirror of the soul, but also as connected to other senses) to profound reflection. And they argue that these were *all* interconnected epistemic activities, part of one metafabric of knowledge. In the final analysis, the essays collected here examines extreme conditions where the limits and the problems associated with sight come into sharpest focus—for those are the sites where the definitions of vision and its truth-value are ultimately questioned and articulated most clearly.

Notes

1. On Teofilo Gallaccini, see Payne, *Telescope and the Compass*; and for the text of the manuscript on astronomy, Payne, *Teofilo Gallaccini*, 245–405. All translations are my own unless otherwise noted.

2. Payne, *Telescope and the Compass*, prologue and chap. 2.

3. On the scientific applications of perspective, see Camerota, *Prospettiva del Rinascimento*.

4. One publication recently emerging from the latter project is Bredekamp, *Theorie des Bildakts*.

5. A seminal associated publication is Daston and Galison, *Objectivity*.

6. See, for example, the essays in Marc Turner's edited volume *The Artful Mind*; and Stafford, *Echo Objects*.

7. See, for example, the yearbook series edited by Horst Bredekamp and Gabriele Werner, *Bildwelten des Wissens*; and Dürbeck et al., *Wahrnehmung der Natur*.

8. Clark, *Vanities of the Eye*, 1–8.

Bibliography

Bredekamp, Horst. *Theorie des Bildakts*. Berlin: Suhrkamp, 2010.

Camerota, Filippo. *La prospettiva del Rinascimento: Arte, architettura, scienza*. Milan: Electa, 2006.

Clark, Stuart. *Vanities of the Eye: Vision in Early Modern European Culture*. Oxford: Oxford University Press, 2007.

Daston, Lorraine, and Peter Galison. *Objectivity*. New York: Zone Books, 2007.

Dürbeck, Gabriele, Bettina Gockel, Susanne B. Keller, Monika Renneberg, Jutta Schikore, Gerhard Wiesenfeldt, and Anja Wolkenhauer, eds. *Wahrnehmung der Natur, Natur der Wahrnehmung: Studien zur Geschichte visueller Kultur um 1800*. Halle: Verlag der Kunst, 2001.

Fischle, Angela, ed. *Instrumente des Sehens*. Vol. 2 of *Bildwelten des Wissens: Kunsthistorisches Jahrbuch für Bildkritik*, ed. Horst Bredekamp and Gabriele Werner. Berlin: Akademie Verlag, 2004.

Payne, Alina. *The Telescope and the Compass: Teofilo Gallaccini and the Dialogue Between Architecture and Science in the Age of Galileo*. Florence: Leo S. Olschki, 2012.

———, ed. *Teofilo Gallaccini: Selected Writings and Library*. Florence: Leo S. Olschki, 2012.

Stafford, Barbara Maria. *Echo Objects: The Cognitive Work of Images*. Chicago: University of Chicago Press, 2007.

Turner, Marc, ed. *The Artful Mind: Cognitive Science and the Riddle of Human Creativity*. Oxford: Oxford University Press, 2006.

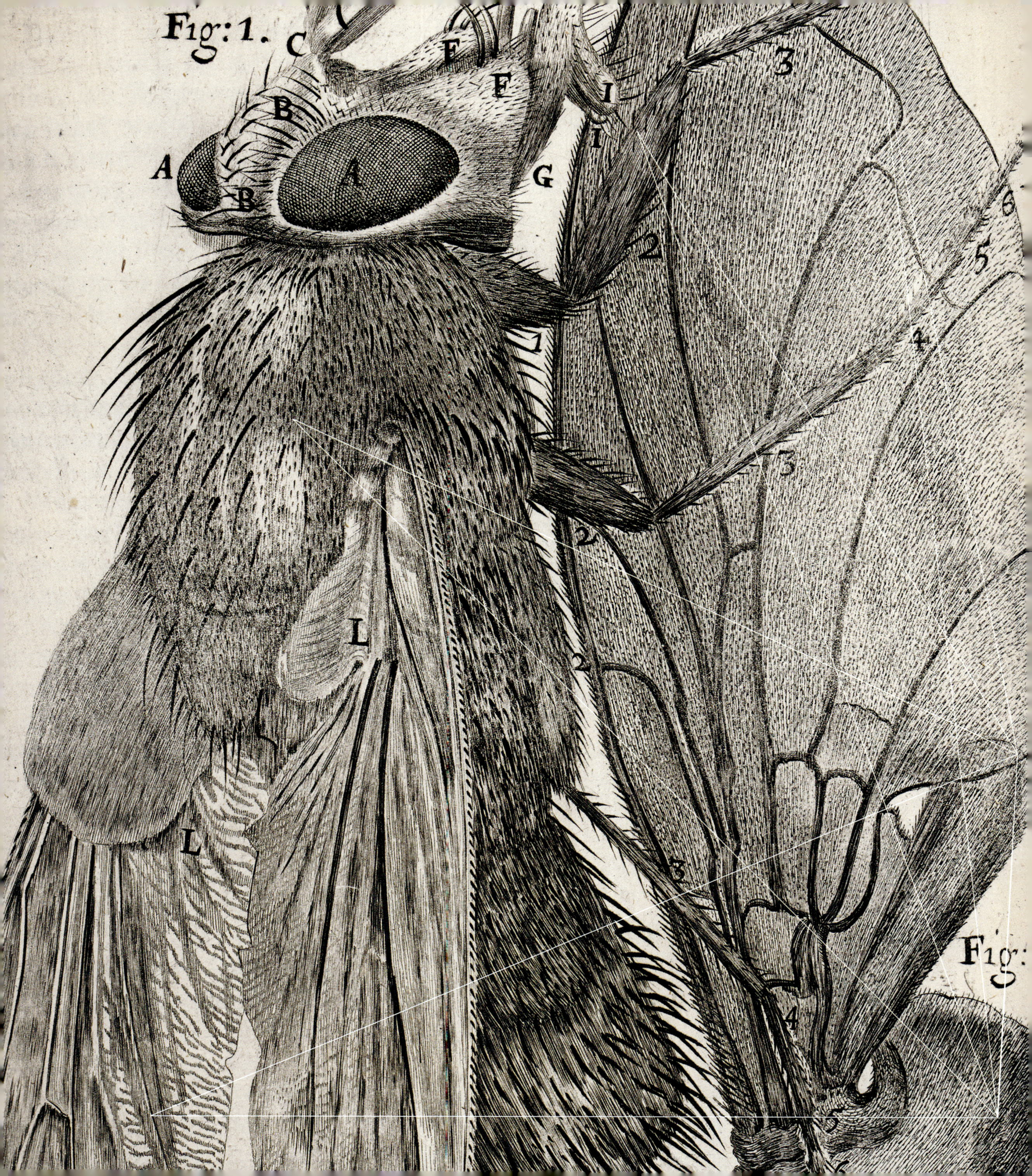
Fig:1.
A
B
C
E
F
G
I
L
1
2
3
4
5
6
Fig:

I.

EPISTEMIC IMAGES

1. Epistemic Images

LORRAINE DASTON

Before Objectivity

Consider two images of the mandrake root, the one from a fifteenth-century manuscript and the other from a printed sixteenth-century herbal (figs. 1.1, 1.2). It is customary among historians of both art and science to pronounce the later image "more objective" than the earlier one. What does this mean? Partly that the sixteenth-century image is simply more accurate than the earlier one, as the author of the work from which it is taken, William Turner, asserted: "The rootes which are conterfited and made like litle puppettes and mammettes, which come to be sold in England in boxes, with heir, & such forme as a man hath, are nothyng elles but folishe feined trifles, & not naturall. For they are so trimmed of crafty theves to mocke the poore people with all & to rob them both of theyr wit and theyr money. I have in my tyme at diverse tymes taken up the rootes of Mandrag out of the grounde but I never saw any such thing upon or in them as are in and upon the peddlers rootes that are comenly to be solde in boxes."[1] But "objectivity" is not simply a synonym for "accuracy." In this case, the accolade "objective" is applied because to modern eyes the earlier image shows signs of subjective projection: anthropomorphic fantasies have

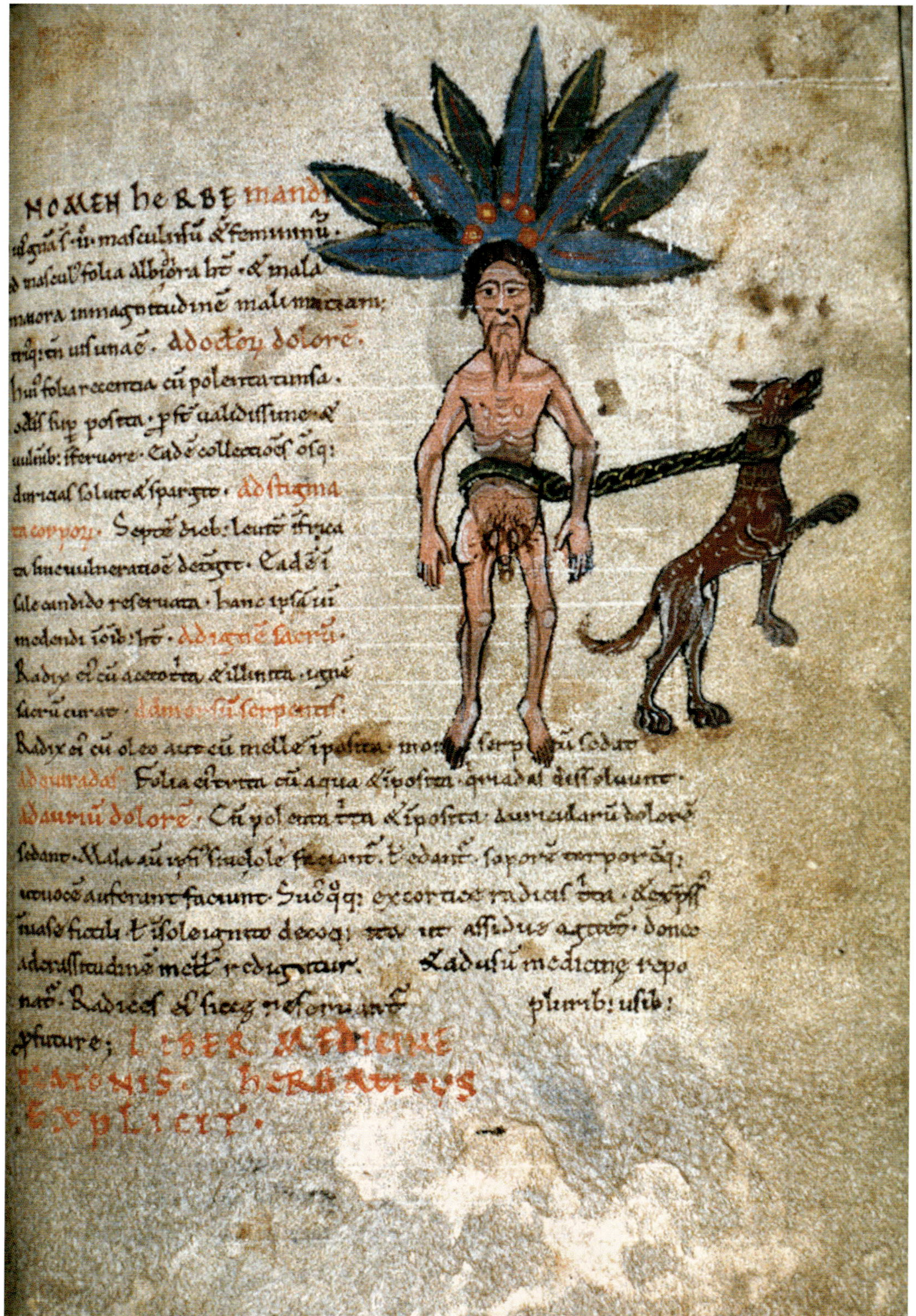

1.1 "Mandragora." In MS Ashmole 1431, fol. 31r. © The Bodleian Libraries, The University of Oxford.

been imposed on the plant. The later image has been purged of these projections; thus it is "more objective."

Had Turner or his readers been confronted with this judgment, they would have been puzzled—not because they doubted the superiority of his image to the earlier one but because they would not have understood the terms in which it is now praised. They would first probably have stumbled over the word itself, which would have had a Latinate, scholastic ring to it in their ears, and, if they recognized it at all, would have conjured up associations with late medieval debates about nominalism and realism. Until Kant in the late eighteenth century redefined the words for his own purposes (and flipped their meanings 180 degrees), the word *objectivus* and its vernacular cognates referred primarily to that which exists as an object of consciousness, as opposed to *subjectivus*, which referred to things in themselves.[2]

We need not, however, make a fetish of the word. Turner and his early modern contemporaries would also have found the concept of objectivity (and its complement, subjectivity) difficult to digest. They would have been familiar with material projections of the imagination, conceived as emanations of animal spirits that could in exceptional cases imprint soft matter with fantasies of fevered brains—most notoriously in the case of fetuses in the womb deformed into monsters by their mothers' errant imaginings.[3] But the notion of psychological projection required very different notions of both the psyche and projection, both of which gained currency only in the early nineteenth century,[4] along with the very category of subjectivity. In terms of early modern theories of perception, projection worked in the other direction: external objects projected impressions onto the sensorium. The magic lantern, probably invented in the early seventeenth century, was understood in explicit analogy to the projection of inverted images righted by means of a lens and projected onto the blank screen of the retina (fig. 1.3).[5] The idea of mental projections outward, capable of distorting

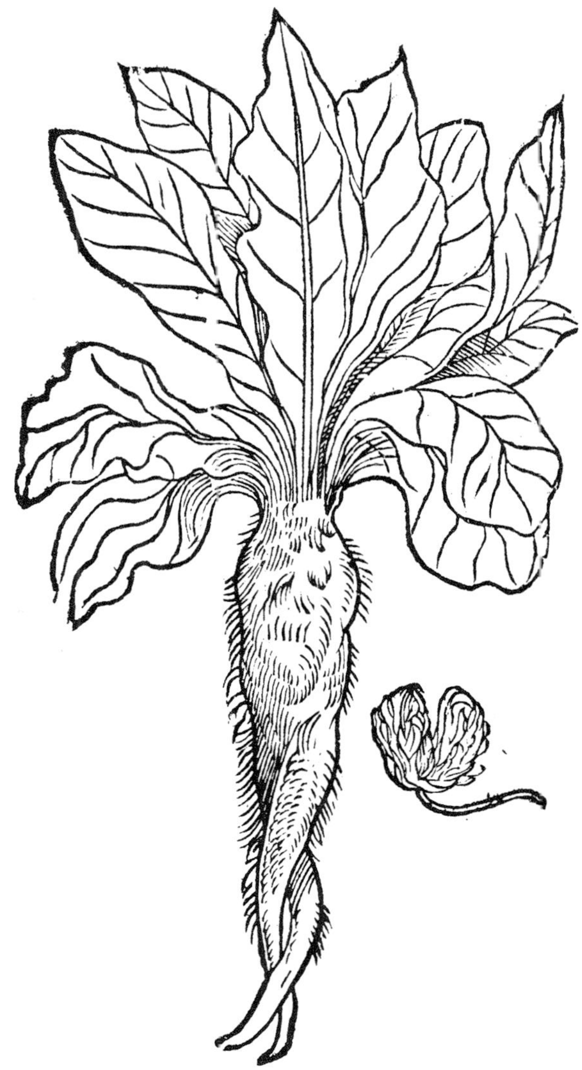

"Mandragore." In William Turner, *A New Herball* (1551), ed. George T. L. Chapman and Marilyn N. Tweddle, 2 vols. (Cambridge: Cambridge University Press, 1989), 2:114. **1.2**

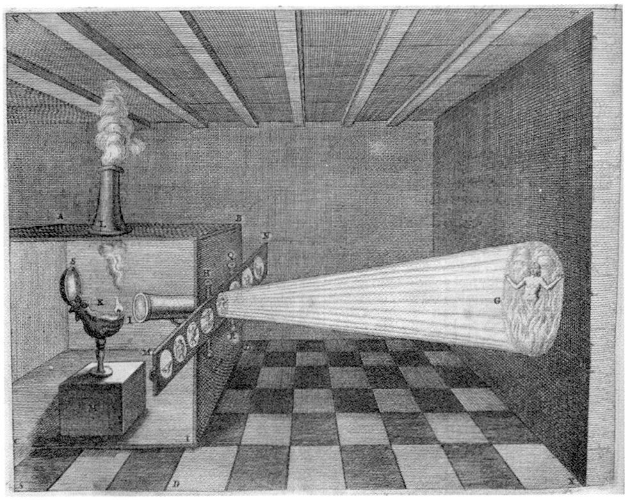

1.3 "Laterna magica." In Athanasius Kircher, *Ars magna lucis et umbrae*, 2nd ed. (Amsterdam, 1671), 768. Maurits Sabbe Library, Faculty of Theology and Religious Studies, KU Louvain.

observation but not altering objects physically, would have seemed an odd reversal. Turner's own explanation for the anthropomorphic mandrake roots was not projection but fraud practiced upon the poor and gullible.

Still harder for early moderns to grasp would have been the desire to eliminate any trace of the naturalist from the presentation of nature, to strive for a "view from nowhere" that depends as little as possible on the "specifics of the individual's makeup and position in the world, or on the character of the particular type of creature he is."[6] They would have protested that the personal qualifications of the naturalist were very much at issue in the evaluation of the scientific authority of an image: Turner, for example, assured his readers that he had traveled widely in foreign parts in order to observe plants there firsthand. Most early modern botanists writing for a learned audience intervened routinely and assertively to correct the images drawn by their artists, in order to prevent them from drawing exactly what they saw.[7] Extreme mimesis that mirrored nature's appearances was sometimes a goal of early modern still life painting but rarely of natural history illustration. And in neither case would any virtue have been found in suppressing the artistry required to achieve the desired effect—quite the contrary.

Does the fact that "objectivity" was not an early modern actor's category disqualify it as a useful term of analysis for historians of art and science who examine images from this period? Not necessarily: it is sometimes illuminating to apply terms and concepts developed in other times and places, especially when comparisons and generalizations are at issue. The meaning of slavery, for example, surely differed in ancient Greece and Rome and in the Americas during the eighteenth and nineteenth centuries, but for the purposes of understanding slavery as a cross-cultural institution, such distinctions can be provisionally set aside. However, the validity of such comparisons and generalizations depends crucially on knowing that the differences exist, what they were, and to what extent they can be overlooked without endangering the assumption of a coherent category.

In the case of terms like "objectivity", which are habitually applied as transhistorical honorifics, still greater caution is called for. This is a matter of philosophical exactitude as well as historical integrity. Objectivity occupies such a commanding position in our modern catechism of scientific values that it is equated with science *tout court*. Objectivity swallows up other values that are not only historically and conceptually distinct from it but also sometimes in conflict with it. Certainty, precision, and truth to nature are also estimable scientific values, and they are not synonymous and occasionally not even compatible with objectivity. Even accuracy can, in some circumstances, collide with rather than reinforce objectivity. To ignore these alternatives to objectivity is to miss something essential about the nature of science, even if one does not share the historian's horror of anachronism.

For historians, anachronism is a professional sin because it erases the past-ness of the past. In the case of early modern ways of seeing and depicting nature, to deploy a critical vocabulary of modern terms such as "objectivity" is the equivalent of a right-handed insult: in an attempt to pay early modern naturalists and artists the compliment of identifying "objective" (and therefore, it is complacently assumed, genuinely scientific) elements in their work, historians obscure what is in fact novel and original in their efforts to reform natural knowledge in both word and image. To claim that early moderns deserve to be called such (i.e., predecessors of modernity) on the basis of glimmerings of objectivity is to misunderstand not only early modern but also modern scientific inquiry. Anachronism blurs the contours of all it touches, enveloping the present as well as the past in a dense fog in which everything looks pretty much like everything else—and all of it drearily familiar.

To rehearse the history of how objectivity emerged as a scientific ideal and practice would take us well beyond the early modern period.[8] My aim in this essay is rather to reconstruct some of the values that guided naturalists *before* objectivity, particularly in the realm of scientific illustration from the mid-sixteenth through the mid-eighteenth century. To that end, I first introduce the notion of the epistemic image and then go on to apply it to the role of outline and color in early modern botanical illustration. I conclude with some reflections about the relationship between the eye of the body and the eye of the mind in early modern vision.

What Is an Epistemic Image?

Since at least the sixteenth century, the sciences have been profligate in images.[9] New media and techniques of reproduction such as the woodcut and engraving, new instruments such as the telescope and microscope, new modes of representation such as perspective and cutaway diagrams, and, above all, new forms of empiricism that emphasized the close and sustained observation of particulars—all these factors combined to elevate the importance of seeing as a way of knowing nature in early modern Europe.[10] During this period, scientific seeing became a cultivated skill that distinguished the savant from mere onlookers: by the late seventeenth century, members of the Paris Académie royale des sciences prided themselves publicly on being among those "who have eyes for these kinds of [natural] things, unlike the majority of people."[11] Scientific illustrations became so integral to publications that they threatened to upstage the texts: already by the mid-sixteenth century, artists were increasingly named and praised in prefaces, especially of anatomical and botanical works, and, in some exceptional cases, even granted a frontispiece portrait on a par with that of the author, as in Leonhart Fuchs's *De historia stirpium commentarii insignes* (1542).[12] By around 1700, the scientific image had established its empire, which has expanded steadily ever since.

Yet not all scientific images, then or now, count as epistemic images. An epistemic image is one made with the intent not only of depicting the object of scientific inquiry but also of replacing it. A successful epistemic image becomes a working object of science, a stand-in for the too plentiful and too various objects of nature, and one that can be shared by a dispersed community of naturalists who do not all have direct access to the same flora and fauna. All scientific images are more or less informative, more or less accurate, more or less competently executed. All serve the aim of representing some aspect of nature faithfully, whether naturalistically or schematically, as woodcut or as digital photo, life-sized or scaled. But epistemic images aspire to more: in the eyes of their makers and users, they are more than just representations of nature; they *are* nature—or rather, nature selected, distilled, refined, and perfected. Even when epistemic images look exactly like what the eye of the body sees, they are nonetheless creations of the mind's eye: the results of deliberation and choice based on long and expert acquaintance with the phenomena in question and informed by strongly held views about how best to seek the truth about them. An epistemic image earns its name by translating abstract epistemological priorities into concrete pictures.

These epistemological priorities are various and not always convergent. Moreover, each has its own history, distinct from that of the others. All epistemology is about how to achieve knowledge—or rather, to avoid error. There may be only one truth, but there are many ways to fail to find it. Depending on which kind of error naturalists fear most, they set their epistemological priorities accordingly. Investigators who most dread the infirmities of the human senses will attempt to replace them with instruments or abandon perception in favor of more formal approaches, such as mathematical modeling. Those anxious about being duped will elevate certainty to the place of honor. Their colleagues preoccupied by the dangers of conflating artifacts with reality will invoke truth to nature. Scientists wary of subjective projections onto the data will insist upon objectivity. Those alarmed by the risks of missing small but significant differences will rally to the cause of precision. Advocates of public over private knowledge will raise the standard of replicability. All will claim, sincerely and rightly, that they are being faithful to nature—but each in a different fashion.

Are certainty, objectivity, precision, accuracy, truth to nature, replicability, and other epistemic virtues really so different? Don't they all, after all, stem from the same impulse, the desire to clear away the obstacles along the path to truth? One way to show that the differences among these epistemic virtues exist and are furthermore consequential is to examine their impact on specific scientific practices such as image making. An astronomer persuaded that a smooth trajectory is more likely to represent the true path of a comet than a jaggedly discontinuous one will jettison outlying values in the name of truth to nature; a colleague worried about how theoretical preconceptions might color judgment will plot the curve objectively, outliers and all. A botanist may prefer a woodcut to an exquisitely mimetic portrait of a plant because the latter renders both accidental and essential traits indiscriminately. The most precise observations and experiments are often also the most difficult to replicate, because they demand virtuosity and special equipment. As in the case of moral virtues, it is not always possible to serve different epistemic virtues simultaneously. Accuracy and objectivity may, for example, diverge in scientific illustrations, just as justice and mercy may point toward different verdicts in the courtroom.

Epistemic images are images that incarnate particular epistemic virtues, sometimes at the expense of others. These choices are reflected in the way the images are made, how they look, and what they are used for. Although there is no firm correlation between particular epistemic virtues and medium (e.g., engraving versus lithograph) or format (e.g., folio versus octavo), the choices among the various technical possibilities for making epistemic images are always made advisedly and the artists chosen and supervised with unusual care. This is because epistemic images are intended to serve as images of reference for whole communities of observers. Collected in scientific atlases and handbooks, these images are designed to standardize the objects of inquiry and calibrate the eyes of the inquirers. What does a certain kind of plant or animal or bodily organ look like? Since the sixteenth century, the scientific novice who poses this question has been instructed to go to nature with atlas in hand and to triangulate among individual specimens, text (or *viva voce* instruction), and the epistemic image.[13] By reading both book and the book of nature simultaneously, the novice gradually develops an expert eye for the essential, the normal, and the typical. Amid the profusion of natural variability, the epistemic image picks out what is scientifically significant. Out of the clamor of individual ways of seeing, the epistemic image creates a visual consensus among initiates.

Epistemic images make collective empiricism possible. The printing press and a widespread interest in natural phenomena were by themselves insufficient to teach far-flung observers to see in unison. Shared working objects, usually in the form of published images, were also necessary. Collections of authoritative images certified by authoritative observers circulated even among those unable to study anatomy with Andreas Vesalius in Padua, or botany with Conrad Gessner in Zurich. The standardizing impact was ontological as well as visual. Although epistemic images were supposed to be studied in tandem with actual specimens, the images were in an important sense more real than any individual exemplar. Each such image, even if it represented an individual, was intended to be representative in the larger sense of standing in for a universal—a species, an anatomical structure, even a kind of pathology or anomaly. One need not embrace the Platonist metaphysics of eternal ideas in order to respect the ontological weight of epistemic images. An Aristotelian view of universal forms abstracted from the particulars of perception would also have endorsed the synthetic image over the individual specimen. Even nominalists skeptical of the reality of anything except individuals acknowledged the necessity of abstractions and classifications in natural history and natural philosophy.

Epistemic images mobilize both senses of the word "representation": to copy and to stand for. Most epistemic images do resemble the objects they represent in the first sense of the word: that is, they are recognizable copies of originals, though not facsimiles. If herbal images bore little or no resemblance to individual plant specimens, they would be useless for field identification. But epistemic images also represent in the second sense of "to stand for." Just as members of a legislature or a committee "represent" a much larger group of constituents, an epistemic image stands in for a whole class of individuals: in the case of botany, an entire species or genus. Epistemic images are almost always the result of synthetic perception, of composites forged by memory and judgment. In some cases (and botany is one of them), the two senses of representation can tug in opposite directions: the more a copy mirrors an original specimen, the less it captures the essentials of the taxon. It is possible that this tension was less pronounced before the middle decades of the nineteenth century, when the development of various chemical techniques for fixing camera obscura images through photography made it possible to capture extraordinary quantities of detail almost instantaneously: apparently infinite detail (and thus extreme individuation) became synonymous with the perfect copy. Yet even in the sixteenth century, botanists argued the merits of more mimetic but less typical figures, most famously in the case of Hans Weiditz's illustrations for Otto Brunfels's *Herbarum vivae eicones* (1530, 1536), which included images of plants with withered flowers and insect-gnawed leaves (fig. 1.4).[14]

The history of epistemic images parallels that of epistemic virtues in the sciences: when, for example, a new epistemic virtue like objectivity emerges, new atlases are made with the explicit aim of producing epistemic images that do justice to new ideals and practices, often in critical contradistinction to older ones. But new epistemic virtues do not necessarily replace old ones, although competition among them sometimes ensues. Nor is there any necessary correlation between kind of medium and kind of epistemic image: photography, for example, has served the epistemic virtues of both objectivity and truth to nature. Nor is style an infallible indicator: naturalism and realism can sometimes pull in different directions. In order to discern whether a scientific image is an epistemic image and, if so, which epistemic virtue it realizes, more is needed than either just reading or just looking. Text, image, and their interaction must be examined together, and so must the context in which they were produced. The next section provides an example from early modern botany.

Outline and Color, or The Multiplication of Entities

In his 1763 treatise on plant families, the French botanist Michel Adanson revived an ancient objection to illustrations of plants: citing Pliny, Adanson agreed that botanical figures could never really do justice to nuances of color and texture or depict the plant in all of its seasonal stages of development. But Adanson did not follow Pliny in rejecting illustrations altogether. Rather, he insisted upon a division of labor between text and image in botany, in which the verbal descriptions indicated "the colors, the polish, the furriness [*velu*], the smoothness or roughness of the surfaces" of plants, while the figures showed "their figure, the situation and disposition of the parts."[15] The debate over the relative merits of images and texts in botanical works was a venerable one, and early modern authors had gone back and forth on the subject.[16] Adanson was not, however, simply giving voice to the latest swing of the pendulum between botanical iconoclasm and iconophilia. In contrast to sixteenth-century botanists such as Fuchs and Gessner, Adanson was much less concerned with the role of images in the identification of known plants than he was with the needless multiplication of species. Of the some seventy thousand figures of plants that had been published since the advent of printing in the late fifteenth century, Adanson estimated that only about ten thousand corresponded to genuine species and that, of these, perhaps only fifteen hundred to two thousand portrayed plants as completely as might be desired.[17] The principal problem was no longer taking inventory of previously unknown domestic and foreign plants, as it had been for much of the sixteenth and seventeenth centuries, or even in imposing a classificatory order upon the swelling catalogue of known genera and species, as it had been for late seventeenth-century botanists such as Joseph Pitton de Tournefort and John Ray.[18] Rather, it was to defend the reality of the classificatory categories themselves, especially that of the species, in the face of skeptics who, like the French naturalist Georges Leclerc de Buffon, pointed to the teeming variety and continuity of natural forms.[19]

The crisis over the multiplication of botanical species had been a long time in the making. Over the course of the sixteenth and seventeenth centuries, the number of plant species described by botanists increased by an order of magnitude, from several hundred to several thousand.[20] Contributing further to this *embarras de richesses* was the booming horticultural market, which was avid for novelty and extravagant in its expenditures—most famously in the Dutch tulip craze of the early seventeenth century.[21] Gardeners who catered to wealthy collectors had every motive to breed new varieties of the flowers most in demand and to emphasize their distinctiveness; expensive, lavishly illustrated florilegia further pandered to the taste for the unusual and the exotic.[22] Botanists eager to secure glory for the discovery of new species (the name of which might immortalize their own with a Latin tag) were also tempted to exaggerate the differences among varieties of the same species.

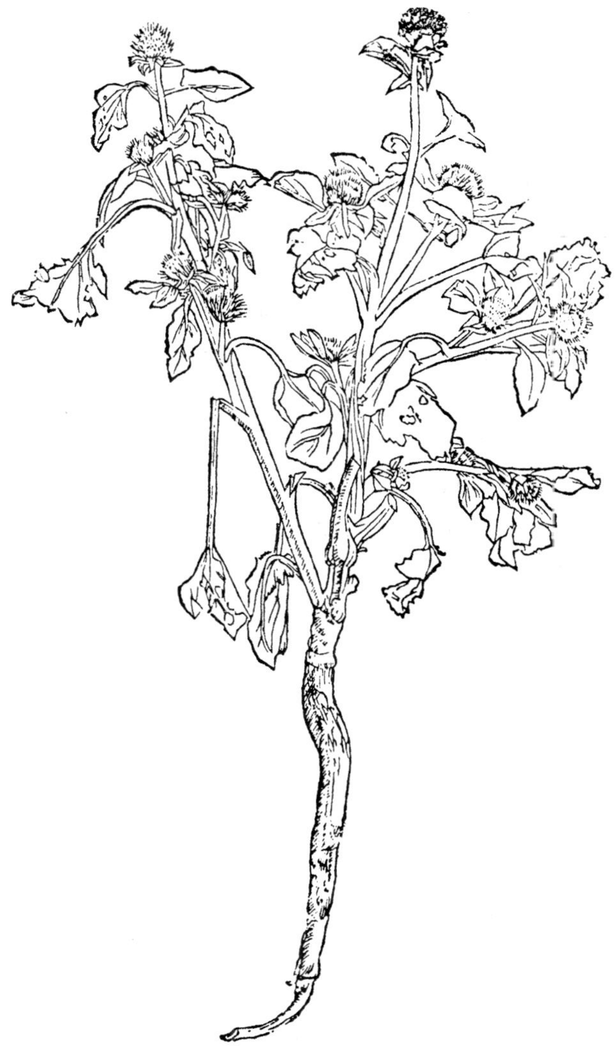

Hans Weiditz, "Lappa." In Otto Brunfels, *Herbarum vivae eicones ad naturae imitationem* (1530), 208. **1.4**

Adanson was by no means the first to flag the alarming proliferation of plant species catalogued by botanists. In his *Critica botanica* (1737), the Swedish naturalist Carl von Linnaeus had some three decades earlier railed against defective drawings that featured misleading classificatory characters, especially those that used the highly variable criterion of color, so easily manipulated by "the florists" and so alluring to undisciplined botanists: "Tournefort enumerates 93 Tulips (where there is only one) and 63 Hyacinths (where there are but two), and others have been no less extravagant."[23] It is no accident that Linnaeus's examples of egregious splitting were the tulip and the hyacinth, both objects of intense horticultural interest and experimentation. As a vigorous defender of the reality of species, divinely ordained divisions among creatures, Linnaeus was scathingly critical of "monstrous" varieties introduced by the florist's art and relentless in his criticism of botanists who, like Tournefort, mistook such monsters for true species: "The species of the Botanists come from the All-wise hand of the Almighty, the varieties of the Florists have proceeded from the Sport of Nature, especially under the auspices of the Gardeners."[24]

Like Adanson, Linnaeus called for the reform (not the elimination) of botanical illustrations. They should restrict themselves to the four "certain and fundamental characteristics, to wit, Number, Shape, Position, and Proportion": "For all the features which are represented in pictures rest on the aforesaid four bases."[25] Linnaeus's own *Hortus Cliffortianus*, published the same year as the *Critica botanica*, was copiously, indeed lavishly, illustrated. Because for this volume Linnaeus was in the fortunate position of having a wealthy patron, the Amsterdam banker George Clifford, whose Hartekamp garden and herbarium provided the specimens, no expense was spared in realizing the figures, under Linnaeus's close supervision and to his exacting specifications.[26] The plants were drawn by Georg Dionysius Ehret and engraved by Jan Wandelaar, both artists of distinction. Already in the subtitle, the generalizing intent of the work was proclaimed: varieties would be related to species, species to genera, and genera to classes. Both text and images distilled descriptions to those essential characteristics required to define what is common to all members of the species (the *descriptio*) and those that distinguished this species from all others in the genus (the *differentia*). In the case of the *Gladiolus foliis linearibus* pictured in Linnaeus's *Hortus Cliffortianus*, the differential characteristic is the long, straight leaf, featured prominently at the center of the plate and corresponding to type number seven of Linnaeus's schematic table of kinds of leaves.[27] The description does specify color ("red blossoms inset with rhomboid-shaped purple spots"), but the figure, a full-page engraving, was fully shaded, suggesting that hand-coloring (as was sometimes commissioned privately by well-heeled purchasers of illustrated botanical works) was not intended (fig. 1.5).[28]

The opposition between outline and color in botanical images may remind art historians of a celebrated thesis advanced by Erwin Panofsky in his study of Albrecht Dürer. Drawing upon Erasmus's praise of the artistry of Dürer's prints as so consummate as to render color superfluous, Panofsky claimed that the cross-hatching techniques introduced by Dürer's teacher Michael Wolgemut and perfected by Dürer in his prints obviated the need for color in subsequent prints.[29] This argument resonated strongly with traditional art-historical assertions, since Vasari, of the superiority of *disegno* over *colore*.[30] But in the case of the botanical images, not only color but also texture was suspect—even if conveyed entirely in black and white. Although cross-hatching techniques complemented the use of color to suggest volume, especially in engravings, any shading that might obscure the distinctness of significant details or muddy the overall outline of the plant was unwelcome. Some botanical artists, such as Ehret, made a distinction between drawn ("Ehret delineavit") and painted images ("Ehret pinxit") in their signatures, suggesting that they too had internalized the distinction (fig. 1.6).[31]

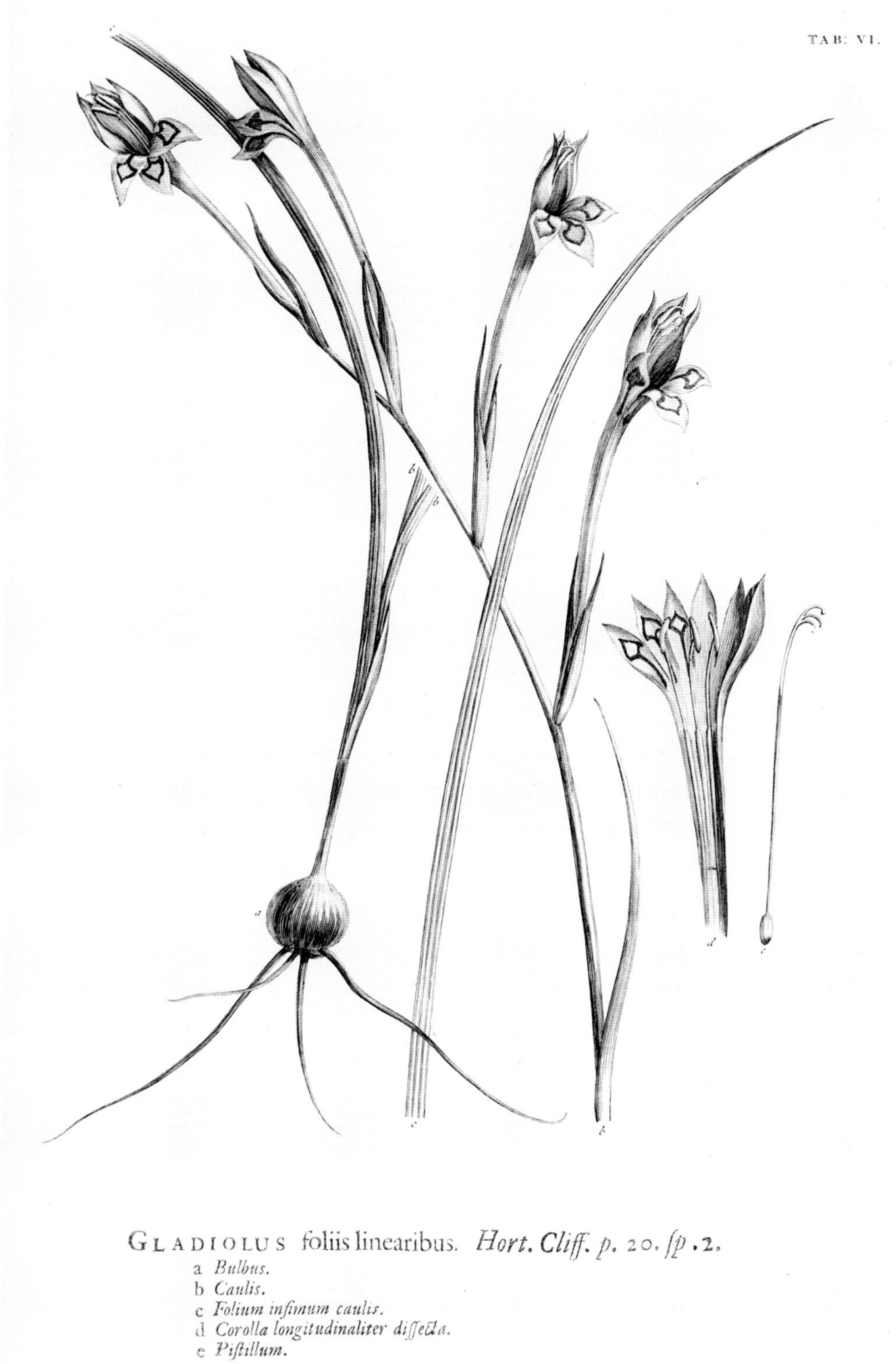

1.5 Georg Dionysius Ehret, *Gladiolus foliis linearibus*. In Carl von Linnaeus, *Hortus Cliffortianus* (Amsterdam, 1737), table 6.

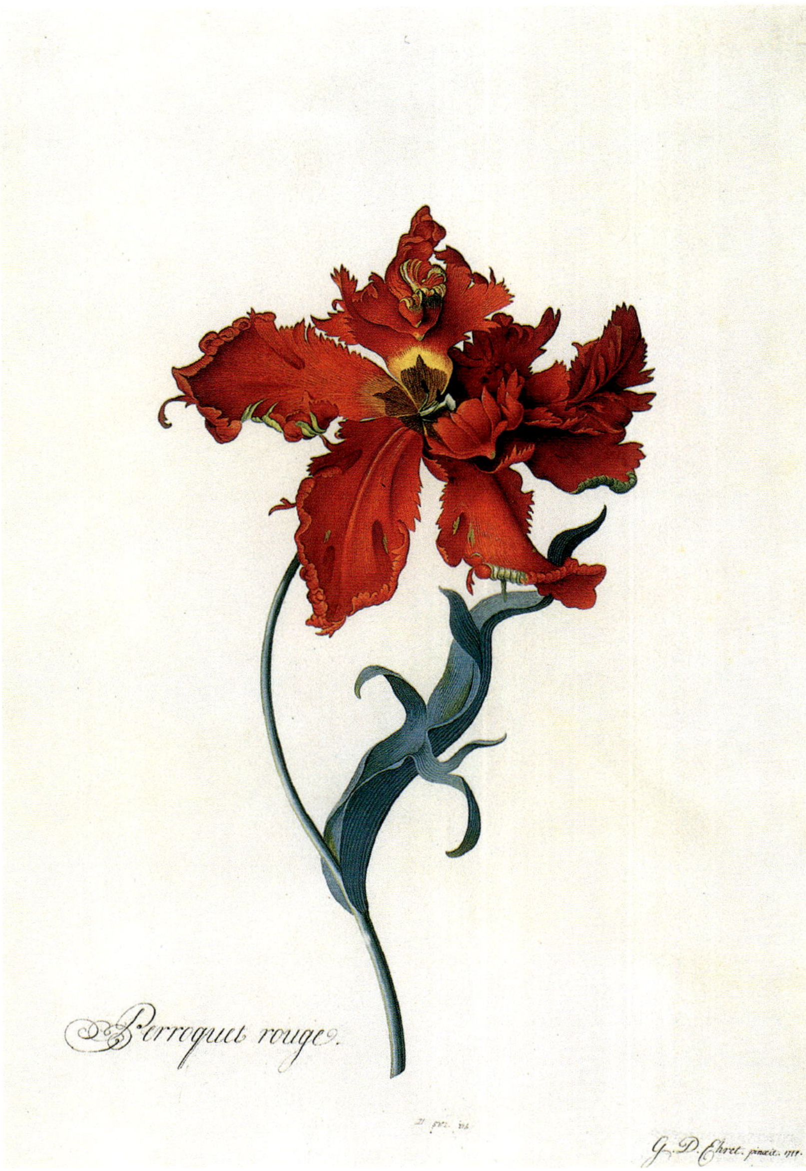

1.6 Georg Dionysius Ehret, *Tulipa gesneriana*, ca. 1744. Published in C. J. Trew, *Hortus nitidissimus* (1750–92). © Victoria and Albert Museum, London.

The issues of color, outline, and shadow increasingly preoccupied botanists in the late seventeenth and early eighteenth centuries. Although deftly applied color and shading undeniably rendered plants, especially those in bloom, more naturalistically, botanists objected that these features obscured or upstaged the essential characteristics used to identify and classify plants. Adanson lumped the two together and called for restraint: "Shadows have the same effect in engravings that colors do in paintings of plants: when they are a bit dark, they obscure and confuse most of their veinings, their outlines, or their smallest parts. Hence it is necessary either to greatly diminish their intensity or to suppress them entirely, marking by a simple stroke the contour of the plant that renders the part or the whole distinctly."[32] Moreover, the very success of the horticulturalists in breeding new strains of garden flowers in rainbow hues unsettled some botanists: How, inquired Linnaeus, could such a variable trait as color reliably indicate species and genus?[33] Standardization of color hues and tints posed further problems.[34] Color did not disappear from botanical texts, as Linnaeus's own descriptions show, but for both financial and scientific reasons, it was relegated to texts rather than to images.[35]

Historians of botanical illustration have remarked on how, in the course of the seventeenth century, the gap widened between the images of botanical works, which were executed in black and white, usually engraved, starkly outlined, and constructed to show all of the parts of the plant clearly (including those, such as flower and fruit, that would never appear simultaneously in nature), and those produced for florilegia, which were opulent color portraits, elegantly composed, softly outlined, and enhanced by the full array of illusionist techniques.[36] The color leached from the botanical treatises, it is suggested, flowed into the florilegia. There is some truth to this claim, to which the botanists themselves attested. Linnaeus thundered his reproaches at the purveyors of luscious color plates: "How many volumes have you written of specific names taken from color? What tons of copper have you destroyed in making unnecessary plates? What vast sums of money have you enticed fraudulently, as it appears, from other men's pockets, the purchasers to wit, on the strength of color alone?"[37]

But such harangues must be taken with a grain of salt. Even if its handsome plates were not colored, Linnaeus's own *Hortus Cliffortianus* was undeniably an *édition de luxe*, privately commissioned and printed by the patron whose garden it documented and celebrated with numerous copperplate engravings, as were many of the most lavish florilegia. Linnaeus himself decorated his bedroom with colored engravings by Ehret for C. J. Trew's *Hortus nitidissimus* (1750–92), and hand-colored editions of the *Hortus Cliffortianus* were not unknown.[38] More significantly, many of the same artists and even some of the same images did double duty in both the decorative arts and natural history illustration. The most celebrated examples were the *vélins* (i.e., vellums, very white, smooth calfskin parchment) of plants and animals collected in the Paris Jardin du Roi (later the Muséum national d'histoire naturelle) from the mid-seventeenth century onward. These color paintings in watercolor or gouache, mostly rimmed in gilt, were originally commissioned in 1631 by Gaston d'Orléans, the brother of Louis XIII, and were executed by artists such as Nicolas Robert and Madeleine Basseporte, who often bore the title of royal miniature painter and worked extensively in the decorative arts that drew heavily on floral motifs, such as embroidery, silk weaving, and porcelain painting (fig. 1.7).[39] Robert's *vélins*, of which he painted more than seven hundred, served as the basis of both the major botanical publication of the Paris Académie royale des sciences and his influential pattern book of floral patterns for the decorative arts. One of Robert's imitators was the German artist and naturalist Maria Sibylla Merian, whose own three-part *Blumenbuch* (1675–80) borrowed heavily from the images in Robert's several flower books, which first appeared in 1640 and were still being pillaged and reprinted as copybooks in Amsterdam as late as 1794.[40]

Both Merian's and Robert's careers were typical of those of many seventeenth- and eighteenth-century flower painters for whom the distinctions between scientific and ornamental illustration hardly held. Whatever the botanists' complaints about the florilegia, they did not preclude such conflations. The botanical illustrations commissioned by the Académie royale des sciences from Robert, among others, were explicitly designed to combat what were perceived as the excesses of florilegia like the *Hortus Eystettensis* (1613), which, according to botanist Antoine Jussieu, depicted "varieties multiplied without necessity, passed off as species, all drawn with little exactitude and correction, and engraved without that taste and art that allows the discernment of different parts."[41] Yet some of the same richly colored, gilt-edged *vélins* that graced Robert's pattern books also appeared as engravings in the Académie's botanical magnum opus. Color gradations were at least suggested by shading: "brown-green and light green will be treated differently [by the engravers], [as will] white flowers and those of a dark color."[42] No simple opposition between art and science, wealthy connoisseurs and serious botanists, the beautiful and the functional does justice to how late seventeenth- and early eighteenth-century botanists judged the merit of illustrations.

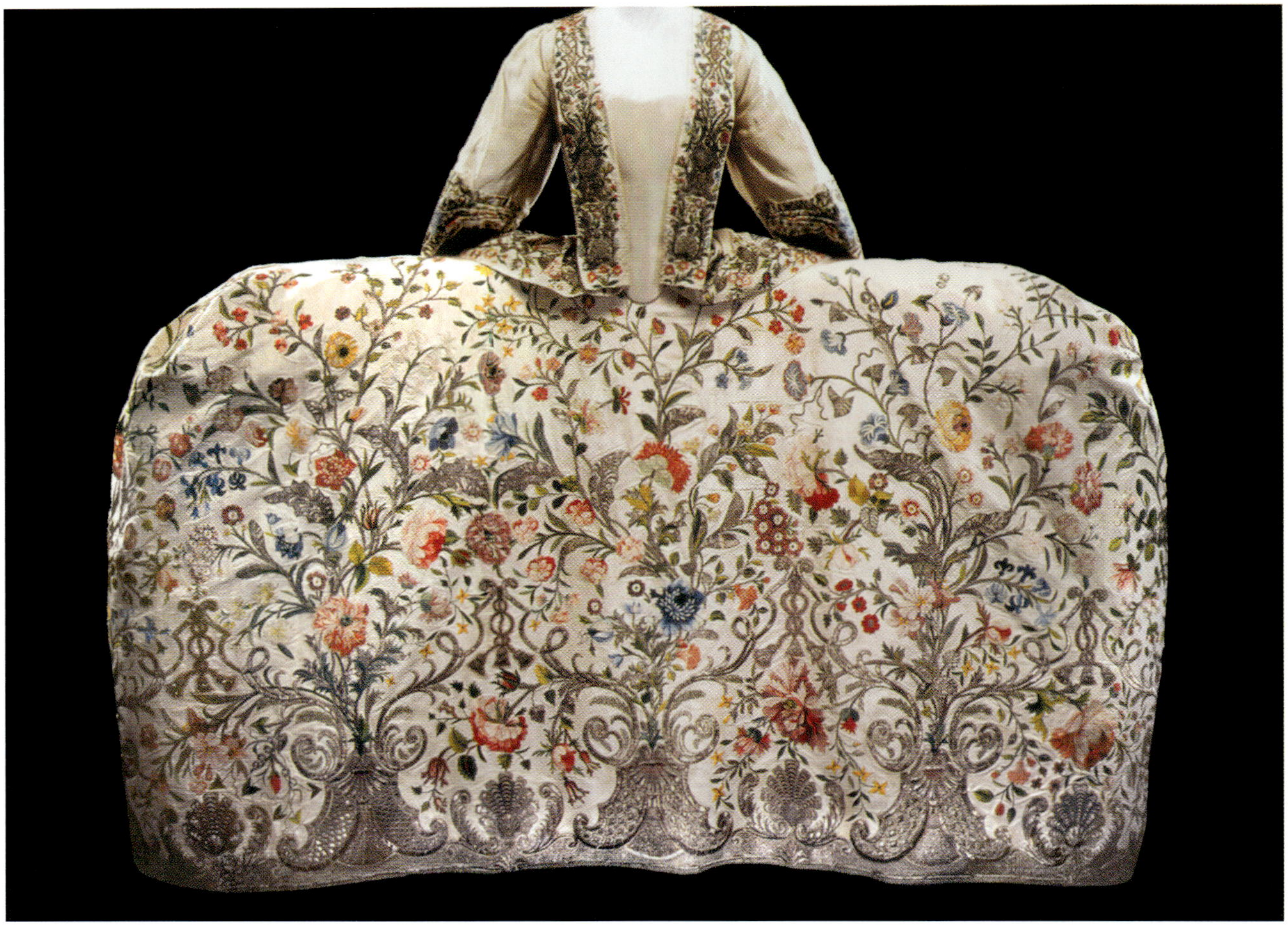

1.7 Mantua, or court dress. Silk embroidered with colored silk and silver thread (1740–45). © Victoria and Albert Museum, London.

Even when botanical images were ultimately destined for highly finished luxury collections, like the *vélins* of the Paris Jardin du Roi, field conditions forced artists to develop expedients and conventions that replaced color with shading. Claude Aubriet, the artist who accompanied Tournefort on his botanizing expedition to the Levant in 1700–1702,[43] usually made drawings of plants on the spot, often in trying circumstances. Eager to outdo Dioscorides's descriptions of plants in this region, Tournefort pressed on to remote locales, regardless of bad weather or rough terrain. Writing to French botanist Denis Dodart from Crete, Tournefort claimed that "the most beautiful plants" were to be found on the most desolate mountain cliffs, adding, "I begin to be strongly persuaded that that good man Dioscorides spoke only of the most common plants."[44] In a letter written from

Constantinople to his fellow artist Jean Joubert back in Paris, Aubriet excused the unfinished state of the drawings being sent to Joubert to be painted for the royal collection: "it is not possible to be able to color all the plants, when there are several to do at the same time; when this happens, M. Tournefort has the goodness to note the principal ones, which I color, and [as] for the others, if time permits, I write down the colors, which will serve when [I] return from the voyage. You need paint, please, only the most finished drawings for the Cabinet du Roi."[45]

What made the images of Ehret, Robert, Aubriet, and other favored artists *epistemic* in the eyes of the botanists was what they showed, how they showed it, and, above all, the degree to which they conveyed the universal in the particular. The optic of botanists during this period was at once analytical, decontextualized, and generalizing. Plant parts deemed important were rendered prominently and with preternatural distinctness through the use of clear-cut outlines, sometimes further emphasized by enlarged detail drawings on the same plate, as in the case of the leaf in the gladiolus Ehret drew for Linnaeus (fig. 1.5). The plant is displayed by itself, floating on a blank white page, with no indication of the ecological context in which it is usually found. If the specimens at the botanist's disposal permitted, the figure, like the text, was a composite of several individuals, aiming at maximum typicality and generality. When Dodart included descriptions of individual plants in his 1676 account of the Académie's botanical project, he felt obliged to explain this deviation from accepted practice: "The description of well-known plants will be ordinarily given as [that] of the entire species; but we will give only an individual description of those foreign plants that are so rare that we will not have been able to observe them over the course of several years."[46] As Dodart's explanation suggests, the ideal description and image were implicitly comparative, the synthesis of multiple observations made over several seasons, following the plant through its full developmental cycle. Even images intended for purely decorative purposes, like those in Merian's *Neues Blumenbuch*, might be composed on the basis of several specimens, a process Merian herself called "ordination."[47] By the late seventeenth century, the practice of consulting both herbarium and garden specimens had become entrenched among botanists, allowing them to hone their ability to detect the *differentiae* that picked out this species from all others in its genus—and ignore or minimize other features.[48]

Botany was not the only science that sought to tame variability in both text and image around the turn of the eighteenth century. The same turn toward "description by omission" in the name of generality occurred in other branches of natural history, as well as in disciplines such as chemistry and anatomy.[49] In reaction to an equally concerted and widespread epistemic program that had emphasized the novel, the rare, and the anomalous in all of their particularity,[50] naturalists of Linnaeus's and Adanson's stamp focused instead on the typical, the universal, and the regular. Their tirades against the monstrosities of the florists and the dubious species of their predecessors underscored the dangers of paying too much attention to natural variability. By deliberately pruning texts and stripping images of copious detail, they achieved on the printed page the generality they sought in nature. Translated into visual terms, this strategy played up the distinct articulation of key plant parts through the use of emphatic contours and magnified details, while downplaying color and context.

Because they are responses to the preoccupations of scientists in a specific historical context, epistemic images rarely go unchallenged for long. By the early nineteenth century, naturalists with other epistemic priorities had become critical of precisely this analytical style: Alexander von Humboldt, keenly interested in local ecology and the geographic distribution of plants, urged botanists to stop peering at the stamens and pistils of isolated specimens and instead widen their gaze to embrace the "spatial and climatic spread of organic types" and the regional clustering of certain species of plants and animals in differing proportions.[51] But the older epistemic values never disappeared, and they held their own whenever fears of the needless and heedless multiplication of species were rekindled. The German botanist Ludolph Treviranus, writing a decade after Humboldt and several decades after the introduction of both lithography and photography, nonetheless still yearned for the stark, bold outlines of the woodcut: "Our [classificatory] systems would be less overburdened with untenable species. . . . For in order to render the allegedly characteristic [traits], far less is required for the copper engraving, the lithograph, or the description than for the woodcut, which represents only definite forms and that which is decisively distinctive in the [plant's] habitus."[52] For botanists haunted by the specter of classificatory clutter, the nuanced shading of the engraving and the subtle textures of the lithograph were the visible sign of blurred species boundaries.

Conclusion: Disinterested, Detached, or Demented?

I have argued that epistemic values have histories and leave their traces in certain kinds of scientific images, especially those intended to serve as a discipline's working objects. Because these epistemic values are both diverse and historical, some care must be exercised in employing them as analytical terms. An objective image is not always an accurate one; a naturalistic image is not always true to nature. Moreover, each of these epistemic values comes into being in specific historical circumstances: truth per se may be as ancient an epistemic goal as justice is a moral one, but the ways of achieving it are multiple and born in different centuries. Fairness, for example, is an old means to the end of justice; human rights, a far more recent path beat to the same destination. Similarly, certainty is a venerable epistemic value, while objectivity emerges only in the nineteenth century. In the case of both epistemic and moral values, new and old coexist, sometimes converging and at other times diverging. And in both cases, despite the lofty abstractness of "truth" or "justice," the choices among rival values can be consequential for matters as concrete and specific as how to draw a plant or the boundaries of a school district. To conflate competing values with one another or to telescope their longer or shorter histories into the flat plane of the present is to obscure the existence of these choices and their effects.

But even those who acknowledge the variety and historicity of epistemic values may still entertain doubts. Is it not, they may ask, the very essence of the transformation of natural knowledge that took place in early modern Europe that systematic precautions were taken to hinder the mind from mistaking itself for the world—in other words, to drive a wedge between objectivity and subjectivity? More specifically, didn't the critique of older botanical images as false or fanciful by early modern naturalists such as William Turner represent a cool detachment that might fairly be called objective, even if *avant la lettre?*

These queries make sense only in the context of a historical narrative that has framed the remarkable innovations in the study of nature in early modern Europe as just that—as the early but decisive steps along the way to modern science. In the twentieth century, this narrative about the history of European science was reinforced by a history of European modernization, more philosophical and usually inflected by central European accents, that interpreted the rise of science since the seventeenth century (and the rise of industrialized technology since the nineteenth century) as the triumph of reason at the expense of desire. In the works of Max Weber, Sigmund Freud, Edmund Husserl, and others, to embrace science and rationality was to abandon the childish but comforting illusion of an enchanted world, animated by forces that mimicked the workings of human thoughts and emotions—a humanized world in which humans were at home.[53] On this account, to abandon the allegedly cozy *Lebenswelt* of premodern Europe was an emotionally costly if intellectually necessary sacrifice. The Russian-French historian of science Alexandre Koyré, who had studied with Husserl in Göttingen, fused these two narratives about science and modernity into an eloquent elegy for a neighborly "closed world" exploded into the loneliness of "the infinite universe."[54]

These are compelling and influential narratives, which have shaped at least a half century of cultural criticism and history of science. There is only one thing wrong with them: there is little or no evidence that they describe how early modern Europeans experienced the extraordinary innovations in the sciences. Learned scholars and laymen alike were acutely aware of the avalanche of novelties that had cascaded over them, from the discovery of new lands and peoples, to inventions such as oil painting and the printing press, to the emergence of new religions, new cosmologies, and new philosophies.[55] But although some feared and others welcomed these novelties, very few interpreted them as an eviction notice from an anthropocentric cosmos.

Quite the contrary: insofar as anthropomorphism became suspect among early modern naturalists and philosophers, it was largely as a perceived threat to anthropocentrism and orthodox religion. The English natural philosopher Robert Boyle worried that endowing nature with too much wisdom and foresight might both encourage the idolatrous worship of the sun, moon, and stars and also discourage the exploitation of nature for human gain: "the veneration wherewith men are imbued for what they call nature has been a discouraging impediment to the empire of man over the inferior creatures of God."[56] Boyle endorsed the mechanical philosophy not because he had stoically shouldered the burden of living in a disenchanted world but rather because he thought that viewing nature as nothing but brute, passive matter would monopolize intelligence and action for God and man alone. The ban on anthropomorphism in early modern natural history and philosophy was proclaimed in *defense* of anthropocentrism—nowadays considered an even more egregious sin against scientific objectivity. Later—much later—romantics would mourn a lost world of tongues in trees and books in brooks and excoriate the primary-secondary quality distinction so central to seventeenth-century natural philosophy for having purged the world of all color, texture, and life. But there are vanishingly few early modern witnesses to such melancholy views.

In fact, late seventeenth- and early eighteenth-century naturalists, heirs to all these allegedly devastating victories of cold reason over warm sentiment, immersed themselves in the minutiae of plants, insects, shells, fish, and a myriad of other *naturalia* so deeply and delightedly that moralists feared for their character and physicians for their health and sanity. They chided the naturalists for a passionate attachment to trifles and for neglecting religious, civic, and familial duties in favor of studying the anatomy of caterpillars or the membranes of leaves. In his long satirical poem *The Dunciad* (1728/1742), Alexander Pope had the Goddess of Dullness chide the naturalists who had brought her offerings of "a Nest, a Toad, a Fungus, or a Flow'r": "O! would the Sons of men once think their Eyes / And Reason giv'n them, but to study *Flies*!"[57] The ridicule heaped upon the new-style naturalists was directed not so much at natural philosophy per se as at the disproportion between the time, resources, and passion consumed and their objects.

At best, this disproportion made savants pedantic, unable to converse on anything except their own specialty. As a result, complained English man of letters Samuel Butler, their conversation was at once boring and boorish, "for as Country-Gentlemen use to talk of their Dogs to those that hate Hunting, because they love it themselves; so will he of his Arts and Sciences to those that neither know, nor care to know any Thing of them."[58] Scientific monomania rendered its victims oblivious to the social cues of age, rank, sex, vocation, and education that skilled conversationalists registered effortlessly and to which they adapted their themes and manner accordingly.[59] At worst, savants, fixated on lunar perturbations or the habits of turtles, squandered the funds set aside for their sons' education or their daughters' dowries on books and instruments. The French naturalist Henri-Louis Duhamel du Monceau, for example, managed to run through not only his own fortune but also that of his brother and nephews by conducting expensive agricultural experiments.[60] At the very worst, they went mad, victims of *idées fixes*. As the British moral philosopher David Hartley explained in his influential 1749 treatise on the association of ideas, scholars who immersed themselves too deeply and too long in one area of study first began to spin crackpot theories and then fell prey to "Fear, Anxiety, Despondency, Peevishness, Jealousy, and the rest of the Tribe of Hypochondriacal Passions."[61]

What was the cause of this excess, so deleterious to sociability, fortune, and health? Although the naturalists and their critics disagreed about the value of such intense research, they were surprisingly consonant about the motivations. Neither a concern for public utility nor a pious desire to glorify God by studying his creation sufficed to explain the fervor of the scientist, oblivious even to food and sleep. The scientist who forgot to dress or lost track of mealtimes was in the grip of a devouring passion, as strong as love. Classically educated savants in early modern Europe would still have been heavily influenced by Aristotelian and neo-Stoic as well as Christian ethics. Within this moral framework, excess of any kind was deplorable, and all vehement passions were suspect. Nothing was at once so ludicrous and so mournful as the spectacle of a philosopher, including a natural philosopher, who succumbed to passionate excess, against all the precepts of wisdom and morality.

The activities of the dedicated naturalist knew no moderation. The program of vigilant observation imposed on the observer a strict discipline that was scarcely compatible with any other activity. The French naturalist René Antoine Ferchault de Réaumur, for example, counted the number of bees leaving the hive in one day, arriving at a sum of more than eighty-four thousand departures in fourteen hours (around a hundred per minute).[62] Yet despite the rigorous demands that attentive observation made on mind and body, the naturalists testified unanimously to its pleasures. Réaumur forgot the broiling summer heat while happily watching bees in a glassed-in hive (fig. 1.8).[63] Linnaeus adapted the biblical language of the Psalms to praise natural history as a "foretaste of celestial pleasure . . . the summit of human happiness."[64] Eighteenth-century manuals of scientific observation exhorted readers to regard their objects of study as "a lover who avidly contemplates the object of his love."[65] There was nothing cool or distanced in the early modern naturalists' attitude toward their subject matter.

This is why it is inaccurate as well as anachronistic to describe early modern scientific images as "objective." Only when viewed through the interpretative lens of postromantic narratives of the history of science and modernity do early modern pursuits become candidates for the stalwart suppression of the emotions and the resolute removal of human beings from the center of the universe that came to be linked to objectivity. In the nineteenth century, scientific abhorrence for anthropomorphism did indeed fuse with the rejection of anthropocentrism, and ascetic self-abnegation with scientific probity. But both connections would have appeared counterintuitive, even paradoxical, to early modern naturalists. Their scientific pursuits may have been disinterested in the sense of flouting conventional social rewards such as power or wealth, but they were hardly detached, and they struck some contemporaries, including their families, as downright demented.

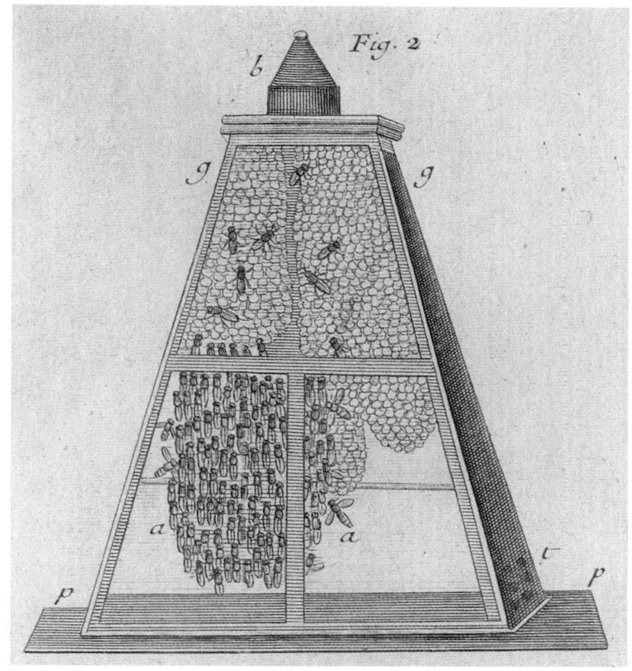

1.8 Glassed-in beehive. In René Antoine Ferchault de Réaumur, *Mémoires pour servir à l'histoire des insectes*, 5:280, plate 24. Max-Planck-Institut für Wissenschaftsgeschichte Library, Berlin.

The images produced under the supervision of botanists like Linnaeus were epistemic, but they served epistemic values other and older than objectivity. Art historians have long distinguished between mimesis and realism.[66] Analogous distinctions are urgently needed in the history of scientific images, especially in history that is epistemologically inflected. In the case of late seventeenth- and early eighteenth-century botanical images, especially those intended as working objects, the depictions were the outcome of collaborations between naturalist and artist and between the eye of the body and that of the mind. Botanists insisted that as many specimens as possible of a genus or species be minutely examined before the image was drawn, the work of the eye of the body, and they were equally insistent that the finished drawing portray a type more real than any one individual specimen, the work of the eye of the mind. The material and cognitive practices that made such titrations possible included the notebook and the herbarium, focused attention and retentive memory.

The vision of naturalists in this tradition was both synthetic and analytic: the image of record was usually a composite of many individuals, but its key parts were visually excised and often magnified. Its visual signature was the dark, crisp outline, softened by neither color nor shading. Although the media in which this vision was first realized, the woodcut and the engraving, have long since been superseded by other techniques, the vision itself and the epistemic value of truth to nature it articulates survive wherever the real must be extracted from the tangle of the merely visible.

Notes

1. Turner, *New Herball*, 2:114. Turner's figures were borrowed from Leonhart Fuchs's *Historia stirpium* (1542), apparently with Fuchs's permission; see Chapman and Tweddle's introduction to the *New Herball*, 1:16–17. However, the fact that Turner's figures are mirror images suggests that the blocks were recut on the basis of the printed illustrations from Fuchs, not printed from the original blocks. Fuchs's figures (and sometimes the original blocks) were still being used well into the eighteenth century; see Arber, *Herbals*, 70, 219. Early modern botanical images were often reused, with or without permission. See Baumann, *Pflanzenabbildungen in alten Kräuterbüchern*, 62–162.

2. See "Objective," *OED* (especially definition 2, in relation to Scholastic philosophy). For the subsequent history of the word in philosophy and science, see Karskens, "Development of the Opposition"; and Daston and Galison, *Objectivity*, 29–32 and passim.

3. On the long history of European ideas about the maternal imagination, see Huet, *Monstrous Imagination*. For early modern views on the powers of the imagination, see Fienus, *De viribus imaginationis*.

4. See Müller-Tamm, *Abstraktion als Einfühlung*.

5. See, for example, Hooke, "Contrivance to Make the Picture of Any Thing Appear on a Wall."

6. Nagel, *View from Nowhere*, 5.

7. See, for example, Charles de l'Escluse's remarks concerning the need to admonish his artists, quoted in Nissen, *Botanische Buchillustration*, 1:63n1.

8. For a detailed history, see Daston and Galison, *Objectivity*.

9. Kusukawa's *Picturing the Book of Nature* offers an excellent account of two disciplines transformed by image making.

10. On the early modern print, see Landau and Parshall, *Renaissance Print;* on the telescope, van Helden, "Telescope in the Seventeenth Century"; and on the microscope, Wilson, *Invisible World*. On visualization techniques in scientific illustration, see Lefèvre, Renn, and Schöpflin, *Power of Images;* and on empiricism, Pomata and Siraisi, *Historia*.

11. Perrault, *Mémoires pour servir*, viii. All translations are my own unless otherwise noted.

12. On the background of the artists and Fuchs's relationship to them, see Nissen, *Botanische Buchillustration*, 1:44–45; also Ivins, *Prints and Visual Communication*, 44, concerning the novelty of giving illustrators such prominent credit for their work.

13. Kusukawa, *Picturing the Book of Nature*, 218–27.

14. There is a large historical literature on this case. See Arber, *Herbals*, 52–70, 204–219; Nissen, *Botanische Buchillustration*, 1:40–45; Rytz, *Pflanzenaquarelle des Hans Weiditz*, 24–31; Reeds, "Renaissance Humanism and Botany"; and Ogilvie, *Science of Describing*, 192–99.

15. Adanson, *Familles des plantes*, 1:clxxxiv.

16. See Kusukawa, "Leonhart Fuchs on the Importance of Pictures"; and Ogilvie, "Image and Text in Natural History."

17. Adanson, *Familles des plantes*, 1:cxlii.

18. On Tournefort, see Becker, *Tournefort;* and on Ray, see Raven, *John Ray: Naturalist*.

19. Adanson, *Familles des plantes*, 1:clxviii.

20. Ogilvie, *Science of Describing*, 222.

21. See Goldgar, *Tulipmania*.

22. Schnapper, *Histoire et histoire naturelle*, 58–59; Sanders, *Picturing Plants*, 41–48.

23. Linnaeus, *"Critica botanica" of Linnaeus*, aphorisms 256 (par. 116) and 259 (par. 122).

24. Ibid., aphorism 310 (par. 197). On Linnaeus's species concept, see Eriksson, "Linnaeus the Botanist."

25. Linnaeus, *"Critica botanica" of Linnaeus*, aphorism 282 (pars. 161–62).

26. On the preparation of the *Hortus Cliffortianus*, see Blunt, *Compleat Naturalist*, 102–26; also Rendle, "George Clifford's Herbarium."

27. Linnaeus, *Hortus Cliffortianus*, table 1, "Folia Simplicia." On the global trade background of Clifford's botanical garden, see Müller-

Wille, "Introduction," in Linnaeus, *Musa Cliffortiana.*

28. Whether engraved botanical illustrations were meant to be hand-colored after printing can be in part inferred from the density and depth of the incised lines used to shade parts of the plant to produce a three-dimensional effect. Nickelsen, "Challenge of Colour," 10n35. On the expense of hand-coloring botanical illustrations, see Arber, *Herbals,* 314–18; Ogilvie, "Image and Text in Natural History," 155.

29. Panofsky, *Life and Work of Albrecht Dürer,* 19. Doubt has recently been cast on Panofsky's claims; see Dackerman, "Painted Prints in Germany and the Netherlands."

30. See Kemp, "Disegno."

31. Ludwig, *Nürnberger naturgeschichtliche Malerei,* 154.

32. Adanson, *Familles des plantes,* 1:clxxxv.

33. Linnaeus, *"Critica botanica" of Linnaeus,* aphorism 266 (par. 139).

34. Lack, "Recording Color"; Nickelsen, "Challenge of Colour."

35. Brian Ogilvie has argued that the number of botanical images actually declined in the late seventeenth century, replaced by an increasingly refined descriptive vocabulary: "they [illustrations] introduced beginners to an unfamiliar subject, but they could be dispensed with by skilled practitioners who could see in their mind's eye objects that were described in technical languages." "Image and Text in Natural History," 162.

36. See, for example, Sanders, *Picturing Plants,* 54–55, which emphasizes the difficulty of distinguishing between florilegia and pattern books for the decorative arts; see also Nissen, *Botanische Buchillustration,* 1:3, 12–13, 66–91.

37. Linnaeus, *"Critica botanica" of Linnaeus,* aphorism 266 (par. 139).

38. I am grateful to Dr. Staffan Müller-Wille for this information.

39. On the history of the *vélins* and links to the decorative arts, see Vezin, *Artistes au Jardin du Roi,* 20–28; also Pinnault, *Peintre et l'histoire naturelle,* 18–26, 136–50.

40. Segal, "Maria Sibylla Merian als Blumenmalerin." On Robert's career and publications, see also Nissen, *Botanische Buchillustration,* 2:152–53.

41. Quoted in Laissus, "Plantes du Roi," 195. On the convoluted publishing history of the project, see Bréchot, Laissus, and Guidès, "Note bibliographique sur les 'Plantes du Roi.'"

42. Dodart, *Mémoires pour servir à l'histoire des plantes,* 7.

43. Tournefort, *Relation d'un voyage du Levant.*

44. Tournefort to Dodart, 23 June 1700, Bibliothèque central Muséum national d'histoire naturelle, Paris, MS 995, 163.

45. Aubriet to Joubert, April 1701, ibid., 402.

46. Dodart, *Mémoires pour servir à l'histoire des plantes,* 2. On botanical projects carried out by the Académie royale des sciences during its first decades, see Stroup, *Company of Scientists.*

47. Ludwig, *Nürnberger naturgeschichtliche Malerei,* 72.

48. Ogilvie, *Science of Describing,* 150, 181–85.

49. Daston, "Description by Omission."

50. Daston and Park, *Wonders and the Order of Nature,* 215–54.

51. Humboldt, *Kosmos,* 18, 25.

52. Treviranus, *Anwendung des Holzschnittes,* 71.

53. Classic expressions of this narrative include Weber, "Wissenschaft als Beruf"; Freud, *Unbehagen in der Kultur;* and Husserl, *Krisis der europäischen Wissenschaften.*

54. See Koyré, *From the Closed World to the Infinite Universe.*

55. For an overview, see Park and Daston, "Introduction: The Age of the New."

56. Boyle, *Free Enquiry,* 15.

57. Pope, *New Dunciad,* bk. 4, lines 390, 430–31, pp. 25, 28.

58. Butler, *Characters,* 2:185.

59. Burke, *Art of Conversation,* 102.

60. Condorcet, "Eloge de M. Du Hamel."

61. Hartley, *Observations on Man,* 1:397.

62. Réaumur, *Mémoires pour servir,* 5:433. See also Torlais, "Réaumur et l'histoire des abeilles."

63. Réaumur, *Mémoires pour servir,* 5:251.

64. Linnaeus, "Curiositas naturalis," 143.

65. Senebier, *Art d'observer,* 1:5.

66. Still seminal is Nochlin, *Realism.*

Bibliography

Adanson, Michel. *Familles des plantes.* 2 vols. Paris: Chez Vincent, 1763.

Arber, Agnes. *Herbals: Their Origin and Evolution; A Chapter in the History of Botany, 1460–1670.* 3rd ed. Cambridge: Cambridge University Press, 1986.

Baumann, Susanne. *Pflanzenabbildungen in alten Kräuterbüchern: Die Umbelliferen in der Herbarien- und Kräuterbuchliteratur der frühen Neuzeit.* Stuttgart: Wissenschaftliche Verlagsgesellschaft, 1998.

Becker, G., ed. *Tournefort.* Paris: Muséum National d'Histoire Naturelle, 1957.

Blunt, Wilfrid, with the assistance of William T. Stearn. *The Compleat Naturalist: A Life of Linnaeus.* New York: Viking, 1971.

Boyle, Robert. *A Free Enquiry into the Vulgarly Received Notion of Nature.* 1686. Edited by Edward B. Davis and Michael Hunter. Cambridge: Cambridge University Press, 1996.

Bréchot, Jean-François, Yves Laissus, and Michel Guidès. "Note bibliographique sur les 'Plantes du Roi.'" *Revue Française de l'Histoire du Livre* 8 (1974): 3–14.

Burke, Peter. *The Art of Conversation.* Ithaca: Cornell University Press, 1993.

Butler, Samuel. *Characters*. In *The Genuine Remains in Verse and Prose of Mr. Samuel Butler.* Edited by Robert Thyer. 2 vols. London: J. and R. Tonson, 1759.

Condorcet, Marie Jean Antoine Nicolas de Caritat, marquis de. "Eloge de M. Du Hamel." In *Histoire et mémoires de l'Académie royale des sciences: Année 1783*, 131–55. Paris: Imprimerie Royale, 1785.

Dackerman, Susan. "Painted Prints in Germany and the Netherlands." In *Painted Prints: The Revelation of Color in Northern Renaissance and Baroque Engravings, Etchings, and Woodcuts*, ed. Susan Dackerman, 9–47. University Park: Pennsylvania State University Press, 2002.

Daston, Lorraine. "Description by Omission: Nature Enlightened and Obscured." In *Regimes of Description: In the Archive of the Eighteenth Century*, ed. John Bender and Michael Marrinan, 11–24. Stanford: Stanford University Press, 2005.

Daston, Lorraine, and Peter Galison. *Objectivity*. New York: Zone Books, 2007.

Daston, Lorraine, and Katharine Park. *Wonders and the Order of Nature, 1150–1750*. New York: Zone Books, 1998.

Dodart, Denis. *Mémoires pour servir à l'histoire des plantes*. Paris: Imprimerie Royale, 1676.

Eriksson, Gunnar. "Linnaeus the Botanist." In *Linnaeus: The Man and His Work*, ed. Tore Frängsmyr, 63–109. Canton, Mass.: Science History Publications, 1994.

Fienus, Thomas. *De viribus imaginationis*. Louvain: Gerardi Rivii, 1608.

Freud, Sigmund. *Das Unbehagen in der Kultur.* Vienna: Psychoanalytischer Verlag, 1930.

Goldgar, Anne. *Tulipmania: Money, Honor, and Knowledge in the Dutch Golden Age*. Chicago: University of Chicago Press, 2007.

Hartley, David. *Observations on Man, His Frame, His Duty, and His Expectations*. 1749. Facsimile with introduction by Theodore L. Huguelet. 2 vols. Gainesville, Fla.: Scholars' Facsimiles and Reprints, 1966.

Hooke, Robert. "A Contrivance to Make the Picture of Any Thing Appear on a Wall, Cub-Board, or within a Picture-Frame &c. in the Midst of a Light Room in the Day-Time; Or in the Night-Time in Any Room That is Enlightened with a Considerable Number of Candles." *Philosophical Transactions of the Royal Society of London* 3 (1668): 741–43.

Huet, Marie-Hélène. *Monstrous Imagination*. Cambridge: Harvard University Press, 1993.

Humboldt, Alexander von. *Kosmos: Entwurf einer physischen Weltbeschreibung*. 1844. Edited by Ottmar Ette and Oliver Lubrich. Frankfurt am Main: Eichborn, 2004.

Husserl, Edmund. *Die Krisis der europäischen Wissenschaften und die transzendentale Phänomenologie*. The Hague: M. Nijhoff, 1954.

Ivins, Walter, Jr. *Prints and Visual Communication*. 1953. Cambridge: MIT Press, 1992.

Karskens, Michael. "The Development of the Opposition Subjective Versus Objective in the Eighteenth Century." *Archiv für Begriffsgeschichte* 36 (1993): 214–56.

Kemp, Wolfgang. "Disegno: Beiträge zur Geschichte des Begriffs zwischen 1547 und 1607." *Marburger Jahrbuch für Kunstwissenschaft* 19 (1974): 219–40.

Koyré, Alexandre. *From the Closed World to the Infinite Universe*. Baltimore: Johns Hopkins University Press, 1957.

Kusukawa, Sachiko. "Leonhart Fuchs on the Importance of Pictures." *Journal of the History of Ideas* 58 (1997): 403–27.

———. *Picturing the Book of Nature: Image, Text, and Argument in Sixteenth-Century Human Anatomy and Medical Botany*. Chicago: University of Chicago Press, 2012.

Lack, H. W. "Recording Color in Late Eighteenth-Century Botanical Drawings: Sydney Parkinson, Ferdinand Bauer, and Thaddeus Haenke." *Curtis's Botanical Magazine* 6 (1997): 87–100.

Laissus, Yves. "Les Plantes du Roi: Notes sur un grand ouvrage de botanique préparé au XVIIe siècle par l'Académie royale des sciences." *Revue d'Histoire des Sciences* 22 (1969): 193–236.

Landau, David, and Peter Parshall. *The Renaissance Print, 1470–1550*. New Haven: Yale University Press, 1994.

Lefèvre, Wolfgang, Jürgen Renn, and Urs Schöpflin, eds. *The Power of Images in Early Modern Science*. Basel: Birkhäuser, 2003.

Linnaeus, Carl von. *The "Critica botanica" of Linnaeus*. 1737. Translated by Arthur Hort, revised by M. L. Green. London: Ray Society, 1938.

———. "Curiositas naturalis." 1748. In *L'equilibre de la nature*, trans. Bernard Jasmin, ed. Camille Limoges, 125–43. Paris: Vrin, 1972.

———. *Hortus Cliffortianus*. Amsterdam, 1737.

Ludwig, Heidrun. *Nürnberger naturgeschichtliche Malerei im 17. und 18. Jahrhundert*. Marburg an der Lahn: Basilisken-Presse, 1998.

Müller-Tamm, Jutta. *Abstraktion als Einfühlung: Zur Denkfigur der Projektion in Psychophysiologie, Kulturtheorie, Ästhetik, und Literatur der frühen Moderne*. Fribourg: Rombach, 2005.

Müller-Wille, Staffan. "Introduction." In Carl von Linnaeus, *Musa Cliffortiana: Clifford's Banana Plant*, trans. Stephen Freer, 15–67. Ruggell, Lichtenstein: A.R.G. Gantner Verlag, 2007.

Nagel, Thomas. *The View from Nowhere*. New York: Oxford University Press, 1986.

Nickelsen, Kärin. "The Challenge of Colour: Eighteenth-Century Botanists and the Hand-Colouring of Illustrations." *Annals of Science* 63 (2006): 3–23.

Nissen, Claus. *Die botanische Buchillustration: Ihre Geschichte und Bibliographie*. 2nd ed. 3 vols. Stuttgart: Hiersemann, 1966.

Nochlin, Linda. *Realism*. Harmondsworth: Penguin, 1971.

Ogilvie, Brian. "Image and Text in Natural History, 1500–1700." In *The Power of Images in Early Modern Science*, ed. Wolfgang Lefèvre, Jürgen Renn, and Urs Schöpflin, 141–80. Basel: Birkhäuser, 2003.

———. *The Science of Describing: Natural History in Renaissance Europe.* Chicago: University of Chicago Press, 2006.

Panofsky, Erwin. *The Life and Work of Albrecht Dürer.* Princeton: Princeton University Press, 1943.

Park, Katharine, and Lorraine Daston. "Introduction: The Age of the New." In *The Cambridge History of Early Modern Science*, ed. Katharine Park and Lorraine Daston, 1–17. New York: Cambridge University Press, 2006.

Perrault, Claude. *Mémoires pour server à l'histoire naturelle des animaux.* Paris: Imprimerie Royale, 1671.

Pinnault, Madeleine. *Le peintre et l'histoire naturelle.* Paris: Flammarion, 1990.

Pomata, Gianna, and Nancy G. Siraisi. *Historia: Empiricism and Erudition in Early Modern Europe.* Cambridge: MIT Press, 2005.

Pope, Alexander. *The New Dunciad: As It Was Found in the Year 1741.* London: T. Cooper, 1742.

Raven, C. E. *John Ray: Naturalist.* 1950. Cambridge: Cambridge University Press, 1986.

Réaumur, René Antoine Ferchault de. *Mémoires pour servir à l'histoire des insectes.* 6 vols. Paris: Imprimerie Royale, 1734–42.

Reeds, Karen. "Renaissance Humanism and Botany." *Annals of Science* 33 (1976): 519–42.

Rendle, A. B. "George Clifford's Herbarium, and the *Hortus Cliffortianus.*" *Journal of Botany* 61 (1923): 114–16.

Rytz, Walther. *Pflanzenaquarelle des Hans Weiditz aus dem Jahre 1529.* Bern: Verlag Paul Haupt, 1936.

Sanders, Gill. *Picturing Plants: An Analytical History of Botanical Illustration.* Berkeley: University of California Press, 1995.

Schnapper, Antoine. *Histoire et histoire naturelle.* Vol. 1 of *Le géant, la licorne et la tulipe: Collections et collectionneurs dans la France au XVIIe siècle.* Paris: Flammarion, 1988.

Segal, Sam. "Maria Sibylla Merian als Blumenmalerin." In *Maria Sibylla Merian (1647–1717): Künstlerin und Naturforscherin*, ed. Kurt Wettengl, 68–87. Frankfurt am Main: Verlag Gerd Hatje, 1997.

Senebier, Jean. *L'art d'observer.* 2 vols. Geneva: Chez Cl. Philibert et Bart Chirol, 1775.

Stroup, Alice. *A Company of Scientists: Botany, Patronage, and Community at the Seventeenth-Century Parisian Royal Academy of Sciences.* Berkeley: University of California Press, 1990.

Torlais, Jean. "Réaumur et l'histoire des abeilles." *Revue d'Histoire des Sciences* 11 (1958): 51–67.

Tournefort, Joseph Pitton de. *Relation d'un voyage du Levant.* Lyon: Anisson et Posuel, 1717.

Treviranus, Ludolph. *Die Anwendung des Holzschnittes zur bildlichen Darstellung von Pflanzen.* 1855. Utrecht: W. de Haan, 1949.

Turner, William. *A New Herball.* 1551. Edited by George T. L. Chapman and Marilyn N. Tweddle. 2 vols. Cambridge: Cambridge University Press, 1989.

van Helden, Albert. "The Telescope in the Seventeenth Century." *Isis* 65 (1974): 38–58.

Vezin, Luc. *Les artistes au Jardin du Roi.* Paris: Éditions Herscher, 1990.

Voyage de Tournefort. MS 995, Bibliothèque centrale, Muséum national d'histoire naturelle, Paris.

Weber, Max. "Wissenschaft als Beruf." 1917/1919. In *Max Weber Gesamtausgabe*, Abteilung I, *Schriften und Reden*, ed. Wolfgang J. Mommsen and Wolfgang Schluchter, with Birgit Morgenbrod, 24 vols., 17:71–111. Tübingen: J. C. B. Mohr, 1992.

Wilson, Catherine. *The Invisible World: Early Modern Philosophy and the Invention of the Microscope.* Princeton: Princeton University Press, 1995.

2.

Drawing as an Instrument of Knowledge

The Case of Conrad Gessner

SACHIKO KUSUKAWA

Picturing, like many other tools of visualization, such as mapping, modeling, tabulating, and drawing diagrams, is a complex, selective, active process of making something visible, and it can take a variety of forms—a sketch, an oil painting, a print, and so on. In this chapter, I focus on the versatility of a particular kind of drawing—a line sketch with watercolor on a sheet of paper—as an instrument of knowledge.[1] I discuss the case of Conrad Gessner (1516–1565), who left about fifteen hundred drawings of plants for his unfinished work, *Historia plantarum*.[2] Indeed, the recent discovery of the original watercolors for his *Historia animalium* leaves no doubt that drawings were an integral part of Gessner's study of the natural world as a whole.[3] He was, of course, not alone in collecting images for study: other contemporaries, such as Ulisse Aldrovandi (1522–1605) and Anselmus de Boodt (1550–1632),

An earlier version of this chapter was presented in 2008 and became the basis of chapter 7 of my *Picturing the Book of Nature: Image, Text, and Argument in Sixteenth-Century Human Anatomy and Medical Botany* (Chicago: University of Chicago Press, 2012). This chapter has been revised to address the key themes of this volume and incorporates scholarship that has appeared since I completed that book.

also amassed drawings of *naturalia*.[4] Gessner's drawings are particularly useful in understanding how pictures of plants in diverse states and of different provenance were pressed into the service of establishing a *scientia* of plants.

Managing Correspondents and Artists

Like many other contemporary physicians, Gessner understood the study of plants as primarily medicinal—plants were part of the naturally occurring materials for medicine, the *materia medica*.[5] And as with his other publishing projects, such as *Bibliotheca universalis* or *Historia animalium*, he wanted to create something comprehensive.[6] In order to be comprehensive, Gessner mobilized his extensive network of correspondents, but he needed to ensure that he did not end up with the commonest plants from all over Europe.[7] Gessner frequently reminded his correspondents that he only wanted pictures of plants that were rare or unknown to him, or whose pictures were not yet available in printed books.[8] He usually requested a catalogue or list of plant names in order to determine whether his correspondents had plants that he did not.[9] When they did, he would request that the plant itself be sent, preferably with roots intact (and placed in a wooden box with a bit of earth) so that he could grow the plant himself in his garden and have its picture drawn when it grew.[10] If a plant or seed was not available, Gessner requested a picture. He thus hoped to obtain from his correspondents either the rare plant itself or a drawing of it.

In return, Gessner offered his correspondents plants, seeds, pictures, books, and the chance to have a plant named after them. He proposed to name a plant unknown to antiquity after its "discoverer," not in the sense of the first person ever to find it, but the first person to communicate its existence to Gessner. Thus he offered to name the *Schelmenkrut*, a species of *Gentiana*, after Benedikt Marti Aretius, because Aretius was the first to tell Gessner its German name and explain its effectiveness against pestilence.[11] Gessner also suggested to Joachim Camerarius the Younger that he should pick a plant unknown to the ancients, and Gessner would name it after him.[12] Gessner's notes on his pictures—for example, "this plant first found by M. Jacob Baumann"—could then be used in the future for naming plants after people.[13] Naming plants after people was part of a gift exchange by which Gessner hoped to give his correspondents an incentive to send him a rare plant.

Controlling the quality of pictures his correspondents sent in was a tricky task. Gessner was disappointed that some of the pictures he received were of fictitious plants or had not been made "ad vivum."[14] Improving the quality of drawings sent to him implied that graphic craftsmen had to be employed, and Gessner was prepared to bear the costs.[15] Gessner wrote to Jean Bauhin that he was prepared to pay four *plapparts* for each picture of a plant, but asked Bauhin to be in touch with him before commissioning a picture so that he could determine whether the plant really needed to be depicted.[16] Johannes Fabricius Montanus, the pastor in Chur, had similarly hired a painter, but Gessner was not impressed with that painter's skills. Gessner complained that his own painters could draw a dried plant better and more vividly than Fabricius's painter could draw a fresh one; thus Gessner requested that Fabricius send a fresh plant or a dried one with its flowers and leaves pressed in a book rather than use that painter. Eight months later, Gessner had to repeat more bluntly that he would rather save the cost of paying the painter.[17]

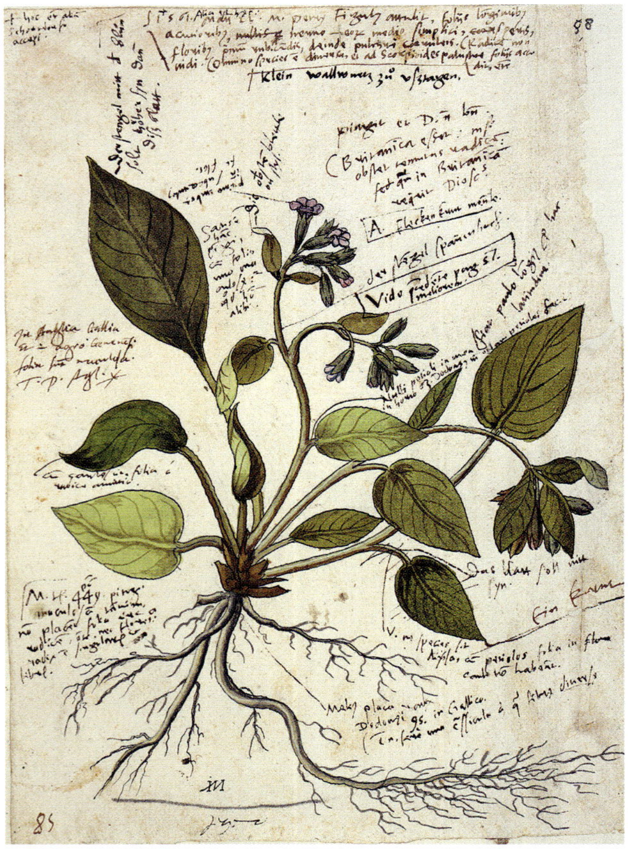

2.1 Watercolor of *pulmonaria gallica* by Jos Murer, with annotations and corrections by Gessner. University Library of Erlangen, MS 2386, fol. 58r.

Drawings were also made of those plants that were sent to Gessner in a box or grown in his garden from seeds he had been given. These were done by a number of artists, and he recorded payment to them on the drawings as a job lot or by the hour.[18] It has been difficult to identify the artists who worked for Gessner on the plant drawings, but one painter we can be sure of is Jos Murer (1530–1580), who was also a glass painter and a map engraver.[19] But painters posed problems for Gessner, too. On Murer's drawing of the *pulmonaria gallica*[20] (fig. 2.1), Gessner wrote, "This leaf shouldn't be [here]"; "The stem with the flower should be higher than this leaf"; "[The leaf] should be somewhat longer than its width."[21] Although Gessner paid for them, painters' drawings did not always show the details he wanted to see, and he had to correct their drawings.[22] Josias Simler remembered that Gessner was constantly present when the painter was drawing pictures of plants, assiduously preventing the artist from displaying his own virtuosity and ensuring instead that he imitated nature as closely as possible.[23] Here is a division of labor that Daston and Galison have dubbed "four-eyed sight"—in which the naturalist had to rely on the artist to picture natural objects but also had to supervise him as closely as possible—a practice that continued well into the eighteenth century.[24] The corrections noted on the drawings are traces of how Gessner sought to align the artist's eye with his own.

Gessner's own pictures of plants, most dating from the last three years of his life, suggest a convergence of his style and his artists'—they drew the entire plant in clear outlines, partially colored with watercolor so as not to blot out the outlines. Gessner was particularly keen on capturing the lines and shapes that made up the external appearance of an object.[25] He asked others to pay attention, too, as he impressed on Theodor Zwinger the Elder that the flower of the *Populus alba* must be drawn "in its color and with all its lines."[26] The emphasis on clear outlines may well have been one reason why Gessner used Jos Murer, a glass painter who was required to draw well-defined continuous outlines.[27] Gessner's insistence on clear lines was underpinned by his belief, expressed toward the end of his life, that God had impressed shapes and outlines onto nature like hieroglyphic marks.[28] It was thus important for Gessner to align the style of drawing by his artists and correspondents.

Icon absoluta

Gessner, furthermore, was looking for a particular kind of picture that captured a particular kind of object. He often requested a "complete" or "perfect" picture of a plant. Gessner asked Adolf Occo III for a complete picture (*iconem absolutam*) of Theophrastus's *coccigrya*—Gessner already had its dried branch with leaves, fruit, and fluffy seeds, but he wanted it to be more complete (*integriorem*) with its flowers. He also asked for a "perfect picture" (*picturam perfectam*) of the *sagapenum*.[29] A complete picture would thus record all parts of a plant: root, stem, shoots, leaves, flowers, fruits, and seedpods. Such a picture would show no trace of individual blemishes, no torn leaf or insect hole. Moreover, Gessner's *dianthus* showed two stages of the flower at once—blooming in red and withering in white (fig. 2.2). The picture of the *leucoia* (fig. 2.3) showed five varieties on a single bush—flowers in pink, gray, red, and white, all with four petals, and one with white flowers in five petals. An *icon absoluta* thus depicted an object in its entirety in form, growth, and variety, which was a pictorial convention deployed earlier by Leonhart Fuchs.[30]

This quest for an *icon absoluta* meant that Gessner's drawings of plants had to be completed over time, often across several years: in the case of a *parvus verbascum*, the violet flowers that grew in his garden were drawn in 1562 and its roots in 1564, probably when he was prepared to uproot the plant.[31] This also meant that a plant had to be put together from different states with different provenances: the plant called *trageriganus* was drawn partly

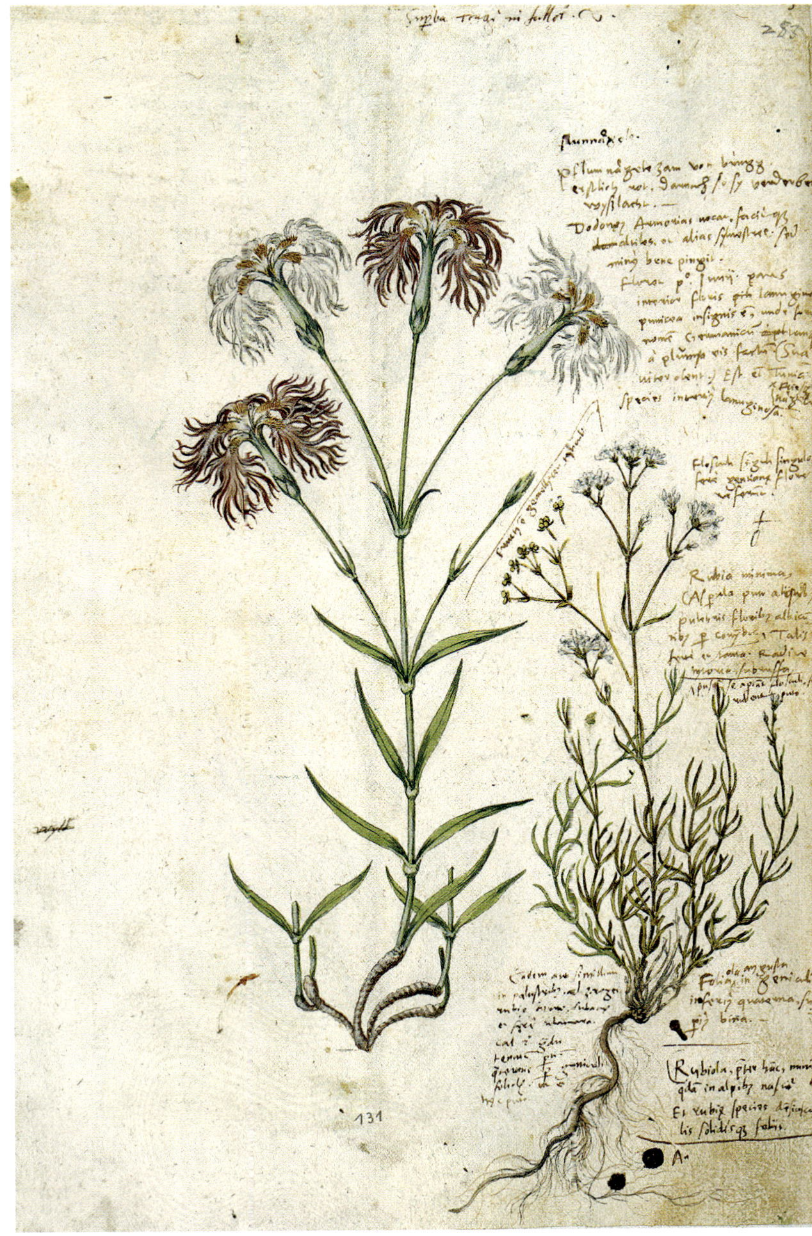

2.2 Watercolor of *dianthus*, with annotations by Gessner. University Library of Erlangen, MS 2386, fol. 283r.

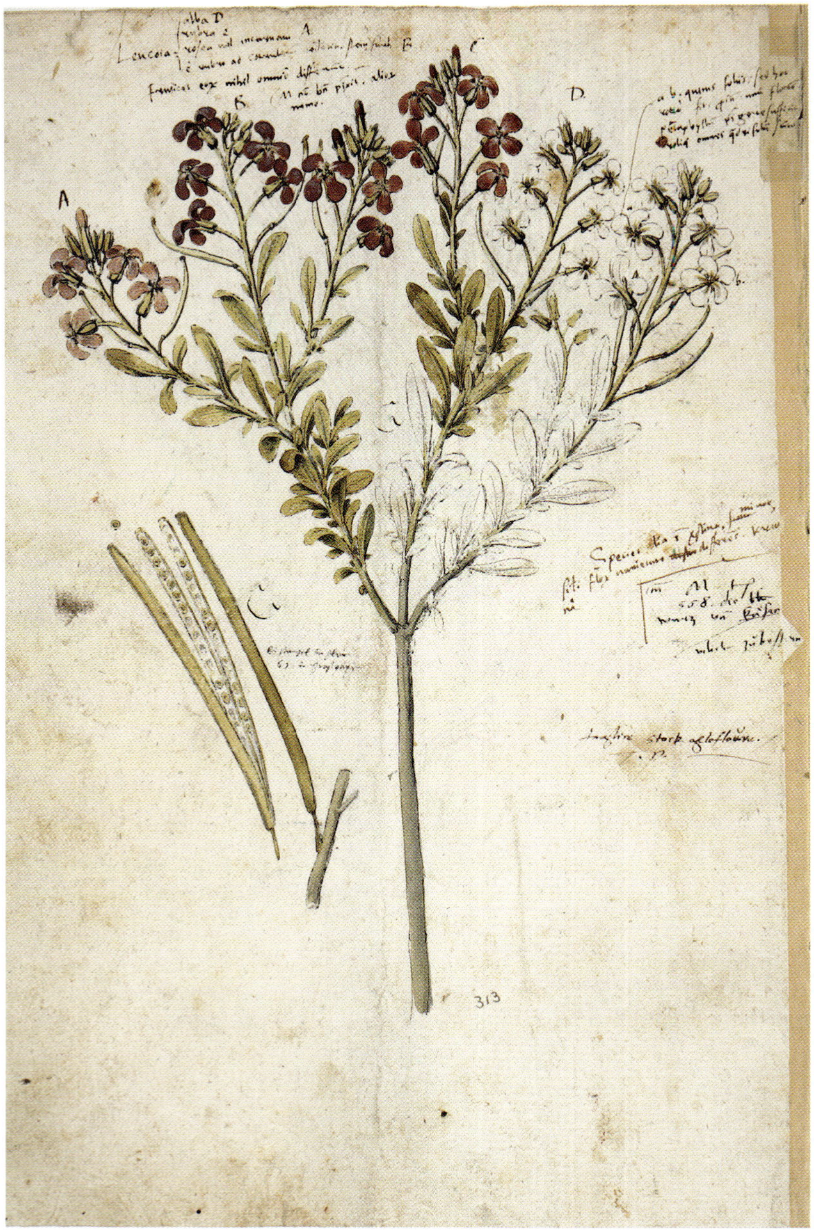

2.3 Watercolor of *leucoia,* with annotations by Gessner. University Library of Erlangen, MS 2386, fol. 377v.

from a dried plant that Giacomo Antonio Cortuso sent from Padua, and partly from a growing plant that had sprouted in Gessner's garden the previous year, but he didn't have its flower yet.[32] A "complete" plant thus had to be put together from a variety of sources and states over several years.

Time could be also stopped on a drawing. Gessner kept dried specimens in a *herbarium*, glued onto large paper, and his drawings contain references to the *herbarium*.[33] Gessner noted often that dried specimens did not retain the color of the fresh plant very well. The *anthora* was "dried and wrinkled and brown, but I think [its color] is so when fresh"; for the *Rote Fluhblümlein*, "the dried flowers are so [i.e., dark purple]. The fresh ones turn red beautifully"; "Tobias Egli sent me blue flowers of [the *Feuerkraut*], dried, and they have probably changed into such a color in dryness."[34] He was also quite specific about the right shade of color, as when he noted that the color of the flower of a *Gentianella* was that of a liquid pigment called "Lettmaß Blau."[35] Gessner thus added notes next to the pictures of plants in order to retain and fix the vivid but all too transient colors and shapes of fresh plants.

Jottings on Gessner's drawings also recorded how local climate and soil affected the variety of species available. He observed that the garden of Joachim Kreich (d. 1575), court apothecary at Torgau, had many rare things that could not be cultivated in Zurich, which was too close to the snowy mountains.[36] The Fugger gardens in Augsburg grew several exotic plants that Gessner had not seen before.[37] He also compared notes with other residents of Zurich: the *Lilium croceum* flowered in the garden of the surgeon and stonecutter Petrus Figulus on 12 June 1564, while it flowered on 26 May in his own garden, although it was cold and raining.[38] The *Scorzonera* grew with natural loam in a dry place and flowered in a sunny spot in the garden of Jacob Baumann, a surgeon who also lived in Zurich, while Gessner himself could not get the plant to grow.[39] Pictures of plants enabled Gessner to overcome local conditions that hampered the growth of certain plants.

Drawings were thus the primary means through which Gessner could gather the fragments of knowledge about plants that he or his friends and acquaintances possessed. These could be in the form of fresh, dried, or pictured plants, and were brought together on paper over time and across space and made into an *icon absoluta.* The *icon absoluta* embodied an entire plant, in all its stages and varieties; it fixed the colors of plants beyond the life span of any individual sample, and was not affected by local conditions such as weather or soil. Though drawn carefully and naturalistically, the depicted plant became something that could not be pinned down to a particular time or place. Such a generalized object of study was one way in which physicians could claim the higher status of *scientia* for a *historia* of plants.[40]

Identifying an Unknown Plant

Gessner frequently annotated his drawings, and his notes indicate the process by which he came to "know" a plant. The plant pictured in fig. 2.4 was first recorded as unknown to him—*herba mihi ignota*—grown from a seed sent to him from either France or Italy in 1554.[41] A morphological description of it followed, similar to his descriptions of a combination of known plants:

> Leaves like the *lapathum,* soft, succulent, somewhat sharp—also pungent, almost like the *noli me tangere* as it probably has some poison. The first taste is somewhat sharp—(see if it is *zinziber caninum*) and smooth. The stem is rather round, though not without some protruding edges, green, moderately hairy, fatty (as is also the flower-cup, such that fingers stick to it because of its stickiness), a foot or higher, solid, but the marrow inside is spongy. The flowers are sallow or of a diluted yellow color (or in shape and division similar to the flowers of the *arthritica*), the pretty petals enclosed almost to its middle part by green cups are detachable, and when they fall off, the seedpods grow out of it and are round-ish like the *avellana,* in which seeds [are contained].[42]

Gessner further recorded that this plant flowered in July in his garden and went to seed in August, that its root was useless, and that the whole plant was soft and limp, had no smell, and grew to three cubits in a rich soil. The seeds yielded no heat, despite the fact that he chewed on them frequently and for a long time.[43] This was a somewhat unnerving habit of Gessner's. Normally, in this period, possible poisons and antidotes were tried out on animals.[44] Gessner was unusual for his time in that he experimented on himself—he is known to have tasted parts of some noxious plants (e.g., *Astrantia nigra, Eupatorium aquaticum, Polygala amara, Hellebore*) and applied medicines to his own body.[45] The drawing recorded Gessner's full experience of the plant, including its taste, smell, and texture.

Gessner then gave the unknown plant a temporary name, presumably because of the similarity of leaf shape and its possible poisonous quality, the *lapathum apocynon* (literally, "sorrel poisonous against dogs"). He then wrote, "It is *priapeia* to Rondelet" ("Priapeia Rondeletio").[46] Though we appear not to have Rondelet's own explanation of the properties of a plant by the name of *priapeia*, it is likely that Gessner had acquired this information directly from Rondelet—he knew Rondelet from the time he had spent at Montpellier at the end of 1540 and had subsequently corresponded with him.[47] Gessner also wrote, "Is it a plant related to Thevet's petun?"[48] André Thévet's (1502–1590) *Les singularitez de la France antarctique* (1558) described an herb similar to the bugloss and called *petun* by New World inhabitants, who rolled up its leaves, set it alight, and inhaled its fumes to cure various ailments.[49] In his copy of Thévet's *Singularitez*, Gessner wrote a similar question on the page on the *petun:* "Is it the *priapeia* of certain people?"[50] Gessner added to his drawing of the "unknown plant," "Dodoens has this too."[51] In his copy of Dodoens's *Histoire des plantes* (1557), next to a picture of a plant called *hyoscyamus luteus*, or the *jusquiame iaulne*, Gessner wrote, "Priapeia of others . . . Baumann said that in Augsburg it is also called the *lunaria*."[52] It must be after this point that Gessner had struck out "herba ignota mihi" and written over it "hyoscyamus luteus," and also deleted the temporary name of *lapathum apocynon*. A little while later, Gessner went back to the drawing and wrote, "look at Mattioli, 460b."[53] The 1563 edition of Mattioli's commentary on the *De materia medica* included a woodcut of this plant on the verso of folio 460 as *hyosciamus III*.[54] We now call Gessner's "unknown plant" tobacco, *Nicotiana rustica*.[55]

Thus, from about 1554 to 1563, Gessner tried to make sense of this new, unknown plant through the printed books of other learned authors, a tendency common in Renaissance scholars and described by Gianna Pomata and Nancy Siraisi as "learned empiricism."[56] The textual notes added to the picture of this "unknown" plant recorded Gessner's sensory experience of the plant as well as his learned empiricism. It would be anachronistic to separate the textual and pictorial elements when discussing Gessner's drawings, as both image and text were necessary for the identification of an "unknown" part. Looking and reading are, after all, both works of the eye, and both were required for establishing knowledge about a plant.

Pictorial Commonplaces

Even for plants better known than the tobacco plant, Gessner's drawings were accompanied by similar "research questions." Next to such questions he would later write down his answers, or cross out the original question: "See if Fuchs has this—he does, but ours is another species"; "See if it is the *scorodoprason*—this is bigger"; "Is this *pulmonaria* to some people?"—crossed out (no).[57] These questions, directed to himself, were usually answered with reference to the description and pictures in other printed books, such as Fuchs's *De historia stirpium*, Dodoens's *Histoire des plantes*, Mattioli's *Commentarii in libros Dioscoridis*, and Hieronymus Tragus's *De stirpium historia*, in addition to the standard works of the classical authors Pliny the Elder, Theophrastus, and Dioscorides. In his own copies of these books, Gessner wrote down further queries and made reference to his drawings and to other books.[58] This web of cross-references indicates how Gessner's drawings were integrated into his study of books, which is perhaps not surprising for the compiler of the *Bibliotheca universalis*. In order to compose the universal history of plants, Gessner tried to read all the books about plants by ancient and modern authors—by the end of his life he claimed to have incorporated 260 books into his study of plants.[59] Yet it is worth noting that the drawings were a focal point in Gessner's study of plants. The research questions started from the drawing, and the answers—found elsewhere—were entered back onto the drawing.

The drawings were made on loose sheets of paper and distributed on two hundred shelves.[60] This allowed ready access to each picture that Gessner would study over a number of years, which were meant to serve as the building blocks of his publication. Earlier, in the second volume of his *Bibliotheca universalis*, Gessner had explained his method of compiling an index, by excerpting texts, cutting them out of books, sorting the strips of text by general themes and then by more specialized topics, and placing them in boxes. This meant that they could be readily brought together later on, and not just for compiling an index:

> I know many learned men who are pleased to apply this convenience to almost all their studies; whether it is something to be written or something to be taught orally in public, they collect the subject matter of their discourse roughly and arrange it in the following manner. Both material that was recently compiled, and that which had long since been acquired is being prepared for use on separate slips (not mounted) so that when needed for whatever subject is to be treated, they can produce them and select from many slips those that serve best for the present purpose; the slips are fastened together with small pins, and are then arranged in whatever useful order one wishes to have them for the purpose of a discourse; what seems to be appropriate is noted down or left out at will, the slips then being put back again in their place.[61]

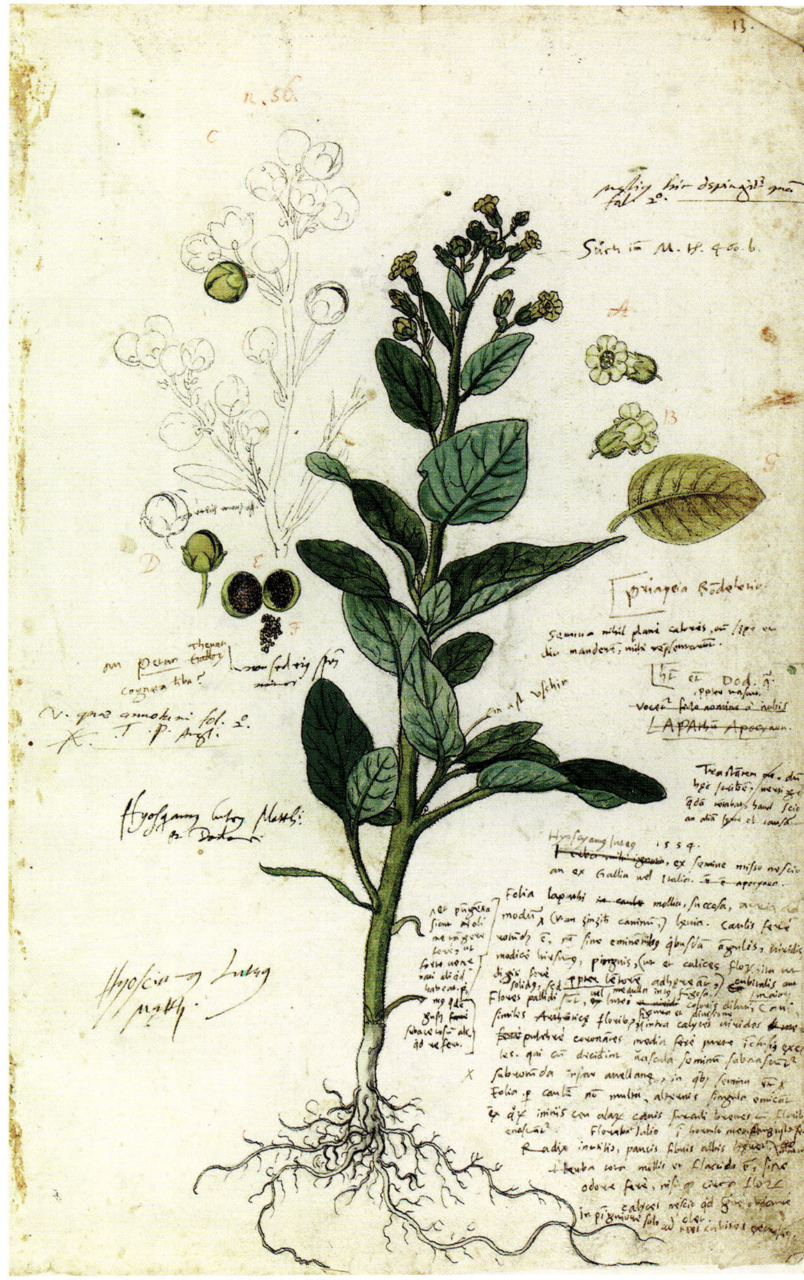

2.4 Watercolor of the tobacco plant, with study notes by Gessner. University Library of Erlangen, MS 2386, fol. 13r.

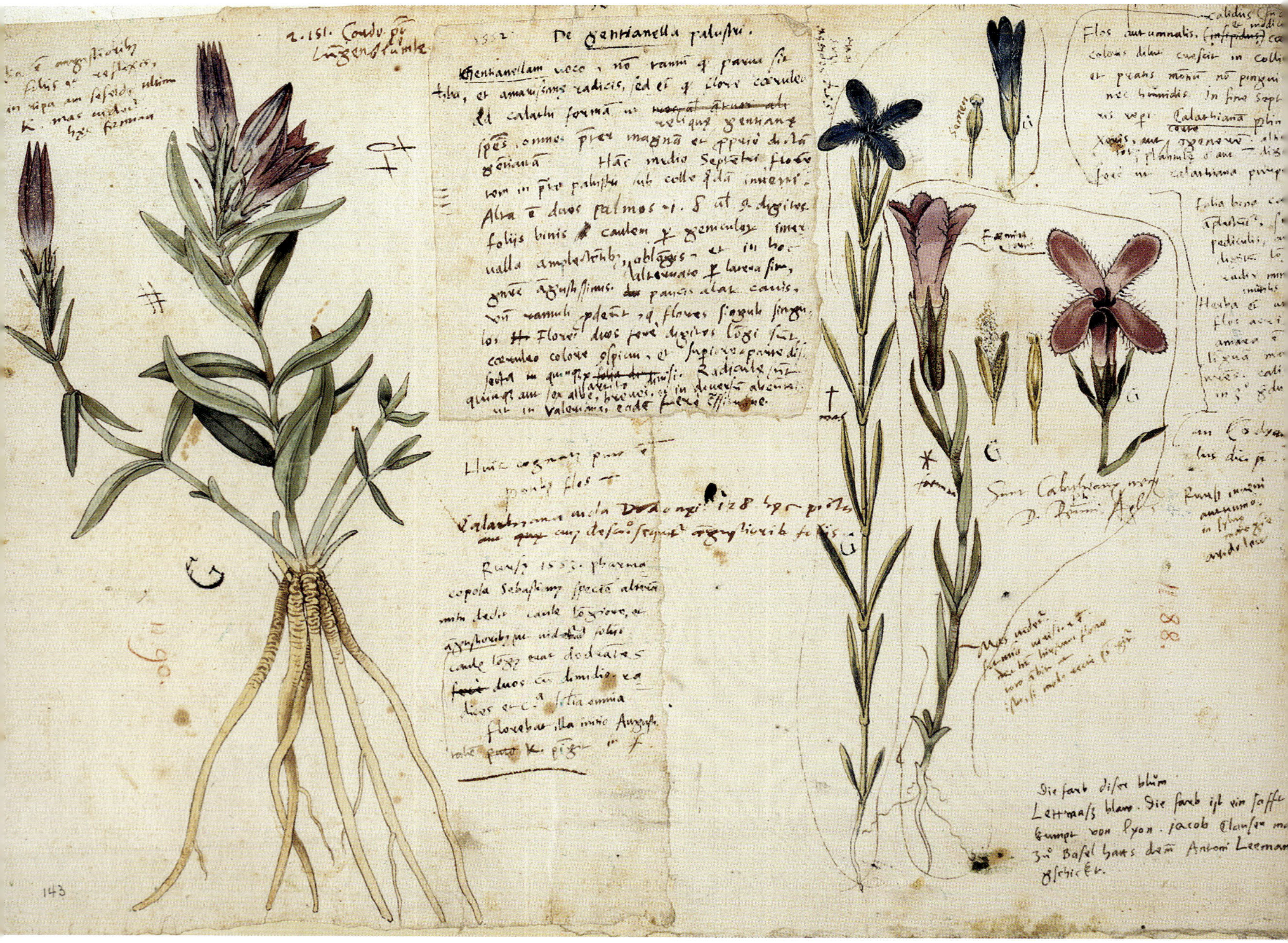

2.5 Assembly of notes and drawings of *Gentianella.* University Library of Erlangen, MS 2386, fol. 143v.

It was important to Gessner that this form of excerpting and ordering in order to compile an index could also be used for purposes of composition. Indeed, he regarded this method as Erasmus's way of compiling commonplaces.[62] As noted recently by Ann Blair, Gessner practiced this method, using slips of paper to work and rework his compositions. This method of quite literally cutting and pasting excerpts from printed pages and handwritten notes became an effective way of parsing and collating information for reference books in this period.[63]

Gessner applied his cut-and-paste method to his drawings as well. A drawing of *Gentianella* (fig. 2.5) made in 1553 and another from 1564 were glued together with a description from 1552.[64] We also know that he cut out or copied woodcuts in printed books and pasted some of them onto another drawing.[65] We do not know for sure what format Gessner intended for his *Historia plantarum*, but we can infer the role of drawings for publication from his edition of Valerius Cordus's *Annotationes* on Dioscorides's *De materia medica* (1561). The woodcut of *chamaenerion* in this work is copied from one of Gessner's drawings of the plant; its morphological descriptions are copied from two other sheets; and further descriptions were added from elsewhere.[66] One entry was thus made up of information from several of his drawings. These drawings could and did function as pictorial commonplaces from which Gessner hoped to compile his own book on the universal history of plants.

Conclusion

Gessner's drawings record the process by which fragments of knowledge of diverse origins became components of a learned, universal history of plants. Through the patient coaxing of correspondents and the constant supervision of artists, with sensory experience and learned empiricism, the depicted plant became visible as a "complete" object with its identity embedded in the learned world of books. Such an object was necessary for a study that aspired not just to comprehensive coverage but also to general claims about medicinal efficacy that would be valid for all members of a given species, wherever they grew. In turn, these pictures were also envisaged, in the form of commonplaces, as the basis for compiling a learned, scholarly book on the universal history of plants. We may note that Gessner had to start with incomplete and fragmented pieces of knowledge in different states—these were seen by many eyes and drawn by different hands. Gessner had to align the many visions and the different styles as well as overcome particularities and variable conditions in order to establish an object worthy of *scientia*. It is on the drawing sheets that much of this work took place, and those very sheets also became the building blocks for the printed book that would transmit this universal history. In the sense that drawings of disparate origin and provenance were refined and transformed into resources that aided in the formation and transmission of the "science of describing," Gessner's drawings were an important tool indeed for knowledge.

Notes

1. On the effectiveness of a drawing on paper as an instrument of knowledge, see Latour, "Drawing Things Together."

2. A selection of the botanical drawings surviving at the University of Erlangen Library was transcribed with botanical and artistic commentary in Gessner, *Historia plantarum: Faksimilieausgabe* (hereafter *HP*, followed by volume number); the full collection is listed in Gessner, *Historia plantarum: Gesamtausgabe*. This chapter is indebted to the transcription, translation, and commentary in these works.

3. See Egmond, "Collection Within a Collection."

4. Aldrovandi's picture albums are now available at http://www.filosofia.unibo.it/aldrovandi/. For a comparison of Aldrovandi and Gessner, see Fischel, *Natur im Bild*. For de Boodt, see Maselis, Balis, and Marijnissen, *Albums of Anselmus De Boodt*.

5. For the study of plants in Renaissance Europe, see Reeds, *Botany in Medieval and Renaissance Universities*; Palmer, "Medical Botany in Northern Italy"; and Ogilvie, *Science of Describing*.

6. For Gessner's publications, see Wellisch, "Conrad Gessner: A Bio-Bibliography."

7. For a study of Gessner's letters, see Dürling, "Conrad Gessner's Briefwechsel"; and Delisle, "Letter: Private Text or Public Place?"

8. Gessner, *Epistolarum medicinalium*, fol. 122v.

9. Ibid., fols. 10r–11r; Bauhin, *De plantis*, 114. Some of these catalogues are noted in Cordus, *Annotationes in Pedacii Dioscorides*, fol. 243r–v.

10. Gessner, *Epistolarum medicinalium*, fol. 107r–v.

11. Ibid., fol. 120v.

12. Rath, "Briefe Konrad Geßners," 152.

13. Gessner, *HP* 3, table 23: "herba inventa primum a M. Iacobo Baumanno." All translations are my own unless otherwise noted.

14. Bauhin, *De plantis*, 101; Gessner, *Epistolarum medicinalium*, fol. 64v. For the concept of "ad vivum," see Kusukawa, "Conrad Gessner on an 'ad vivum' Image."

15. For example, he asked Camerarius to jot down the cost on the drawings. Rath, "Briefe Konrad Geßners," 164.

16. Bauhin, *De plantis*, 123.

17. Gessner, *Epistolarum medicinalium*, fol. 89r–v. See also an earlier letter to the same, fols. 88v–89r.

18. For the artists who worked for Gessner, see Nissen, *Botanische Buchillustration*, 58–59. For records of payment, see *HP* 5, table 9; *HP* 2, tables 12, 17, 19; *HP* 3, table 3.

19. For Murer, see Anderes, "Glasmalerei in reformierten Zürich," 18; Dürst, "Map-View of Zurich." Murer also drew Gessner's portrait in the third volume of the *Historia animalium* (1555).

20. I have used the historical plant names given by Gessner and his contemporaries; they are not the same as the post-Linnean classification names, and in several cases, their identification is difficult to establish securely.

21. "Das blatt soll nitt syn"; "Der stengel mitt der blùm solt hoeher syn diß blatt"; "Fiat Paulo longius pro hac latitudine." *HP* 4, table 3.

22. *HP* 6, table 10.

23. Simler, *Vita C. Gesneri*, fol. 15r.

24. Daston and Galison, *Objectivity*, 84–98.

25. For more on Gessner's interest in outlines and color, see Kusukawa, "Conrad Gessner on an 'ad vivum' Image."

26. "Populus alba cum florebit apud vos, quod forte iam fit, cura quaeso, ut flos mihi pingatur suo colore et lineamentis omnibus bene expressus, et florem unum chartula involutum mitte." Gessner to Zwinger, 22 March 1563, in Gessner, *Epistolarum medicinalium*, fol. 111r.

27. For drawing clear outlines as a skill required of glass painters and woodblock cutters, see Butts, "Albrecht Dürer and the Modernization of Stained Glass," 342.

28. See Kusukawa, "Conrad Gessner on an 'ad vivum' Image."

29. Gessner, *Epistolarum medicinalium*, fol. 65r.

30. See Kusukawa, *Picturing the Book of Nature*, 114–19.

31. *HP* 6, table 12.

32. Bauhin, *De plantis*, 123.

33. Gessner's *herbarium* appears not to have survived.

34. "Sicca et corugata. Prun est. sed recens puto sic est." *HP* 6, table 22. "Flores aridi tales sunt [dark purple]. Recentes vero pulchre rubent." *HP* 5, table 15 (my interpolation). "D. Thobias Egli misit flores subcoeruleos huius siccos; et forte siccitate mutant ita colorem." *HP* 7, table 10.

35. *HP* 4, table 18.

36. Gessner, *Epistolarum medicinalium . . . liber quartus*, B4r.

37. For images from the Fugger gardens, see *HP* 6, tables 3–8.

38. *HP* 7, table 15.

39. *HP* 5, table 1.

40. For the status of *historia* as a discipline, see the essays in Pomata, Gianna, and Siraisi, *Historia*.

41. "Herba mihi ignota, 1554 ex semine misso nescio an Gallia vel Italia." *HP* 1, table 17.

42. "Folia lapathi mollia, succosa, acria admodum—et pungentia sicut Noli me tangere fere, ut forte venenati aliquid habeat. Primus quidem gustus subacetosum aliquid refert—(vide an zinziber caninum), laevia. Caulis fere rotundus est, non sine eminentibus quibusdam angulis, viridis, modice hirsutus, pinguis (ut calices florum, ita digiti fere propter lentorem adhaereant), cubitalis aut maior, solidus, sed medulla intus fungosa. Flores pallidi sunt vel lutei coloris diluti, (aut similes Arthriticae floribus figura et divisione), intra calyces virides pulchre coronantes media fere parte inclusi exemptiles, qui cum deciderint, vascula seminum subnascuntur subrotunda instar anellanae, in quibus semina etc." Ibid. (my interpolation in translation).

43. Ibid.

44. Palmer, "Medical Botany in Northern Italy," 155.

45. The examples are noted in Dürling, "Conrad Gessners Briefwechsel," 106–7. See also Simler, *Vita C. Gesneri*, fol. 14v.

46. *HP* 1, table 17.

47. For Rondelet's teaching at Montpellier, see Lewis, "Clusius in Montpellier," 70–94. For the explanation of the name *priapeia* by Leonhart Fuchs (who obtained the seeds from Gessner), see Meyer, Trueblood, and Heller, *Great Herbal of Leonhart Fuchs*, 1:631.

48. "An Petun Theveti cognata herba?" *HP* 1, table 17.

49. Cf. Thévet, *Singularitez de la France antarctique*, fol. 59r.

50. "V[ide] an priapeia quorundam?" Basel University Library, Hx VI 30. Noted also in *HP* 1, 89. Gessner owned the Paris and Antwerp editions of Thévet's *Singularitez;* only the Antwerp copy survives. See Leu, Keller, and Weidmann, *Conrad Gessner's Private Library*, nos. 361–62.

51. "Habet etiam Dodonaeus." *HP* 1, table 17. For Gessner's copies of Dodoens's herbal, see Leu, Keller, and Weidmann, *Conrad Gessner's Private Library*, nos. 115–16.

52. "Priapeia aliorum. . . . Bummanus dicebat Augustae etiam Lunariam vocari." Annotation in Gessner's copy of Dodoens, *Histoire des plantes*, 305. Also noted in *HP* 1, 89. For Dodoens, see Ockenden, "Note on Hyoscyamus Luteus," 273–74. The comparison between the tobacco plant and henbane goes back to Oviedo (1535); see Brooks, *Tobacco*, 1:204.

53. "Such imm Matthiolo 460b." *HP* 1, table 17. Gessner's copy of the 1563 edition appears to be lost. Leu, Keller, and Weidmann, *Conrad Gessner's Private Library*, no. A41.

54. *HP* 1, 89.

55. For Gessner's study of New World flora and fauna, see Leu, "Konrad Gessner und die neue Welt."

56. Siraisi and Pomata, *Historia*, 17. Simler estimates that Gessner identified at least fifty plants unknown to the ancients. Simler, *Vita C. Gesneri*, fol. 15r.

57. "V. ne Fuchsius habebat. Habet 86 sed alia species noster." *HP* 3, table 6. "V. an scorodoprason—hoc maius est." *HP* 2, table 10. "[An pulmonaria quibusdam?]" *HP* 6, table 14.

58. Not all of his library has survived, but see Leu, Keller, and Weidmann, *Conrad Gessner's Private Library*, for instance, nos. 115–16, 241, 294, 322, 359–60.

59. Salzmann, "Francesco Calzolari," 102.

60. Leemann van Elck, *Buchschmuck*, 24.

61. "De indicibus librorum." Gessner, *Pandectarum sive partitionum*, fols. 19v–20v, translated in Wellisch, "How to Make an Index," 12.

62. Gessner, *Pandectarum sive partitionum*, fol. 24r. For Erasmus, see Moss, *Printed Commonplace-Books*, 111–15.

63. Blair, *Too Much to Know*, 173–229, esp. 213–25.

64. *HP* 4, table 18.

65. See, for example, *HP* 3, tables 17–18.

66. Cordus, *Annotationes in Pedacii Dioscorides*, fols. 213v–216v. See the commentary at *HP* 1, 90, 96 (commentary on *HP* 1, table 26); *HP* 4, table 5; *HP* 4, 71–74.

Bibliography

Primary Sources

Bauhin, Jean. *De plantis a divinis sanctisve nomen habentibus additae sunt Conradi Gesneri medici a Casparo Bauhino.* Basel: J. Waldkirch, 1591.

Cordus, Valerius. *Annotationes in Pedacii Dioscorides Anazarbei de medica materia libros V.* Edited by Conrad Gessner. Strasbourg: J. Rihel, 1561.

Dodoens, Rembert. *Histoire des plantes.* Antwerp: J. van der Loë, 1557.

Gessner, Conrad. *Epistolarum medicinalium.* Zurich: C. Froschauer, 1577.

———. *Epistolarum medicinalium . . . liber quartus.* Wittenberg: S. Gronenbergij, 1584.

———. *Historia animalium.* 4 vols. Zurich: C. Froschauer, 1551–58.

———. *Historia plantarum: Faksimilieausgabe; Aquarelle aus dem botanischen Nachlass von Conrad Gessner (1516–1565) in der Universitätsbibliothek Erlangen.* Edited by Heinrich Zoller, Martin Steinmann, and Karls Schmid. 8 vols. Zurich: Urs Graf, 1972–80.

———. *Historia plantarum: Gesamtausgabe.* 2 vols. Dietikon-Zurich: Urs Graf, 1987–91.

———. *Pandectarum sive partitionum universalium libri xxi.* Zurich: C. Froschauer, 1548–49.

Simler, Josias. *Vita C. Gesneri philosophi et medici clari.* Zurich: C. Froschauer, 1566.

Thévet, André. *Les singularitez de la France antarctique.* Antwerp: C. Plantin, 1558.

Secondary Sources

Anderes, Bernhard. "Glasmalerei in reformierten Zürich." In *Zürcher Kunst nach der Reformation: Hans Asper und seine Zeit*, ed. Marianne Naegeli, 15–20. Zurich: Schweizerisches Institut für Kunstwissenschaft, 1981.

Blair, Ann. *Too Much to Know: Managing Scholarly Information Before the Modern Age.* New Haven: Yale University Press, 2010.

Brooks, Jerome E. *Tobacco, Its History Illustrated by the Books, Manuscripts, and Engravings in the Library of George Arnets, Jr.* 5 vols. New York: Rosenbach, 1937–52.

Butts, Barbara. "Albrecht Dürer and the Modernization of Stained Glass." *Master Drawings* 41, no. 4 (2003): 341–58.

Daston, Lorraine, and Peter Galison. *Objectivity.* New York: Zone Books, 2007.

Delisle, Candice. "The Letter: Private Text or Public Place? The Mattioli-Gesner Controversy About the *aconitum primum*." *Gesnerus* 61 (2004): 161–76.

Dürling, Richard J. "Conrad Gessners Briefwechsel." In *Humanismus und Naturwissenschaften*, ed. R. Schmitz and F. Krafft, 101–11. Boppard am Rhein: Boldt, 1980.

Dürst, Arthur. "The Map-View of Zurich by Jos Murer, 1576." *Cartographica Helvetica* 15 (1997): 23–37.

Egmond, Florike. "A Collection Within a Collection: Rediscovered Animal Drawings from the Collections of Conrad Gessner and Felix Platter." *Journal of the History of Collections* 24 (2012): 1–22.

Fischel, Angela. *Natur im Bild: Zeichnung und Naturerkenntnis bei Conrad Gessner und Ulisse Aldrovandi*. Berlin: Gebr. Mann, 2009.

Kusukawa, Sachiko. "Conrad Gessner on an 'ad vivum' Image." Forthcoming.

———. *Picturing the Book of Nature: Image, Text, and Argument in Sixteenth-Century Human Anatomy and Medical Botany*. Chicago: University of Chicago Press, 2012.

Latour, Bruno. "Drawing Things Together." In *Representation and Scientific Practice*, ed. M. Lynch and S. Woolgar, 19–68. Cambridge: MIT Press, 1990.

Leemann van Elck, Paul. *Der Buchschmuck in Conrad Gessners naturgeschichtlichen Werken*. Bern: Verlag Paul Haupt, 1935.

Leu, Urs B. "Konrad Gessner und die neue Welt." *Gesnerus* 49 (1992): 279–309.

Leu, Urs B., Raffael Keller, and Sandra Weidmann. *Conrad Gessner's Private Library*. Leiden: E. J. Brill, 2008.

Lewis, Gillian. "Clusius in Montpellier, 1551–1554: A Humanist Education Completed?" In *Carolus Clusius: Towards a Cultural History of a Renaissance Naturalist*, ed. F. Egmond, P. Hoftijzer, and Arnoud S. Q. Visser, 65–98. Amsterdam: Edita, 2007.

Maselis, Marie-Christine, Arnout Balis, and Roger H. Marijnissen. *The Albums of Anselmus de Boodt (1550–1632): Natural History Painting at the Court of Rudolph II in Prague*. Rotthalmuenster: Heribert Tenschert, 1999.

Meyer, Frederick G., Emily Emmart Trueblood, and John L. Heller. *The Great Herbal of Leonhart Fuchs: De historia stirpium commentarii insignes, 1542*. 2 vols. Stanford: Stanford University Press, 1999.

Moss, Ann. *Printed Commonplace-Books and the Structuring of Renaissance Thought*. Oxford: Oxford University Press, 1996.

Nissen, Claus. *Die botanische Buchillustration: Ihre Geschichte und Bibliographie; Supplement*. Stuttgart: Hiersemann, 1966.

Ockenden, R. E. "A Note on Hyoscyamus Luteus." *Isis* 30, no. 2 (1939): 273–76.

Ogilvie, Brian W. *The Science of Describing: Natural History in Renaissance Europe*. Chicago: University of Chicago Press, 2006.

Palmer, Richard. "Medical Botany in Northern Italy in the Renaissance." *Journal of the Royal Society of Medicine* 78 (1985): 149–57.

Pomata, Gianna, and Nancy G. Siraisi, eds. *Historia: Empiricism and Erudition in Early Modern Europe*. Cambridge: MIT Press, 2005.

Rath, Gernot. "Die Briefe Konrad Geßners aus der Trewschen Sammlung." *Gesnerus* 7 (1950): 140–215.

Reeds, Karen Meier. *Botany in Medieval and Renaissance Universities*. New York: Garland, 1991.

Salzmann, Charles. "Francesco Calzolari, der Erforscher der Flora des Monte Baldo, und seine Pflanzensendung an C. Gessner in Zürich." *Gesnerus* 16 (1959): 81–103.

Wellisch, Hans. "Conrad Gessner: A Bio-Bibliography." *Journal of the Society of the Bibliography of Natural History* 7, no. 2 (1975): 151–247.

———. "How to Make an Index—Sixteenth-Century Style: Conrad Gessner on Indexes and Catalogs." *International Classification* 8 (1981): 10–15.

3.

Aesthetic Appreciation of Nature in Early Modern Science

CATHERINE WILSON

Three features of northern European visual art of the early seventeenth century stand out as historically distinctive. One is naturalism, the interest in the faithful depiction of plants and animals. Flowers, crustaceans, game birds, and domestic animals are drawn and painted with affection and care. A second, related feature is ordinariness and intimacy. In Dutch painting of the period, ordinary folk are shown calmly engaged in the activities of daily life, preparing their meals, minding their children, employing their brooms and scales, absorbed in their reading, or practicing their musical instruments. They replace the saints and historical and mythological figures portrayed in dramatic or highly emotional states, connected to supersensible reality.

A third feature is the virtuosic rendering of materials, the transparency of glass and the sheen of metal, but also the minute detailing of the products of human fabrication, such as cloth, carpets, figured porcelain, and silver. Earlier, painters of the English school had abandoned the classical interest in folded monochromatic drapery to depict their aloof subjects in spectacular outfits of figured tapestry, or embroidered with geometrical designs, and with jewelry and trim tending toward the fantastic—webs of gems, lace ruffs and cuffs. The interest in repeated patterns and geometrical forms such as floor tiles and brocades extends to the "mathematical engagement"[1] of still life composition, and in these pictures, both defamiliarization and an intimacy suggestive of Platonic reminiscence seem to remove them to a higher order of experience.

A style of presentation implies a set of choices—what to depict and how to depict it—and choice implies a need for justification, for both the production and the consumption of art objects require investment of training, time, and effort, on one hand, and of money, on the other. The objects depicted must therefore reveal "a dimension of value capable of justifying their representation,"[2] and historians have long sought to explain the values of these modalities, with Svetlana Alpers's *Art of Describing* still the preeminent guide to the period. The choice of natural and humble over sacred and mythological themes reflects material well-being and gratitude for it, and a sense that earthly existence has intrinsic value apart from its role as a proving ground for the next world. The depiction of everyday maintenance tasks and ordinary commercial activity reinforces the moral lesson that the general well-being is sustained by diligence and application. More recently, social historians have pointed to the role of trade, conquest, and the plantation systems of the Dutch, which brought china, textiles, fruits, flowers, the objects of *reflexy-const*, to the center from distant parts of the globe. The shops of Amsterdam were stuffed with Persian rugs, Chinese porcelain, Japanese lacquer, Venetian glass, Spanish taffeta, Italian majolica.[3] The newfound luxury, as morally problematic as it is to modern eyes, was evidently worth displaying, with the sweetness of living sometimes juxtaposed with memoranda of fragility, death, and decay.

New technologies, meanwhile, assisted with the development of illusionistic technique and naturalistic composition. Optical devices, especially convex mirrors and concave lenses, can enhance the appearance of a landscape and heighten its colors, and the use of the camera obscura explains many of the innovative and sometimes peculiar features of mid-seventeenth-century painting.[4] People and furnishings could be arranged into the most natural and elegant-appearing configurations, as on a stage, before being projected, and their outlines could be rapidly traced to establish proportions and therefore perspective. The camera was therefore exceptionally well suited to capturing not only landscapes but also interiors.[5] Magnifying lenses of low power had been available since the late medieval period, and while their use has not been documented to the same extent as that of the camera obscura, use of a lens was essential for painting details of texture that produce the intended effect only when viewed from some distance. David Piper remarks that "the careful imitation of minute detail, from the wiry curl of the individual hair to the tiniest facets of complex jewelry is common to all North European portraiture of the period."[6] That some details in painting can only be seen with magnifying glasses indicates that lenses were used in their composition (fig. 3.1).

The analogy between these visual styles and early micrographical natural history is too striking to be coincidental, and it too has attracted attention.[7] So too has the analogy between curious works of art and curious works of nature.[8] The invention of the microscope, and its increasing deployment in the middle of the seventeenth

3.1 Unknown, *Queen Elizabeth I*, ca. 1585–90. Oil on panel, 953 x 819 mm. © National Portrait Gallery, London.

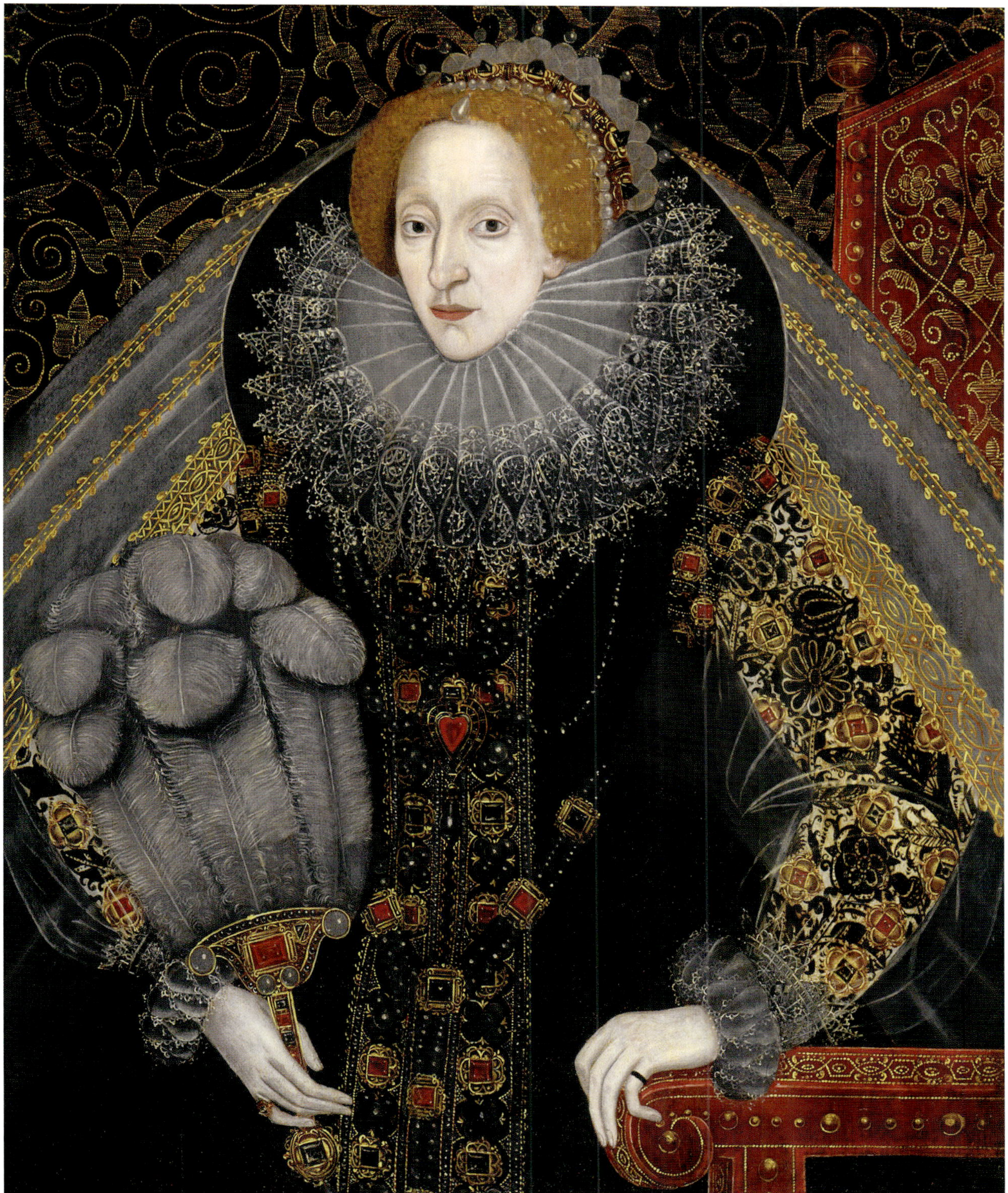

century, brought a new array of objects into the human repertoire of the depictable and the describable.[9] Seeds, grains of sand, parasitic insects, assuredly known to exist by previous generations but scarcely interesting to them, came to be characterized not merely as "curious" and "wonderful" but even as "beautiful." Magnifying lenses brought a new aesthetic object into existence that shared some formal characteristics with the most prized articles of human manufacture—jewels, embroidered goods, and tiny mechanical devices such as watches—that were at the same time both natural and humble. It is perhaps difficult to imagine the figures depicted in seventeenth-century northern European visual art as holding, in the minds behind their composed or expressive faces, an array of curious philosophical theories concerning the forms of crystals, plants and animals, their generation and destruction, but they—or their acquaintances—assuredly did. Scientific commerce and frequent literary exchanges, not only between England and the Netherlands but throughout Europe, promoted new reflections on the relationship between divine, human, and natural artistry.

What qualities and relations justified the verbal description and pictorial representation of these tiny objects—"epistemic images," to use Lorraine Daston's convenient and telling term? What explains the enthusiasm that greeted Robert Hooke's *Micrographia* and that inspired Anton van Leeuwenhoek to fire off his own lengthy reports to the Royal Society? Despite the obvious location of these word pictures and engravings in the category of the "art of describing" of the fine arts, the social history of the accumulation of riches, of self-presentation, and of morality seems at first to provide no useful clues. We seem to be thrown back on nothing more—though also nothing less—profound than the basic human interest in reality. But the authors and readers of these micrographical productions did find external meaning in them, and they understood in various ways the experience of finding geometrical regularity and exquisite patterning in the objects revealed by their microscopes.

Well before the invention of the microscope in the early seventeenth century, Renaissance naturalists throughout Europe had been keenly aware of form and pattern in nature: the veining of mineral deposits and ores, the growth and regrowth of shells, teeth, bones, nails, horns, cysts, tumors, and stony deposits on and in the human body, monstrous births, fossil animals, butterflies, the reticulation of roots and cones, the forms of frost, the honeycombs of bees, the growth of concentric rings in trees and onions. That natural geometry is more apparent in very small natural objects than in larger ones perhaps explains Plato's curious proposal in the *Timaeus* 65b that the world is composed of elemental atoms: pyramids of fire, cubes of earth, octahedrons of air, and icosahedrons of water.

The Pythagorean moment, when an observer comes upon geometry ready-made in nature, not flowing from human art, can induce a sort of awe; nature seems to need no compasses or rulers to judge her lengths or steady her hands. It also induces a sense of affinity, insofar as it implies the presence of a mind with values and interests similar to the human one. Norman Bryson describes the topological character of certain still lifes, seemingly organized along mathematical principles, that reveal a world both near and untouchably remote. To depict these objects as though only they existed, the painter must isolate them from their context and erase every trace of his own effort and involvement in the picture.[10] Yet, invariably, micrographical authors reflect on the implications of beauty and geometrical intricacy in nature for their own situations, as though information could be extracted concerning their own purposes and destinies. Lancelot Law Whyte observes in this connection that "it is characteristic of the human mind, when well-defined forms are observed, sooner or later to ask how they came into existence, and to start by regarding them as the products of a divine activity."[11]

Yet while geometry for the ancients was an exalted discipline, the crafts as such were not. Weaving, carving, and decorating—though painting had in the meantime attained greater dignity—remained servile occupations in early modernity that were not held in great esteem. When René Descartes, in his *Rules for the Direction of the Mind*, recommends that the aspiring scientist attend to "the simplest and least exalted arts, especially those in which order prevails, such as weaving and carpet making, or the more feminine arts of embroidery, in which threads are interwoven in an infinitely varied pattern," his reference to the "simplest and least exalted" arts may surprise the modern reader who considers these arts both esoteric and splendid.[12] But mass textile production has always been the activity of slaves, women, and the poor; and needlework is only an occupation of the feminine elite when there is nothing else as interesting, or more interesting, to do. Order and beauty—indeed, infinite variety—can proceed from such unpretentious activity. Gardeners, too, have a method for laying out elliptical flowerbeds that shows more intuitive understanding of the curve than mathematicians who treat the ellipse as a conic section possess.[13] To the extent that nature can be seen as an intricately woven surface, as fabric, it is analogous to the productions of mere artisans, lowly creatures possessed of a certain skill, in and of itself no more remarkable than that of spiders; the insect world, with its dumb and tireless porters, builders, and weavers, is the model and emblem of human effort. This ambiguous feature of nonhuman workmanship—the way in which it lends itself to being appreciated as stemming from a divine mind, or, alternatively, from blind, unintelligent nature—explains the frequent coexistence of what appear to be two incompatible discourses, sometimes in the same micrographical author.

Analytical and philosophical consideration of geometrical forms and figures in minute nature probably originated in modern times with the English mathematician Thomas Harriot. Harriot studied the efficiencies of certain patterns of stacking and packing, which interested him as an atomist and which he understood by analogy with the decorator's practices of tiling and inlaying.[14]

Harriot's works were not published, but he corresponded with Johannes Kepler about these problems. Kepler's own treatise on the snowflake, *Strena seu de Nive* (1611), presented as a "New Year's Gift" to his patron, Rudolph II, though it begins with literary joking on the themes of smallness, triviality, and humility, is the product of intense reflection on geometrical form in nature.[15] Kepler touches on a number of topics, such as the symmetry of animal bodies and their three-dimensionality, and the tendency of flowers to have five petals, but he addresses two main questions: Why are the honeycombs of bees hexagonal, and why do snowflakes have just six points? What is the agent that stamps these forms from without or builds them from within? Aware that these are two rather different questions—the functionality of hexagonal honeycombs and the beauty of useless snowflakes—he asks whether there might be a "nature . . . in which there is inborn either the idea of beauty inherent in the hexagon or knowledge of the purpose which that form subserves." He gives no definitive answer to the question in the treatise, but he makes several notable points about geometric efficiency. The least boundary material is needed to divide a given space into cells when these take the form of hexagons. Round cells would leave gaps, and the hexagon is the closest figure to a circle of the three regular shapes, including triangles and rectangles, that can cover a plane surface without gaps. "The bee . . . by nature," Kepler decides, "has this instinct as its property, to build in this shape rather than others. This original pattern has been imprinted on it by the Creator." Because snowflakes have no apparent purpose, Kepler must turn to a notion other than efficiency to explain their six points, namely, that of "formative reason," which, he says, "does not act only for a purpose, but also to adorn." Formative reason is responsible as well for the "beauty or special quality" of plants. This formative reason is resident in the earth as well as in water vapor. "It imparts itself to different bodies and cooperates with them. It engrafts itself on to them, and builds now one design, now another, as the inner disposition of each matter or outer conditions allow."[16] Geometrical archetypes and other forms in the mind of God are realized through the operations of world-building spirits.

Copies of Kepler's book were scarce, but Descartes and Gassendi knew of his research on the snowflake, and both Descartes and Thomas Bartholin wrote on the subject, the former in his *Meteorology* of 1637, the latter in his *De Nivis* of 1661. Because Descartes excluded from his natural philosophy all spirits and formative agents, excepting only the human mind, which, he thought, could form only ideas and volitions, and God, who creates only extension and motion and lays down a few simple laws of mechanics, his accounts of the snowflake and the salt crystal, as well as that of the development of the fetus, suffer from vagueness and implausibility. His ambitious and restrictive version of the "mechanical philosophy" was widely agreed to be inadequate to the explanation of the phenomena.

Geometrical regularity in mineral forms in the post-Cartesian period was extensively discussed by Athanasius Kircher in his *Mundus Subterraneus* (1664), and in plant and animal forms by Sir Thomas Browne, most notably in his discourse of 1658, *The Garden of Cyrus, or The Quincunciall, Lozenge, or Net-Work Plantations of the Ancients, Naturally, Artificially, Mystically Considered.* Browne mentions tiling and pavements, filigree, net making, and gratings as analogous to natural patterning. Further commentary on architectural shapes in ancient designs and building is followed by an exhibition of the triangle figure and the quincunx in constellations, the rhomboids of stones, and the "elegant ordination of vegetables." The "Quincunciall order" characterizes the process of vision itself: "All things are seen Quincuncially"; perception may take place by lines that are "magnified, diminished, distorted, and ill placed in the Mathematicks of some brains," whereby they have "irregular apprehensions of things, perverted notions, conceptions, and incurable hallucinations." Browne makes further discoveries of quincuncial patterns in plants, honeycombs, cocoons, snakeskins, the insides of birds' skins, fish scales, the network of veins and arteries, and the quincunx of the human figure with arms and legs akimbo.[17] "Right lines and circles," he finds, "make out the bulk of plants," and the legs of spiders are made after a "sesqui-tertian" proportion. He invokes the theory of signatures:

> Studious Observators may discover more analogies in the orderly book of nature, and cannot escape the Elegancy of her hand in other correspondencies. The Figures of nails and crucifying appurtenances, are but precariously made out in the *Granadilla* or flowers of Christs passion: . . . In young Wallnuts cut athwart, it is not hard to apprehend strange characters; and in those of somewhat elder growth, handsome ornamental draughts about a plain crosse. In the root of *Osmond* or Water-fern, every eye may discern the form of a Half Moon, Rain-bow, or half the character of Pisces. Some find Hebrew, Arabick, Greek, and Latine Characters in Plants.[18]

The sense that these minute natural forms convey a higher truth incapable of articulation is evident in Browne's admiration, expressed in his earlier *Religio Medici*, for "the mystical way of *Pythagoras*, and the secret Magicke of number," his belief that "in this masse of nature there is a set of things, that carry in their front, though not in Capitall letters, yet in stenography, and short Characters, somthing to Divinity, which to wiser reasons serve as Luminaries in the abysse of knowledge."[19]

Browne's investigations led him to assert the individuality of every existing object. "It is a common wonder of all men," he says,

> how among so many millions of faces, there should be none alike. . . . Let a Painter carefully limbe out a Million of faces, and you shall finde them all different, and after all this art there will remaine a sensible distinction from the patterne of every thing in the perfectest of that kind. . . . There was never any thing so like another, as in all points to concurre, there will be ever some reserved difference slip in, to prevent the Identity without which, two severall things would not be alike, but the same, which is impossible.[20]

Nor is there any "*Grotesco* in nature, nor [in] any thing framed to fill up empty cantons, and unnecessary spaces in the most imperfect creatures, such as were not preserved in the Arke." In addition to the beauty of insects generated by the sun from "seeds and principles in the wombe of nature," and filling up all voids, Browne finds "a generall beauty in the works of God, and therefore no deformity in any kind or species of creatures whatsoever: I cannot tell by what Logick we call a Toad, a Beare, or an Elephant, ugly, they being created in those outward shapes and figures which best expresse the actions of their internall formes."[21]

Robert Hooke, writing only a few years later, continues to invoke the figure of playful nature. "That which makes an Enquiry pleasant, are, first, a noble *Inventum* that promises to crown the successful endeavor; and such must certainly the knowledge of the efficient and concurrent causes of all those curious Geometrical Figures be, which had made the Philosophers hitherto to conclude nature in these things to play the Geometrician."[22] In his *Discourse of Earthquakes,* he mentions the "curiously figured Bodies of Salts, Talks, Spars, Crystals, Diamonds, Rubies, Amethysts, Ores and divers other Mineral Substances." Salt forms cubes or parallelepipeds; niter, triangular and hexangular prisms; alum, octahedrons.[23] According to the *Micrographia*, crystals are "the Elemental Figures, or the ABC of Nature's working, the Reason of whose curious Geometrical Forms . . . is very easily Explicable Mechanically." They have the important property of self-similarity, the shape of the large crystal replicating that of the smaller crystals of which it is composed. In the *Micrographia,* he discusses within the space of a few pages the six-sided figures in frozen urine, snow, hoarfrost, and niter, and the leaves of ferns, concluding that "if the Figures of both be well consider'd, one would ghes that there were not much greater need of a *seminal principle* for the production of *Fearn,* than for the production of the branches of *Urine,* or *the Stella martis,* there seeming to be as much form and beauty in one as in the other."[24]

Hooke rejects, however, sporting nature when he comes to his discussion of "figured stones," a category that included fossil forms as well as crystals. He notes that "divers Stones have been found, formed into the Shapes of Fishes, Shells, Fruits, Leaves, Woods, Barks, and other Vegetable and Animal Substances." The forms of plants and animals in stones are, however, to be explained as residues of living creatures. "It seems very difficult to imagine that Nature formed all these curious Bodies for no other End than to play the Mimick in the Mineral Kingdom and to imitate what she had done for some more noble End, and in greater Perfection, in the Vegetable and Animal Kingdoms." Most people, he says, "have recourse to some vegetative or plastic Vertue inherent in the Parts of the Earth where they were made," and he criticizes the astrologer Jacques Gaffarel, who transferred them over to his "Astrological and Magical fancy and thinks that they were produced from some extraordinary Celestial Influence." In his own view, they are either actual animals and vegetables petrified, or else the impressions of actual animals and vegetables imprinted in the stone.[25]

Following Browne's lead in seeing every animal as beautiful and none as grotesque, Henry Power professes to find a bag of spider eggs and a mass of creeping and crawling baby spiders "a very pleasant spectacle":

> I Took a Field-Spider under a stone, 13. of *June,* with a bag of eggs fastned to her tayl, bigger than all the bulk of her body; I opened it, and saw abundance of blewish eggs in it, which in the *Microscope* look'd white and round, like your counterfeit pearl, and I could most clearly see abundance of very minute Spiders, newly hatch'd, no bigger, and just like Mites in Meal, with white hairs and bristles, especially in their tail, creeping and crawling amongst the eggs: The Nett-work of the Purse or Bag seem'd all diaphanous; a very pleasant spectacle, and of curious workmanship.[26]

Hooke, in turn, describes the flea as strong and beautiful, "adorn'd with a curiously polish'd suit of *sable* Armour, neatly jointed, and beset with multitudes of sharp pinns, shap'd almost like Porcupine's Quills, or bright conical Steel-bodkins; the head is on either side beautify'd with a quick and round black eye,"[27] and "several other Flies" resemble butterflies, birds, and Persian carpets (fig. 3.2):

> In several other Flies, there are infinite of small Feathers, which cover both the under and upper sides of this thin film as in almost all the sorts of Butterflies and Moths: and those small parts are not onely shap'd very much like the feathers of Birds, but like those variegated with all the variety of curious bright and vivid colours imaginable; and those feathers are likewise so admirably and delicately rang'd, as to compose very fine flourishings and ornamental paintings, like *Turkie* and *Persian* Carpets, but of far

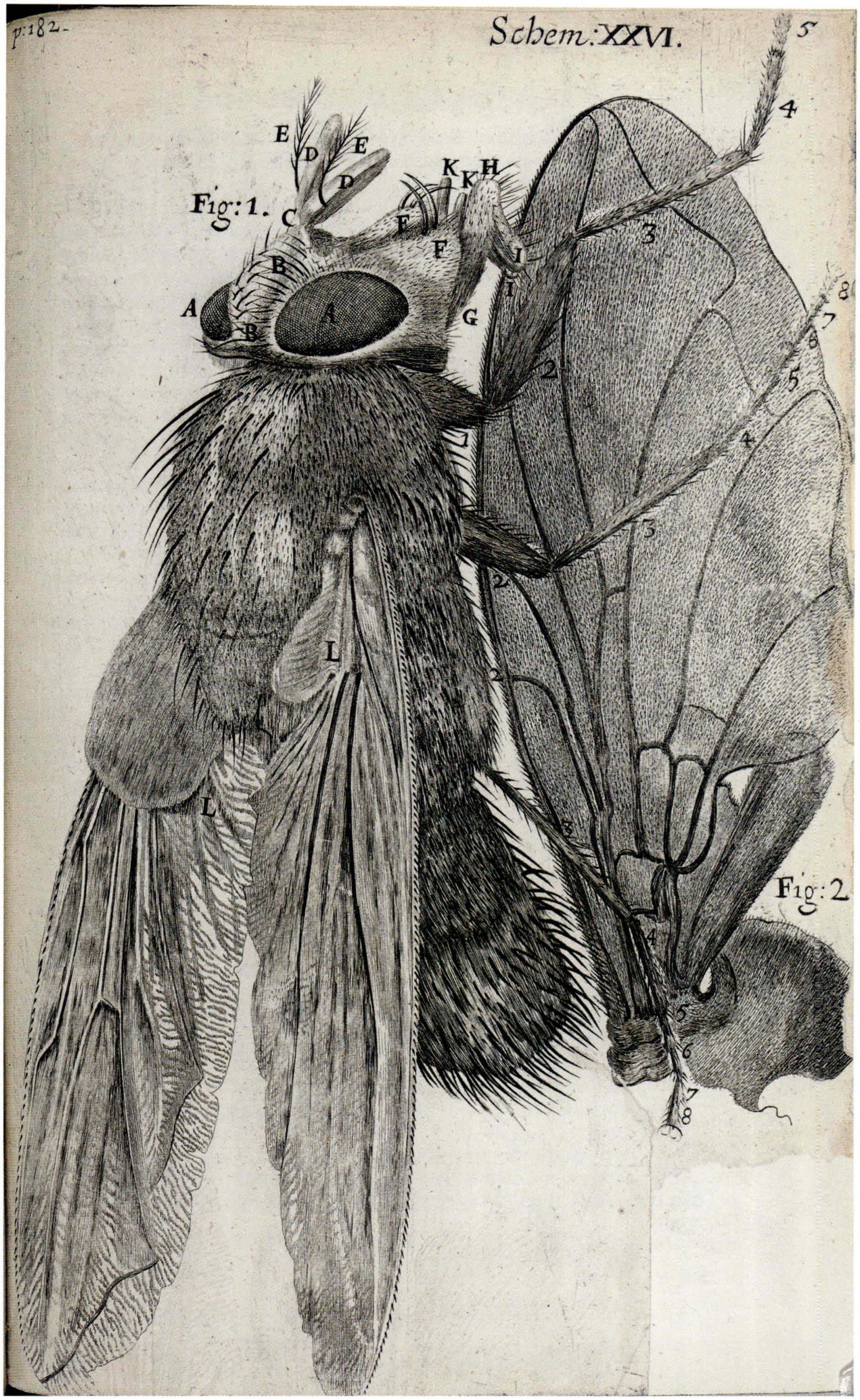

3.2

Fly. Engraving. In Robert Hooke, *Micrographia,* 182. © Trustees of the British Library.

> more surpassing beauty, as is evident enough to the naked eye, in the painted wings of Butterflies, but much more through an ordinary *Microscope*.[28]

Colored silks, peacocks' feathers, and mother-of-pearl are evoked in Hooke's discussion of the microstructure of vivid colors in nature, which, he finds, result from the reflections within the tiny transparent parts of objects. This accounts for the difference between hair and feathers, and Hooke describes the tactile sensuality of the drab but thick fur of northern animals. Their outer coats are not only "adapted to the peculiar use and convenience of that Animal which they inwrap," but "very much also for the ornament and beauty of it, as will be most evident to any one that shall attentively consider the various kinds of cloathings wherewith most creatures are by Nature invested and cover'd." The colors of "Leopards, Cats, Rhein Deer, *&c.*," as well as the shapes of "Horses manes, Cats beards, and . . . the Vestments of Fishes, Birds, Insects," seem purely ornamental.[29]

Even "the humble moss," Hooke tells the reader, is not to be scorned, "for, as to its shape, it may compare for the beauty of it with any Plant that grows." The seeds of plants, with their "delicate, strong, and most convenient Cabinetry," are Nature's repositories, "wherein are laid up its Jewels. . . . Nor indeed is there in any part of the Vegetable so curious carvings, and beautifull adornments, as about the seed." Browne's insistence on the uniqueness of every natural object is repeated by Hooke, who observes that the seeds of the corn violet are also "cover'd with a tough thick and bright reflecting skin very irregularly shrunk and pitted, insomuch that it is almost an impossibility to find two of them wrinkled alike, so great a variety may there be even in this little seed."[30]

Browne's *Religio Medici* was written (and revised and amended) to excuse the physician for the "scandal of his Profession," for doctors were assumed on the basis of their involvement with the body and with death and decay to be theological skeptics who denied the immortality of the human soul and the resurrection of the body promised in scripture. By way of countering this impression, Browne devotes considerable attention to the question of the resurrection, which he thinks we should endeavor to understand "naturally, and as Philosophers." In a burned-up plant reduced to ashes, the form only seems to have "taken his leave forever. . . . To a subtile Artist the formes are not perished, but withdrawn into their combustible part, where they lie secure from the action of that devouring element." This can be demonstrated by experience, whereby the plant is revived from its ashes and cinders.[31] Some of the most curious examples of seventeenth-century experimental philosophy are the attempted palingenetical demonstrations of the disappearance and reemergence of complex forms. So Kenelm Digby recounts an experiment repeated from the chemist Quercetanus, in which he reduces the roots, stalks, leaves, and flowers of nettles to ashes, leaves the ashes outdoors to freeze, and notes "an abundance of Nettles frozen in the ice," adding, "They had not the colour of Nettles. No greennesse accompanied them. They were white. But otherwise it is impossible for any Painter to delineate a throng of Nettles more exactly, then they were designed in the water. As soon as the water was melted, all the Ideall shapes vanished: but as soon as it was congealed again, they presently appeared afresh."[32] Another experiment of Digby's involved boiling crayfishes, separating them from their shells, which were distilled, burning the fishes, and filtering a decoction to extract their salt. In a few days, Digby claimed, young crayfish appeared that could be nourished on ox blood; with good care, "you may bring them on to what bignesse you please."[33]

The question posed at the beginning of this chapter was what features of micrographical description and representation justified the investment of effort and expense for authors and readers. For, as the foregoing has made clear, the impulse to describe—and the willingness to dedicate many pages of print and costly engraving to the mere description of objects found underfoot or in drawers and crevices—is not wholly explained by the resonance of geometrical forms with the human mind, as important and as poorly understood as this phenomenon is, or by the natural human interest in generation, perishing, and survival after death. One might think here of what Roland Barthes refers to, in connection with literary technique, as the sought-after reality effect: "The real is assumed not to need any independent justification, . . . it is powerful enough to negate any notion of 'function,'. . . it can be expressed without there being any need for it to be integrated into a structure. . . . The *having-been-there* of things is a sufficient reason for speaking of them."[34]

Yet such a presumption could hardly go unchallenged in a theological age in which *curiositas* was a form of concupiscence and therefore sin, and too much concern with the material world, as opposed to the spiritual, was considered deplorable.[35] Here, the architecture of the invisible world revealed by the microscope compensated for certain losses, notably the collapse of the traditional appeal to the majesty and utility of the heavens as plain and evident proof of the existence of God, available to the simplest peasant. After the acceptance of Copernicanism, the heavens could not be taken as a sort of roofline feature of our little world, created for the sake of the beings inhabiting its surface and answering to their navigational, calendrical, and prophetic needs. The vast size and arrangement of the Copernican universe indicated that the vault of the heavens was not a spectacle specially arranged for us to marvel at on clear nights; the presumed semiotic function of the stars in spelling out messages was doubtless diminished by the realization that they were distributed in three-dimensional, not two-dimensional space. However, meanings and messages could still be found in the microcosm; and in a host of authors, including Kepler and Browne, though not Hooke, the revelation of nature's minute geometry appears to reinforce the theology of benevolent creation. There is, Browne urges, "a Phistognomy or Physiognomy, not onely of men, but of Plants, and Vegetables. . . . The finger of God hath left an inscription upon all his workes, not graphicall or composed of Letters, but on their severall formes, constitutions, parts and operations." Every essence, "created or uncreated, hath its finall cause, and some positive end both of its essence and operation: This is the cause I grope after in the workes of nature, on this hangs the providence of God; to raise so beautious a structure, as the world and the creatures thereof."[36] Nehemiah Grew describes frost "like a Lavender's Spike or mallow Flower," and "Rivers and Seas as well as Clouds of figured Icicles" in his *Cosmologia Sacra*, taking this artistry as proof that God's government is "Universal. Reaching not only to Celestial, but Terrestrial Worlds; and among them, this we live in. . . . Not only Unto Things of greater Moment: but unto those also, which seem to us, to be the most Casual, and the most Trivial."[37]

Here a very old problem presented itself. Was nature an intelligent and purposeful being, distinct from God and autonomous in her own sphere? Proverbs 8:30–31 describes Wisdom as standing beside God at the creation. "Then was I beside him as his craftsman, and I was his delight day by day, Playing before him all the while, playing on the surface of his earth; and I found delight in the sons of men."[38] So disturbing was this evident intrusion of a demiurge into the signature monotheism of the Bible that commentators strained to insist that this playful Wisdom was identical to, or foreshadowed, Christ. If the magnifying lens revealed how widespread in nature the beauty, detail, and geometrical regularity associated with artistry really were, this could be interpreted in three ways: first, as evidence of the constant operations of the divine mind and hand in nature, as God's personal fashioning and maintenance of every existing individual thing; second, as evidence of the powers of mere blind mechanism to produce objects perceived by the human eye as beautiful; or third, as evidence of some unknown formative power neither purely mechanical nor wholly divine. Hooke concluded along with the Cartesians that the forms and operations of life could be mechanically produced. "As far as I have been able to look into the nature of this Primary kind of life and vegetation [that of mushrooms], I cannot find the least probable argument to perswade me, that there is any other concurrent cause then [*sic*] such as is purely Mechanical." His confidence in spontaneous generation, which he describes as like a watch set in motion by the sun, and his finding of the "first footsteps of *Sensation* and Animate motion" in plant life were indications of an ambitious program to explain the evolution of forms rising from "Fluidity and Orbiculation, through Crystallization and Vegetation, to Animation, Sensation, and Imagination."[39] By contrast, the Cambridge Platonists, denying that the Cartesian laws of mechanical interaction could produce the forms of plants and animals, posited a demiurge—the "plastic nature," a form-building principle under the direction of God. For Henry More, plastic nature builds as well according to human criteria of beauty, indicating its link to the divine mind. Certain plants, he says, "[bear] such a sutablenesse and harmony with the more refined sense and sagacity of the Soul of Man, that he cannot chose . . . but acknowledge that some hidden Cause much a kin to his own nature . . . is the contriver & perfecter of these so pleasant spectacles in the world." The "Pulchritude of things . . . acknowledged by the whole generations of Men to be in Trees, Flowers and Fruits," as well as in "most sorts of birds," "is an ample & undeniable Testimony" of an intellectual principle.[40]

By the end of the eighteenth century, a world of self-replicating organic forms produced entirely by the building forces of nature, rather than by or with the assistance of a God or a quasi-divine laborer, was thinkable. Formative forces were posited as inherent in matter and as distinct from ordinary mechanical forces and the laws of chemical combination. But if the world was not, after all, created to please and instruct human beings by a divinity with similar optical tastes, how was More's "Pulchritude of things" to be explained? Why did the human eye and mind find natural forms, including normally invisible forms, such as those of crystals and plants and plant parts, beautiful? How did one explain the resonance of flowers, shells, and leaf forms with human cognition, and if the beauty of nature did not point to the existence of a benevolent and all-powerful Creator, what exactly did it point to? This question preoccupied Kant. It was inconsistent with his "critical" methodology to posit dogmatically, or as the conclusion of any philosophical argument, a divine Creator, or to assign to such a being, if one existed, any desire to please human beings. He could not assume with Browne that "the world was made to be inhabited by beasts, but studyed and contemplated by man."[41] But the beauty of the flower, he thought, could not be understood as functional in the same manner as the harmony of the limbs and organs of animals.[42] It was gratuitous—seemingly without any purpose. For Margaret Cavendish, our aesthetic experience arises out of nature's desire to be looked at and loved; nature resolved to "Make some things [e.g., human beings] adore us, worship give, / Which now we only to our selves to live."[43] Kant was not quite so explicit. Yet, on his view, the formative forces of nature seem to have conspired to produce—not by chance, but as if purposively—a world visible to and knowable by human beings that gives them aesthetic pleasure.

Kant, like Hooke, was virtually certain that some forms of natural beauty are produced by physicochemical means: "As far as the beauty of flowers, of bird feathers, and seashells is concerned, in terms of both their shape and their color, these can be ascribed to nature and its faculty for forming itself aesthetically and purposively in its freedom, without special ends aimed at that, in accordance with chemical laws, by the deposit of the matter requisite for their organization."[44] The agreement of beautiful objects with our minds, with what Kant calls the "play of our cognitive faculties," is in this sense accidental and admits of no further explanation.[45] It has, however, significance. Through the experience of natural beauty, nature gives us "some trace or . . . a sign that it contains in itself some sort of ground for assuming a lawful correspondence of its products with our satisfaction that is independent of all interest."[46] The experienced harmony between the natural world and our nonutilitarian aesthetic preferences indicates the possibility of a harmony between the social world and our nonutilitarian moral commitments. As Alexander Rueger points out, "What is needed for such conclusions is the experience of a realm that is not our own fabrication."[47] Nature, in other words, is not only our moral enemy and seducer, leading us down the paths of desire and inclination. She seems as well to recognize and reward our indifference to functionality and purpose, which indifference is the essence of moral experience as well.

This chapter has considered three related phenomena: the urge to reproduce in painting the complex patterns and textures of man-made luxury objects—carpets, lace, embroidery, ceramics, and jewelry, and other items of human manufacture, by English and Dutch artists of the sixteenth and seventeenth centuries; the interest in depicting humble, domestic subjects, activities, and interiors—people sweeping, peeling, or reading; and, finally, the urge to describe and to reproduce in scientific drawings and engravings the patterns and textures of common plants, animals, and minerals. Optical instruments played a role in the emergence of all three desires. Without magnifying lenses to re-create the tiny brushstrokes responsible for the surface appearances of transparency, dimensionality, and the gleam of reflected light, the illusionism of *reflexy-const* would have been impossible to bring off. Without projection devices to capture exact geometrical relations and to assist in naturalistic compositions involving human figures, the depiction of domestic interior architecture, with its linked rooms, doorways, and windows, would not have been possible either. And the lens revealed the fabulous intricacy and beauty of a microworld lying beneath the threshold of their God-given optical equipment and long hidden from them, which human beings seemed to have stumbled on by chance.

While painted depictions of elegant manufacture and painted depictions of everyday activities might seem to express opposing values, they were related to each other and to the emergence of a new form of visual attention. Artists turned from the heroic subjects of history and mythology, including Christian mythology, and from portrayals of hunting, warfare, bold, seductive femininity, and pageantry, to portrayals of domestic life in a way that has long been understood to correspond to the introduction of a new set of values. The discovery of the beauty and interest of the close at hand and its expression in painting reinforced and perhaps helped to create a preference for peace and a settled order—the well-wrapped-up female form occupied in inglorious household tasks reinforces the message—and for production, trade, and consumption over honor and triumph. For Royal Society naturalists as well, the observation and depiction of such mundane objects as insects, cork, and hair expressed an orientation to the present, local world, a relaxation of strivings after heaven, on the one hand, and foreign wars and overseas adventure, on the other. Science offered gentlemen an alternative to theology, politics, seduction, and warfare, and gave the lowborn a new route to modest distinction. It removed the observer from the dangerous fray of human emotions. "What anger, envy, hatred or revenge can long torment his breast, whom not only the greatest and noblest objects, but every sand, every pebble, every grass, every earth, every thing can divert?"[48]

For a long time, the interest in diversity and individuality, in understanding the principles of growth and development, that is characteristic of late seventeenth-century micrographical observation received little recognition in the historiography of science. The ruminations of a Leeuwenhoek or a Hooke seemed to contrast with the revolutionary accomplishments in mathematical physics that depended upon abstracting from and ignoring the visible surface of the world, and considering matter or body as such in its idealized relations to other bodies, and earlier historians insisted that the objectivity of modern science originated in this refusal to take the world at face value. The observation of nature was seen, at best, as a preamble to the delayed revolution in biology that allegedly did not take place until the nineteenth century. It was associated with the literary and eccentric side of English letters, and was considered a residue of Renaissance eclecticism. But as Kepler, Hooke, and Descartes show, the deep Pythagoreanism of the inquirers into mathematical astronomy, optics, and mechanics is continuous with their interest in the visual surface and its visible geometry. In their appreciation of organic and inorganic forms, they intimated the unfolding of mathematically describable processes that they were helpless to understand or explain.[49]

Challenging the earlier conceptualization of the origins of objectivity in the mathematicization of nature in the seventeenth century, Daston and Galison have argued that scientific objectivity should be understood not merely as a cognitive technique but as a quasi-moral virtue. They have been reluctant to trace its history back to the seventeenth century.[50] Yet the descriptive styles of the close observers and micrographical writers already reveal a moral attitude toward observation, one of openness to experience and acceptance of whatever is.

This nonjudgmentalism is met with in Margaret Cavendish's declared refusal "to extol my own kind above all the rest, or above nature herself."[51] It is exemplified in Hooke's and Power's readiness to find not only curiosity but beauty in insects, their nests, and their eggs; in Browne's admiration for the sheer variety of faces, beautiful or ugly, surely an advance over the Elizabethan preference for perfection over personality; and in Descartes's declaration that a sick body is just as much a natural thing, just as much one of God's creatures, as a well one. These forms of humility have nothing to do with abjection before God, or the archbishop, or the prince; they indicate a respectful stance toward the complexity and individuality of nonhuman nature and an interest in its preservation. Browne declares himself free of all ordinary prejudices: "I am of a constitution," he declares,

> so generall, that it consorts, and sympathizeth with all things; I have no antipathy, or rather Idiosyncrasie, in dyet, humour, ayre, any thing; wonder not at the *French*, for their dishes of froggs, snailes, and toadstooles; . . . I could digest a Sallad gathered in a Church-yard, as well as in a Garden. I cannot start at the presence of a Serpent, Scorpion, Lizard, or Salamander; at the sight of a Toad, or Viper, I finde in me no desire to take up a stone to destroy them. I feele not in my selfe those common antipathies that I can discover in others.[52]

Notes

1. The term is Bryson's. *Looking at the Overlooked*, 70.
2. Brusati, "Natural Artifice and Material Values," 147.
3. Bryson, *Looking at the Overlooked*, 110.
4. Wheelock, *Perspective, Optics, and Delft Artists*, 160–70. See also Huerta, *Giants of Delft*, 49–51.
5. See Steadman, *Vermeer's Camera*.
6. Piper, *English Face*, 59.
7. Alpers, *Art of Describing*, 73; Huerta, *Giants of Delft*, 79 and passim.
8. Daston and Park, *Wonders and the Order of Nature*, 314.
9. Wilson, *Invisible World*, 75–98.
10. Of Juan Sánchez Cotán's *Quince, Cabbage, Melon and Cucumber*, Bryson says, "We begin at the left with the quince, a pure sphere, revolving on its axis. Moving to the right, the sphere seems to peel off its boundary and disintegrate into a ball of concentric shells, revolving around the same vertical axis. Moving to the melon, the sphere becomes an ellipse. . . . At the right the segmented shapes recover their continuous boundary in the corrugated form of the cucumber." The curve they describe is "precisely logarithmic." *Looking at the Overlooked*, 69.
11. "In seeking to understand the genesis of forms," Whyte continues, we have the sense, "of touching a singularly deep issue, affecting everything that exists. . . . It is as though to ask why forms exist were to probe the meaning of existence itself." Whyte, "Introduction," in Kepler, *On the Six-Pointed Snowflake*, 58.
12. Descartes, *Philosophical Writings*, 1:35. See also Gauvin, "Artisans, Machines, and Descartes's Organon," for a study of Descartes's attitude toward workers and artisans, especially weavers; and Zittel, *Theatrum Philosophicum*, 189–226, on "aesthetic form" in Cartesian science.
13. Descartes, *Oeuvres*, 6:166.
14. For further details on Harriot's life and work, see Shirley, *Thomas Harriot*.
15. Findlen gives a rich history of references to "jokes of nature" and "merry conceits," which peak in the sixteenth and seventeenth centuries and fade in the eighteenth, in "Jokes of Nature and Jokes of Knowledge."
16. Kepler, *On the Six-Pointed Snowflake*, 9, 19, 33–35.
17. Browne, *Garden of Cyrus*, 169ff., 202, 204, 184–89.
18. Ibid., 191.
19. Browne, *Religio Medici*, 20–21.
20. Ibid., 118.
21. Ibid., 24, 28.
22. Hooke, *Micrographia*, 87.
23. Hooke, *Discourse of Earthquakes*, 294, 280, 293.
24. Hooke, *Micrographia*, 90.
25. Hooke, *Discourse of Earthquakes*, 289, 288, 291.
26. Power, *Experimental Philosophy*, 15.
27. Hooke, *Micrographia*, 210.
28. Ibid., 174.
29. Ibid., 169–70, 161.
30. Ibid., 132, 155, 153.

31. Browne, *Religio Medici*, 91.
32. Digby, *Discourse Concerning the Vegetation of Plants*, 76–78.
33. Ibid., 84–85.
34. Barthes, "Reality Effect," 15.
35. For more on this theme, see Wilson, "Some Philosophical Criticisms of Early Microscopy."
36. Browne, *Religio Medici*, 116, 23.
37. Grew, *Cosmologia Sacra*, 86.
38. Bredekamp, in *Antikensehnsucht und Maschinenglauben*, 66, quotes the passage, noting that "playing" is probably a mistranslation of "enjoying."
39. Hooke, *Micrographia*, 130, 151, 127.
40. More, *Antidote Against Atheism*, 259, 271.
41. Browne, *Religio Medici*, 23.
42. Kant, *Critique of Judgment*, 229.
43. Cavendish, *Poems, and Fancies*, 1–2.
44. Kant, *Critique of Judgment*, 349–50.
45. See Rueger, "Kant and the Aesthetics of Nature."
46. Kant, *Critique of Judgment*, 300.
47. Rueger, "Kant and the Aesthetics of Nature," 155.
48. Sprat, *History of the Royal Society*, 45, quoted in Alpers, *Art of Describing*, 108.
49. The symmetries and mathematical principles underlying the forms of living creatures were explored by Ernst Haeckel and by D'Arcy Thompson. More recently, the theory of "cellular automata" and their generation of fractal patterns has been employed by David Young to explain patterns of coloration such as the stripes and spots of animals' coats that captured Hooke's attention.
50. Daston and Galison, *Objectivity*. A "prehistory" of objectivity appears in Daston's "Baconian Facts."
51. Cavendish, *Observations upon Experimental Philosophy*, 112.
52. Browne, *Religio Medici*, 111–12.

Bibliography

Alpers, Svetlana. *The Art of Describing: Dutch Art in the Seventeenth Century*. London: John Murray, 1983.

Barthes, Roland. "The Reality Effect." In *French Literary Theory Today*, ed. Tzvetan Todorov, 11–17. Cambridge: Cambridge University Press, 1992.

Bredekamp, Horst. *Antikensehnsucht und Maschinenglauben*. 2nd ed. Berlin: Wagenbach, 2002.

Browne, Sir Thomas. *The Garden of Cyrus, or The Quincunciall, Lozenge, or Net-Work Plantations of the Ancients, Naturally, Artificially, Mystically Considered*. In Browne, *Selected Writings*. Edited by G. Keynes. Chicago: University of Chicago Press, 1968.

———. *Religio Medici*. London: A. Crooke, 1642.

Brusati, Celeste. "Natural Artifice and Material Values in Dutch Still Life." In *Looking at Seventeenth-Century Dutch Art: Realism Reconsidered*, ed. Wayne Franits, 144–57. Cambridge: Cambridge University Press, 1997.

Bryson, Norman. *Looking at the Overlooked: Four Essays on Still Life Painting*. London: Reaktion Books, 1990.

Cavendish, Margaret. *Observations upon Experimental Philosophy*. 1666. Edited by Eileen O'Neill. Cambridge: Cambridge University Press, 2001.

———. *Poems, and Fancies*. London: Martin and Allestrye, 1653.

Daston, Lorraine. "Baconian Facts, Academic Civility, and the Prehistory of Objectivity." *Annals of Scholarship* 8 (1992): 337–63.

Daston, Lorraine, and Peter Galison. *Objectivity*. New York: Zone Books, 2007.

Daston, Lorraine, and Katharine Park. *Wonders and the Order of Nature, 1150–1750*. New York: Zone Books, 1998.

Descartes, René. *Oeuvres*. Edited by Charles Adam and Paul Tannery. 11 vols. Paris: Vrin, 1996.

———. *Philosophical Writings*. Edited and translated by J. Cottingham, R. Stoothoff, D. Murdoch, and A. Kenny. 3 vols. Cambridge: Cambridge University Press, 1985–91.

Digby, Kenelm. *A Discourse Concerning the Vegetation of Plants*. London: John Dakins, 1661.

Findlen, Paula. "Jokes of Nature and Jokes of Knowledge: The Playfulness of Scientific Discourse in Early Modern Europe." *Renaissance Quarterly* 43 (1990): 292–331.

Gauvin, Jean-François. "Artisans, Machines, and Descartes's Organon." *History of Science* 44 (2006): 187–216.

Grew, Nehemiah. *Cosmologia Sacra, or A Discourse of the Universe as it is the Creature and Kingdom of* God. London, 1701.

Hochstrasser, Julie Berger. *Still Life and Trade in the Dutch Golden Age*. New Haven: Yale University Press, 2007.

Hooke, Robert. *Discourse of Earthquakes*. 1705. In Hooke, *Posthumous Works*, ed. Richard Waller, 279–450. Hildesheim: Olms, 1970.

———. *Micrographia*. London, 1665.

Huerta, Robert D. *Giants of Delft: Johannes Vermeer and the Natural Philosophers; The Parallel Search for Knowledge During the Age of Discovery*. Lewisburg: Bucknell University Press, 2003.

Kant, Immanuel. *Critique of Judgment*. Translated by Werner S. Pluhar. Indianapolis: Hackett, 1987.

Kepler, Johannes. *On the Six-Pointed Snowflake*. Edited by Lancelot Law Whyte, translated by Colin Hardie. Oxford: Clarendon Press, 1966.

———. *Strena seu de Nive Sexangula*. Frankfurt, 1611.

More, Henry. *An Antidote Against Atheism.* Books 1–2. In *The Cambridge Platonists*, ed. C. A. Patrides, 213–87. Cambridge: Cambridge University Press, 1969.

Piper, David. *The English Face.* London: National Portrait Gallery, 1992.

Power, Henry. *Experimental Philosophy.* London: Martin and Allestry, 1664.

Rueger, Alexander. "Kant and the Aesthetics of Nature." *British Journal of Aesthetics* 47 (2007): 138–55.

Shirley, J. W. *Thomas Harriot: A Biography.* Oxford: Clarendon Press, 1983.

Sprat, Thomas. *History of the Royal Society.* London: J. Martin, 1667.

Steadman, Philip. *Vermeer's Camera: Uncovering the Truth Behind the Masterpieces.* Oxford: Oxford University Press, 2001.

Thompson, D'Arcy W. *On Growth and Form.* 2nd ed. New York: Dover, 1992.

Thorndike, Lynn. *A History of Magic and Experimental Science.* 8 vols. New York: Macmillan, 1923–58.

Wheelock, Arthur K., Jr. *Perspective, Optics, and Delft Artists Around 1650.* New York: Garland, 1977.

Wilson, Catherine. *The Invisible World: Early Modern Philosophy and the Invention of the Microscope.* Princeton: Princeton University Press, 2005.

———. "Some Philosophical Criticisms of Early Microscopy." In *Evidentia: Reichweiten visueller Wahrnehmung in der Frühen Neuzeit*, ed. F. Büttner, M. Friedrich, K. Leonhard, and G. Wimböck, 217–36. Münster: LIT Verlag, 2007.

Young, David A. "A Local Activator-Inhibitor Model of Vertebrate Skin Patterns." *Mathematical Biosciences* 72 (1984): 51–58.

Zittel, Claus. *Theatrum Philosophicum: Descartes und die Rolle ästhetischer Formen in der Wissenschaft.* Berlin: Akademie der Wissenschaft, 2009.

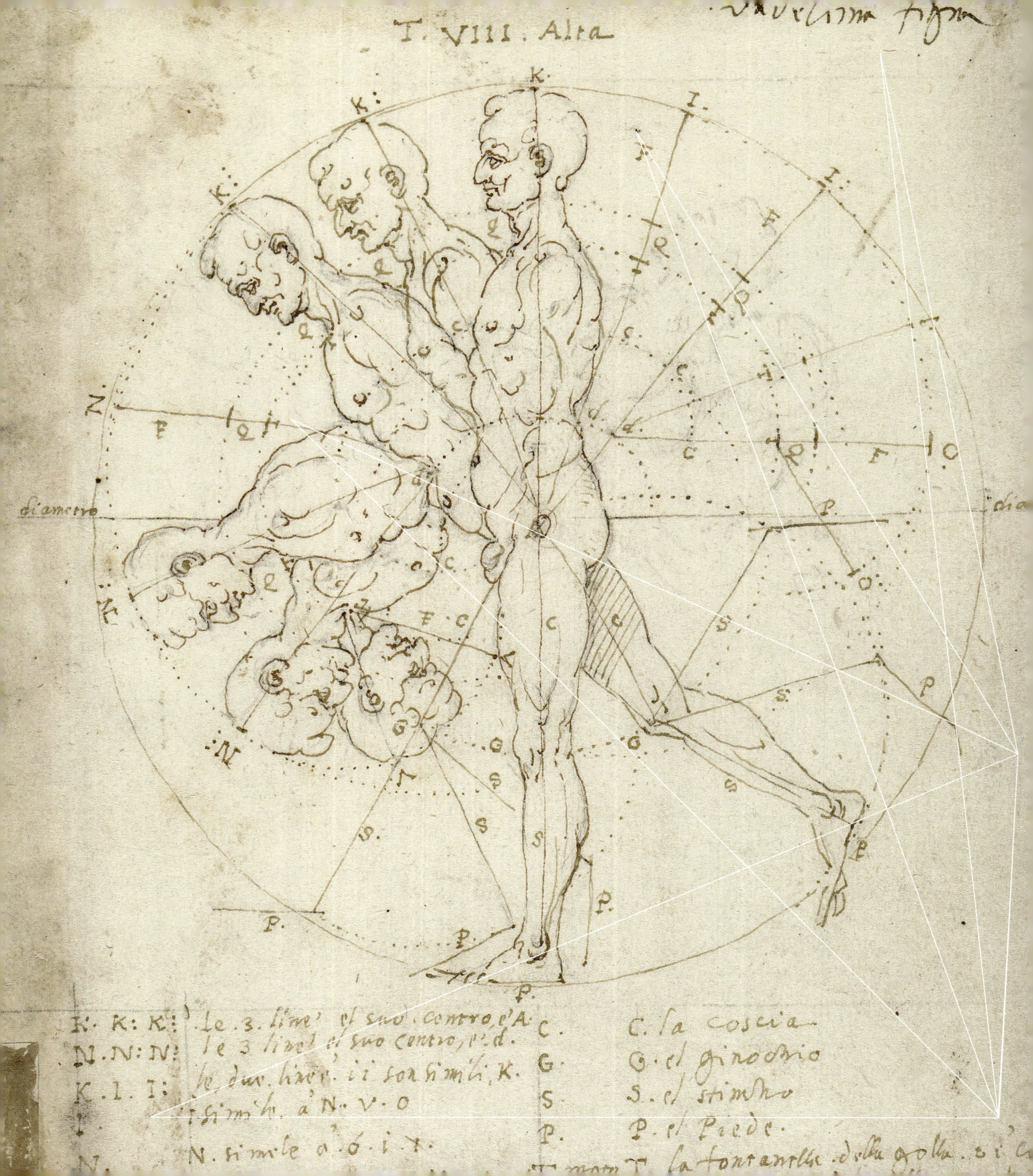
T. VIII. Alta
diametro
K. K: K: le 3. linee el suo centro è A
N. N: N: le 3 linee el suo centro è d.
K. I. I: le due linee ci son simili K.
simili a N. V. O
N. simile a O. I. T.
C. la coscia
G. el ginocchio
S. el stincho
P. el Piede.
T. la fontanella della golla

II.

SEEING THE UNSEEABLE

4. Leonardo's Point

FRANK FEHRENBACH

Emphatic trust and consistent doubt are intricately bound together within the history of visual experience in Western culture after the thirteenth century, and have inevitably led to prolonged reflections upon the organ of vision. Both the invention of increasingly powerful optical devices, which overcame the apparent invisibility of very remote and very small objects, and the development of scientific tools that ultimately made the presence of a perceiving eye obsolete[1] are grounded in a conception of the eye as a natural, but inherently limited, instrument of observation. In this history of faith and mistrust surrounding the visual organ, a specific form of "romance" between man and his eyes, Leonardo da Vinci, the paradigmatic artist-scientist of the Renaissance, plays a prominent role. What still remains largely overlooked, however, is his surprisingly persistent interest, culminating in the

I wish to express my gratitude to Marisa Bass and Aliki Economides for polishing my translation, Michael Cole and Fabio Frosini for their critical comments, Mareike Wübbenhorst and Vivian Michalski for editing the text, and Alina Payne for her infinite patience. All translations are my own unless otherwise noted.

years around 1505, in the interpenetration of the visible and the invisible in the act of vision—and his reflections on the paradoxical nature of physical reality itself. In the following, I rapidly sketch out Leonardo's persistent focus on this paradoxical ontology and its reverberations in some major fields of his intellectual endeavors: optics, cosmology, physics, and art theory.[2]

In the context of the present volume, Leonardo's approach to vision and its instruments holds an important position. It was the artist's emphasis on imitation that led Leonardo to a revision of his representational convictions and to an acknowledgment of the dramatic pervasion of being and nonbeing (zero, nothing, naught)[3] in the majestic performance of nature. Far from becoming an armchair skeptic in a time of very limited research technologies and mathematical understanding—the right man in the wrong era[4]—Leonardo pursued scientific and artistic strategies to comprehend the world as transition, a world that dies and is reborn at any moment in time, in short, a world that *reinvents* itself continuously. Following the line of his major argument, it would have been quite natural for Leonardo to embrace the technical support of optical devices. But Leonardo's reflections upon the inclusion of the invisible in vision also hint at a phenomenology *avant la date*. Despite some recent studies, this aspect of his work still demands further clarification.[5]

Nothing

In the first paragraph of his unabridged, posthumously edited *Treatise on Painting*, written circa 1500–1505,[6] Leonardo claims that painting is a science (scientia) because, like geometry, it is founded on an ultimate principle (ultimo principio): the point. "Therefore, the point is the first principle of geometry, and no other thing can exist either in nature or in the human mind that would be more fundamental than the point [che possa dare principio al punto]. If you were to say that the creation of a point is the final contact made with the point of a stylus on a surface, this is not true; we would say such contact is a surface that surrounds a center, and in that center is the location of the point."[7]

All the points of a surface, even of the world, would not create "more" than a single point, Leonardo continues. The point is not materially part of the picture surface (non è della materia di essa superfitie); nevertheless, it mysteriously creates the image. By comparing the point to zero, Leonardo seems to equate the point with "nothing," but the fact that the "addition" of zero, as Leonardo puts it, changes the value of numbers—from 1 to 10 to 100 ad infinitum—provides an analogy to the dynamic qualities of the point, namely, its extension into line and surface on a picture's plane.[8]

Lines and surfaces are created by the transit of the point (la linia è il transito del punto). Consequently, lines and surfaces also have no extension; they are "something spiritual rather than substantial" (cosa spirituale che sustantia).[9] The point is thus intimately connected to movement and therefore to Leonardo's categories of physics. I will return to this issue.

In the light of the apparently abstract quality of the point, Leonardo's bold assertion that "painting is mental" (la pittura è mentale)[10] comes as little surprise, yet it is, nevertheless, only one side of the coin. Leonardo also emphasizes again and again the importance of visuality, leaning strongly toward an identification of sensory activity as a mental activity per se, thereby circumventing the traditional hierarchy between "outer" and "inner" senses.[11] However, as is well known, Leonardo's remarks on the point as an invisible principle of painting contrast significantly with those of Alberti, Filarete, and Piero della Francesca. All three authors describe the first element of painting as a *visual* point, as the smallest *perceptible* element on the picture surface.[12] Alberti categorically denies the relevance of invisible entities for any discourse on painting: "Nobody would argue against the fact that the things that we cannot see are irrelevant for the painter."[13] Following the Aristotelian tradition, Piero della Francesca calls the point a product of human imagination (inmaginativo).[14]

Leonardo, instead, seems to follow closely the authority of Euclid, who defines the point, at the very beginning of his *Elements*, as "that which has no part."[15] But the paradox in Leonardo's argument is obvious. Identifying the first principle of painting as a mathematical point seems to be a negative answer to the question that Leonardo poses in the aforementioned passage, as to whether or not painting is a science: "That mental discourse is termed science that originates in first principles beyond which nothing else can be found *in nature* as part of this science."[16] Later in the same paragraph, Leonardo once again relates mathematics to nature or visibility, emphasizing that all the sciences have to be transparent to mathematical demonstration. However, only those that do not begin and end in the mind, but are based on sensory experience, are true sciences.[17] The question is, how can an "abstract" quality like the point be part of nature, and therefore belong to sensory experience?

In his brilliant article "Leonardo da Vinci e il 'nulla,'" Fabio Frosini reconstructs the artist's ideas about *nulla*, *punto*, and *zero*, and locates them in classical, medieval, and contemporary discourses on mathematical entities, cosmology, and metaphysics. According to Frosini, Leonardo developed and accepted paradoxical formulas in his definitions of the point for at least two reasons: first, to overcome the categorical difference between mathematics and nature; and second, to depart from this traditional juxtaposition in order to develop an ontology that interprets nature as a transition between being and nonbeing. In his meditations on the paradoxical nature of the point, Leonardo established an interpretive model that permits reflection upon the continuity and, at the same time, discontinuity of bodies. Within this dramatic monologue, the *punto* emerges as a term used to describe the transitional world of objects in contact, movement, and metamorphosis. This chapter focuses on these *dynamic* qualities of the point and its impact on art and science. Leonardo conceives of the point not as a mathematical, nondimensional entity, nor as the smallest visible sign, but rather as an infinitesimally small entity that motors the transition of nonbeing into being, and vice versa.[18]

In a series of fascinating meditations that can be dated—following Carlo Pedretti[19]—around 1505–8, and which are found mostly in the Codex Arundel of the British Museum, Leonardo struggles with the paradoxical properties of point and "nothing." At the beginning, he defines "nothing" as an abstract and static concept that is completely separate from the natural realm of bodies and space: "What is called naught [niente] can only be found in time and in words; in time it is between past and future, while it is not part of the present, and also in the names of things that do not exist or are impossible . . . and its [nothing's] power is not extended to the things of nature."[20] According to this traditional view, *nulla* is just the negation of any continuous and extended object, pure "nothingness," a semantic construction that defines the ontological status of time (past and future) and imagination. This reflects earlier passages in Leonardo's notebooks, for instance, the Codex Trivulzio, fol. 34v: "The point is not part of the line. / The water that you touch in rivers is the last of the one that passed, and the first of the one that comes; similarly the present instant."[21] The point as *niente* does not reside in nature (e la sua potestà non s'astende infra le cose di natura); like the infinite, it merely provides an analytic tool to investigate nature.[22]

The ontologically *positive* role, the actual presence of "nothing" in the midst of "being," emerges only in Leonardo's subsequent reflections. On the same folio of the Codex Arundel, Leonardo writes, "Among the magnitude of things that are among us, the being of nothing is the main principle, and its domain [ofitio] extends to the things that have no being."[23] In this more complex view, the point, and the lines, namely, the surfaces it creates, are "limits," the common "border" of two things or the "border" between an object and "nothing." One of Leonardo's favorite examples in this context is the common surface of water and air. While there is actually "nothing" (materially) between both elements, they are nevertheless "in contact." What differentiates them is neither water nor air; it is "nothing": "Therefore there's one surface as a common border of two bodies that are not continuous, and it does not participate either in one or in the other. Because if this surface would be part of [one or either of them], it would have a divisible extension, and since this [border] is not divisible, the nothing [el nulla] separates those bodies from each other."[24] It is the inclusion of the definite article that transforms, as Frosini has shown, "negative nothing" (as a semantic placeholder for absence), or empty space (the vacuum), into the *positive* nothing, an intrinsic category of physical reality that ensures both separation (autonomy, coherence) and continuity of bodies at the same time.[25]

At the end of this breathtaking intellectual exploration, Leonardo defines the point as a third, liminal entity *between* nothing and something. "Nothing can be called smaller than the point, and it is the common border [termine] of naught and line, it is neither naught nor line, and it does not occupy any space between naught and line. Therefore, the end of nothing [!] and the beginning of the line are in contact, but not connected. And in this contact the point is the divider between the continuity of naught and line."[26] There is not an identity but an intimate relationship between point, the infinitely small, and naught.[27] It is striking that whenever Leonardo tries to establish a complex and dynamic relationship between identity and difference (for instance, among the "four powers" in nature—force, movement, gravity, and percussion—or painting and music), he uses metaphors of kinship. "Therefore, the end of the nothing and the beginning of the line are in contact, but not connected; and in this contact is the point. Naught is the brother of this point."[28]

The point ensures, to put it somewhat differently, continuity and discontinuity *at the same time.* As the "extension" of the point, the surface itself oscillates between nothing (pure absence) and something: "Air is conjoint with water, and the end of the one is shared with the other, in a way that it could be called continuous quantity because they are connected, and discontinuous because they have two different natures."[29]

However, by defining the point as the infinitely divided, smallest unit *in nature,* a unit that constitutes the irreducible core of lines, surfaces, and bodies, and by ultimately reducing all positive data to one single point,

Leonardo had to face the collapse of the part and the whole, of naught and infinity: "all the things that do not occupy any space are similar among themselves . . . and each of them is equal to all of them. Therefore, it follows in this case that the part is equal to the whole, and the whole to the part, and the divisible to the indivisible, and the finite to the infinite. Because of this, the surface, the line, the point are nothing, because they do not occupy any space, and all the nothings [zeros] are equal to all, and all to each, as arithmetic demonstrates. . . . The body is draped in surfaces, the surfaces are surrounded by lines, and the lines are terminated by points."[30] Frosini concludes that this paradox (the identity of the infinite with the whole, and the visible with the invisible) marks the major achievement of Leonardo's intellectual journey,[31] a natural extension, we might conclude, of Neoplatonic theology and its notion of God as "everywhere and nowhere" (ubique et nusquam).[32]

The Dynamics of the Point

Following but also diverging from Frosini's main argument, I would like to emphasize that Leonardo introduced his discussion of dynamics (movement) in an attempt to dialectically reconcile the terms of the paradox that he had previously established. Leonardo conceived of the point as a liminal entity between "something" and "nothing," an entity that not only ensures continuity and discontinuity but actively (in atto) oscillates between the two states. In other words, at the core of physical reality, an immanent principle works constantly against nondimensional "unity" in an effort to achieve "multitude" and also, simultaneously, wears down the distinctions and plurality of the three-dimensional world. The point, in this view, is the chief agent of the continuous drama of a world that "contracts" into the infinitely small, that collapses into the principle of unity, *and* that reemerges at every infinitely small moment of time and at every infinitely small point of transparent space. It is through his identification of the point with the principle of movement that Leonardo reintegrates "nothingness" and nature in her capacity for transformation. In the point, as a principle of every process, nature dies away and is reborn *continuously*. It is hardly surprising that Leonardo develops this argument further after around 1500 in the context of his general focus on antagonism in major fields of his scientific exploration: hydrology, geology, and art (*Battle of Anghiari*).[33]

Without naught, objects in the world would not be differentiated from one another. But without naught, movement would also be impossible. The classic Epicurean argument already emphasized that there would be no "space" for change if "void" did not separate the positive data of objects (atoms).[34] But Leonardo's approach is more complicated and risks even more contradictions. At the beginning of his subsequent investigation into this concept (ca. 1506–8, according to Pedretti), Leonardo clarifies the identity between point and moment, line and *ispazio di tempo*. Since it has no surface, or body, time is not a continuous "quantity" but is instead comparable to point and line and therefore one-dimensional. Time is the result of the movement of "nothing"—the point of pure presence, the instant—comparable to the movement of the point creating the line (Lo instante non ha tenpo. El tenpo è causato dal moto dello instante).[35] In the continuum of time, the point of the instant works as the connecting and separating unity between past and future, which are also "nothing," as Augustine had already demonstrated.[36]

Implicitly critical of his own earlier attempt to ground the *scienza* of painting in the principle of the point as an invisible entity, Leonardo continues, "the point is nothing [nulla], but on the nothing one cannot build up any *scienza*. And to avoid this principle we will say: Nothing can be smaller than the point, and the line is created by the movement of the point, and its ends are two points, and the surface is generated by the transversal movement of the line, . . . and the body is created by movement [e il corpo è fatto dal moto]."[37] But the converse of the argument also holds. In the latter case, privation (i.e., "nothing" in action) *reduces* the dimensionality of bodies, surfaces, and so on.[38] As will soon become clear, this understanding of a gradual negation as a privation of dimensions determines Leonardo's description of major processes in nature as well as his interpretation of the painter's creations.

Therefore, in contrast to the opposition between space that is filled and the empty space within a vacuum, nothing and something are everywhere *gradually* connected by the infinitely small: "Where nothing ends, the thing is born, and where the thing disappears [dove manca la cosa], nothing emerges."[39] The relationship between naught and things does not dissolve in a simple teleology—God's *creatio ex nihilo*, as the most powerful paradigm, or the "almost nothing" (ultima pene nihil; Matthaeus Frigillanus, *De nihilo*, Paris, 1562) of the *prima materia* in Christian and Neoplatonic thought[40]—but Leonardo conceives it as a continuous, oscillating, ambivalent process, open to movement in both directions. Corporeal nature, in other words, is entirely, permanently permeated by nothing, but *nulla* is—through the borderline activity of the point—continuously "emanating" into space and time. It is the point that creates and negates the dimensions at the same time.

Leonardo's point, in other words, is the power of transition itself—a liminal entity connecting and dividing, a paradoxical being, the all-pervasive "One," a motor that works against its own integrity, identical with itself only through permanent self-transcendence: "because movement is of the same nature as line, and the line initiates in a point, therefore movement, too, originates in the point."[41] Consequently, the end of movement in percussion, for instance, is also a point, the liminal entity that marks the completion of a change between two states: "Percussion is the end of movement in indivisible time, because percussion is caused in the point, the end of the line of movement."[42] We will soon see how this argument continues in Leonardo's reflections on the nature of impetus as the fundamental, punctiform entity of his theorization of dynamics.

At this juncture, it is useful to situate, albeit briefly, Leonardo's paradoxical notion of point and naught in the extended history of the idea of nothing. Indeed, it would be especially interesting to outline the implicit nonindifference of naught, the concept of an absence that awaits its transition into presence, or, in other words, a history of the dynamics of the void. Hebrew thought radicalized the nonbeing of nothing (to use a tautology), but already the early Christian conflations of *nihil* and evil (for instance, in Augustine) qualified the void not simply as an absence but as absolute negativity.[43] However, the Christian substitution of the precreational naught with Aristotle's *prima materia* (*hypokeimenon*) and the mystic's conflation of God and *nihil* overcame the odd negativity of radical nonbeing.[44] In Greek cosmology, Plato's mysterious concept of the primordial "nurse of coming-to-be" (*Timaeus* 52d–53a) and his definition of space as receptacle (chora) indicate a teleological understanding of absence as a gap waiting to be filled.[45] Brian Rotman and Robert Kaplan traced the persistence of this idea of zero as a "pregnant void"—the hollow circle waiting to be filled with pebbles on the calculator's table or on the Mancala board; the substitution of zero's circle by the egg; the name *theca*, signifying "container," for zero (see Giovanni da Sacrobosco); but also the sign burned into the skin of criminals;[46] the invention of the pause in Western musical notation;[47] or "'nothing' as slang for vagina in Elizabethan English."[48]

This could be compared, as John Barrow showed, to Hindu meditations on naught (kha) as the lodge of an object, as a figure in mathematics, or directly as receptivity (sunya).[49] Indeed, despite the lack of any positive connections, Hindu thought provides probably the closest affinity to Leonardo's reflections. Among the rich Indian vocabulary for zero and nothing, the concept of *bindu* (or *vindu*)—the point—reflects three decisive features of Leonardo's argument: the motion of the primordial unit, creating lines, planes, and three-dimensional bodies (i.e., the creation of space); the conception of *bindu* as a generational force; and the regression of being into nonbeing as a general trajectory of Hindu thought.

In our context, the dynamics of naught or, better, its impact on the realm of being are of particular importance. Again, paradoxical theological thought provides the most detailed materials for the interrelation of the partners in this dialectic, naught and being (negative theology; God as the one who embraces nonbeing, etc.), as the texts in Colie's and Ossola's works amply demonstrate.[50] Of course, one could add the ethical paradoxes of naught, so wonderfully explored in Shakespeare's *King Lear*.[51] Quite obviously, in both areas the central notion of imperfection includes by definition a lack, an absence, the impact of negativity *in action* (as in every privation). In Pico della Mirandola's short treatise *Of Being and Unity*, naught enters the stage precisely as the force of the immeasurable distance between God and his creation, but also as the necessary category that distinguishes the units of individual bodies from one another.[52] Closer to Leonardo's world, however, are the dynamics of zero in two intimately connected areas, computing and economy. Mathematicians soon recognized the power of *theca vel circulus vel cifra*. While zero passively takes the value of every figure it is added to or subtracted from ($12 + 0 = 12$; $12 - 0 = 12$)—the perfect example of the benevolent indifference of naught—its annihilating powers appear when the figure is involved in multiplication and division. Here, zero sets free exactly those implosive and explosive energies that are mirrored in Leonardo's identification of the infinitely small (the point) with the whole ($12 \times 0 = 0$; $12 : 0 = \infty$). But as a result of calculation in double-entry bookkeeping, the revolutionary invention of the late medieval Italian marketplace, zero is the equivalent of the merchant's sigh of relief, the fragile *balance* of credits and debits that keeps the trader's ship afloat.[53]

One consequence of Leonardo's meditations on *punto* and *nulla* is quite obvious: without the infinitely small (the point, as a transitional element between being and nonbeing), there would be no distinction between objects in nature, but also no multitude in the visual field; objects would be indistinguishable, and therefore invisible. Only "nothing" allows for the possibility of vision.[54] At the same time, it is true that *no thing* is really visible (because of the invisible nature and the lack of extension and quality in points, the final constituents of bodies and time). The major impact of this idea on art and representation in general cannot be underestimated. However, instead of following this line of thought, I would like to rapidly situate Leonardo's insistence on the immanence of the point in the broader horizon of his scientific thought, which extends to optics, hydrogeology, and physics. As I hope will become clear in the following rough outline, in each of these fields the notion of the point is directly linked to dramatic processes on a macroscopic level. The reversible action of the point—creating extensions, and establishing, at the same time, the presence of naught, of an absence in nature—leads to oscillations between being and nonbeing on a cosmological scale. In each case, the point appears as a motor of transition, as the transitional entity through which being (movement, the irregular shape of the earth, images) collapses and is—marvelously—re-created.[55]

Optics

How is the point related to optics? Perspective and its reliance on the transformation of three-dimensional bodies into punctiform marks on the "window pane" of the picture surface played a fundamental role not only in Leonardo's theory of painting, but also in other major fields of research.[56] Leonardo reconceptualized Alberti's main goal—the description of a constructive device for painters—by interpreting perspective as a fundamental phenomenon of nature itself. Since late medieval optics, the linear transmission of punctiform surface properties (rays) was reconciled with the linear transmission of object *forms* (*eidola, spetie, simulacra*) in transparent media.[57] Alhazen's influential concept of *physical* Euclidian rays that transport surface relationships (mainly form and size) reliably to the eye (as long as they are projected perpendicularly on the pupil's convex surface) did not contradict the Western (Epicurean) model of image emissions.[58] In other words, the division of the homogenous optical field in single rays of light provided primarily an *analytical* model, especially for the geometrical reconstruction of reflections (catoptrics) and refractions (dioptrics). But the de facto continuity of punctiform surface emissions still allowed late medieval perspectivists to conceive of these emissions as coherent *images*. Leonardo's optical studies embraced this twin model. In his diagrams, he visualized the optical process by lines (light rays), while his texts refer mostly to *spezie* and (less frequently) *similitudini*.[59]

Alberti remained explicitly undecided in the long and complicated debate about the direction of visual rays.[60] But Leonardo, following the mainstream of late medieval optics, emphasizes that it is the lit object that sends out two-dimensional images of its "true form" through transparent media (air, water, glass, etc.). These image emissions diminish regularly, by straight lines, ensuring the fundamental law of perspective—the inverted ratio of distance and apparent size of objects. The problems of constructing space and the apparent convergence of orthogonals occupied a much lesser rank in Leonardo's thought. This does not mean that he was not interested—theoretically and practically—in Alberti's (or Brunelleschi's) perspective construction. However, constructing an image of an architectural space (with the main problem of the regular diminution of the transversals) became just a special case within the broader framework of Leonardo's optics. Or, as Leonardo stated categorically, "The eye does not see if not through a pyramid. The perspective of painters is unfounded without the pyramid."[61] Instead of developing geometrical models to create fictive space (a tradition that culminates in Piero della Francesca's and Albrecht Dürer's writings), Leonardo highlighted the complementary aspect of perspective—the diminution of object images on their way through space.

The exceptional state of physical light does not consist in its almost instantaneous swiftness and geometric order alone; it is also proved by its unique independence from the natural vertical movement of the four basic elements and the mixtures of them. Light is free to radiate through transparent space in every direction. But why do visual rays converge? Light sources behave differently; they *spread* light in space. One of the most powerful cosmological paradigms of the thirteenth century, Robert Grosseteste's model of creation, describes the emergence of space as the expansion of light from God's *punctus lucis*, a point that marks the transition from naught into being.[62] Indeed, emissions of light, sound, or odor (as "rarefied" substances, Leonardo refers to them sometimes as "spiritual")[63] proceed in this way. In an early passage of his Codex Trivulzio (ca. 1487–90), Leonardo writes, "All the spiritual powers, the more they move away from their first or second cause, the more space they occupy, and the more they diminish in force."[64] The accompanying diagram shows a candle whose flames emit a cone *diverging* in space.

Alhazen explained the convergence of visual rays by the "physiological" necessity of perpendicular rays in perception. A passage in Leonardo's MS A, fol. 27r (ca. 1490) seems to reflect this idea; here, the eye appears as "target and magnet" of species (dalle quali spezie l'occhio si fa berzaglio e calamità). But is the decrease of apparent size in space different from the diminution of other object qualities (color, sound, odor, etc.)? In other words, is the expansive diffusion of surface images in transparent space irreconcilable with the regular diminution of individual *species*? Obviously not, as a diagram in the same MS A demonstrates (fig. 4.1).

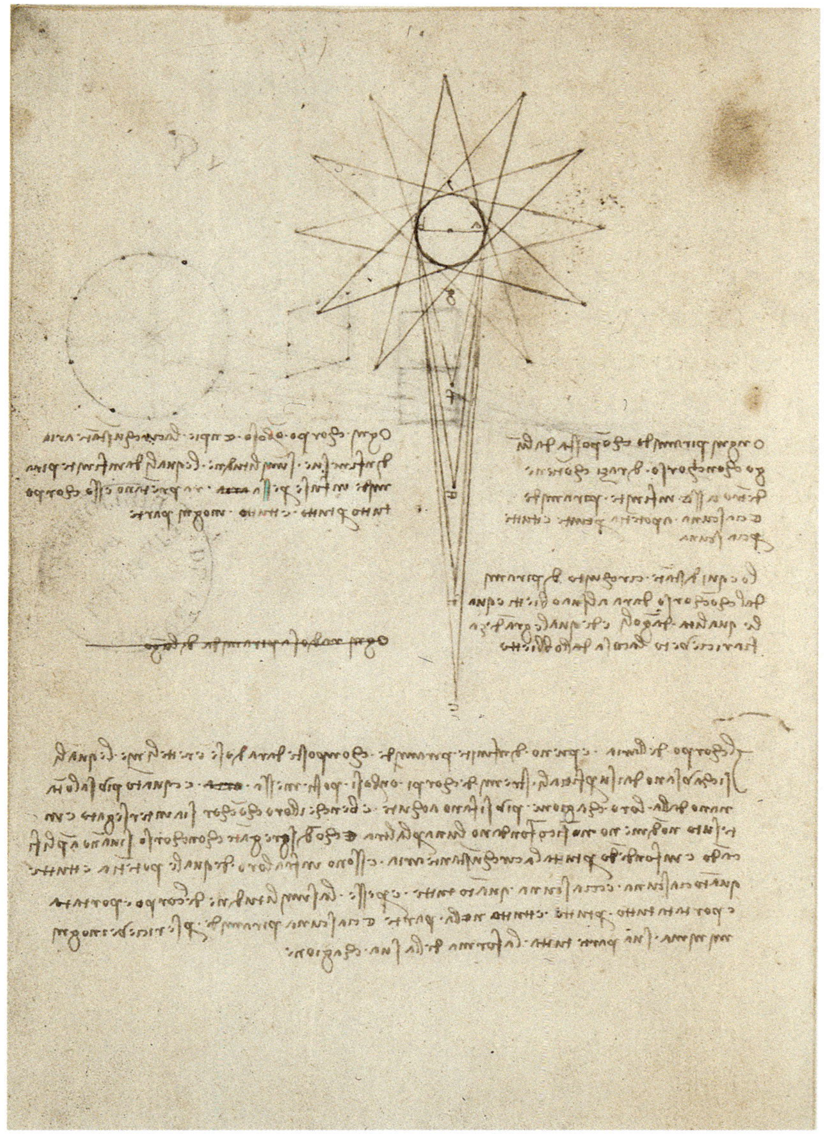

4.1 Leonardo da Vinci, Manuscript A, fol. 86v, Institut de France. Paris.
© Bibliothèque de l'Institut de France, Paris.

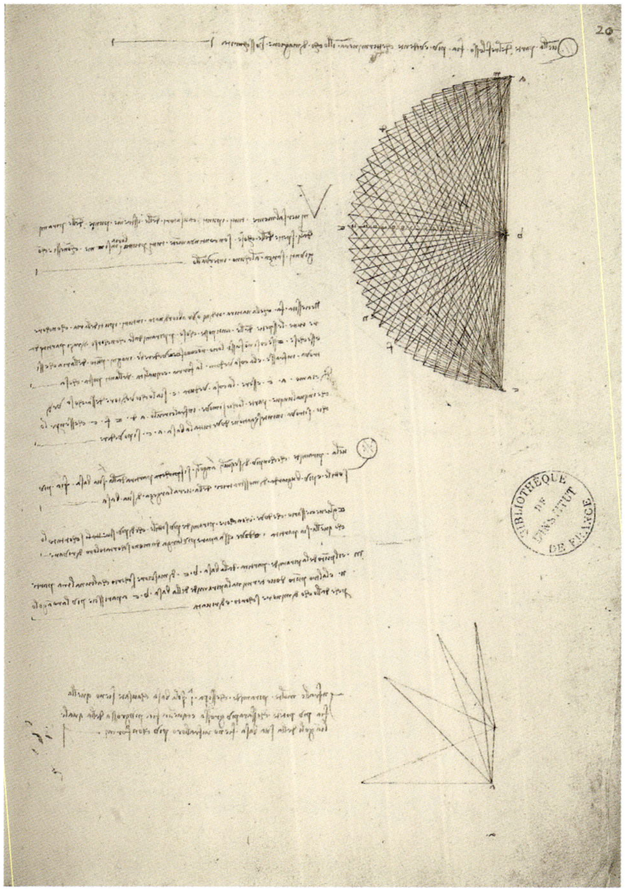

4.2 Leonardo da Vinci, Manuscript C, fol. 20r, Institut de France, Paris. © Bibliothèque de l'Institut de France, Paris.

As two-dimensional images of three-dimensional objects, the simulacra become, in the end, the apex of the visual pyramid and are reduced to points. Undeniably, the whole process has a sublime character. Activated by light, objects emit images of their "true form" that diminish at every point across the transparent medium. The semiosis of nature, as I feel tempted to say—the permanent communication of forms in space, the fundamental mechanism of every lit object to flood space with images of itself—creates countless, regularly diminishing simulacra of surface forms, an ocean of *visible* points.[65] "Pure air is capable of receiving in itself, outside the interval of time, every likeness of a corporeal nature that is hit by the rays of the sun or is illuminated by another agent. . . . Every [minimal] part of the air receives in itself all the images/simulacra of bodies that see it [this part of the air] and that are seen by this part of the air at the same time" (see fig. 4.2).[66]

The particular nature of these points and of their relationship to the optical process concerned Leonardo for many years—and not only because the emission process seems to transform being into nonbeing, thereby extending the topos that large objects (entire mountains, the starry sky, etc.) appear miraculously diminished within the space of the minuscule eye.[67] First of all, Leonardo had to clarify whether perception is located at the apex of the visual pyramid or not. Some artists writing in the fifteenth century answered in the affirmative.[68] Leonardo's early optical texts followed this tradition, in the context of a "correct" reconstruction of space via perspective. Therefore, the center point of the eye became the point of reference for the establishment of the distance between eye and object—i.e., the distance point of perspective—a mobile point hit by image points in space, the place where the shapes and colors of objects are transformed into perception (fig. 4.3).[69]

However, in his later optical studies, conducted after 1500, Leonardo realized that there would be no perception of magnitudes if the place of perception were a nondimensional point. This is an argument that had already motivated Alhazen to speculate about the complex refractions caused by the complicated layers, spheres, and densities of the eye. The site where three-dimensional objects are transformed into perception must therefore necessarily be extended, an argument that blurs the mathematical simplicity of perspective and contributes decisively to Leonardo's sharp critique of perspectivists in the years around 1500.[70]

Leonardo's struggle for an alternative model included the traditional hypothesis that the *virtù visivia* should be located at the surface of the eye or at the front surface of one of its "humors," that is, *before* the rays intersect in a point.[71] However, since nature produces nothing in vain, how could this solution explain the complicated structure of the eye? Aristotle had already posed the same question.[72] Relying strongly on medieval optical texts like those by Roger Bacon and John Pecham (both of whom he referred to directly in his writings),[73] Leonardo had to accept that the eye's apparatus redirected the visual rays (fig. 4.4). During the very same years of his reflections on point and nothing (ca. 1505–8), Leonardo felt more and more convinced that the most important site of perception was the rear part of the organ, the place where the optical nerve penetrates the eyeball. But how could this observation be reconciled with the intersection of the visual rays in the eye's transparent *humores?*

The major problem was the necessarily inverted projection of the species in the rear part of the eye, Alhazen's and Bacon's main argument *against* the intersection of rays.[74] Leonardo was quite inventive in overcoming this apparent dead end (which was only accepted some one hundred years later by Johannes Kepler). He postulated a double intersection of the rays (fig. 4.5), or a plurality of intersections, in order to rectify the projected image, for some time even speculating on a first intersection in front of the eye, caused by the reflection at lash and lid (fig. 4.6).[75]

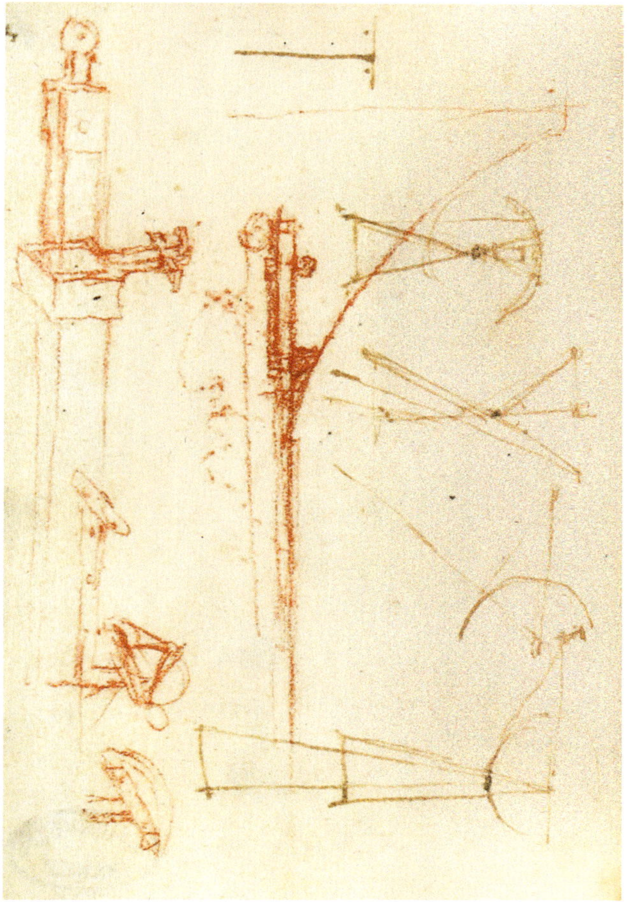

4.3 Leonardo da Vinci, Manuscript I, fol. 97v, Institut de France, Paris. © Bibliothèque de l'Institut de France, Paris.

In all of these observations and thought experiments, however, one thing remained undisputed: the fact that a two-dimensional image is one or several times reduced to a point, the infinitesimally small transition between body and naught. In a text that tries hard to differentiate between natural and mathematical points, a text that can be dated around 1508, Leonardo states, "All the minimal parts of the species penetrate each other without

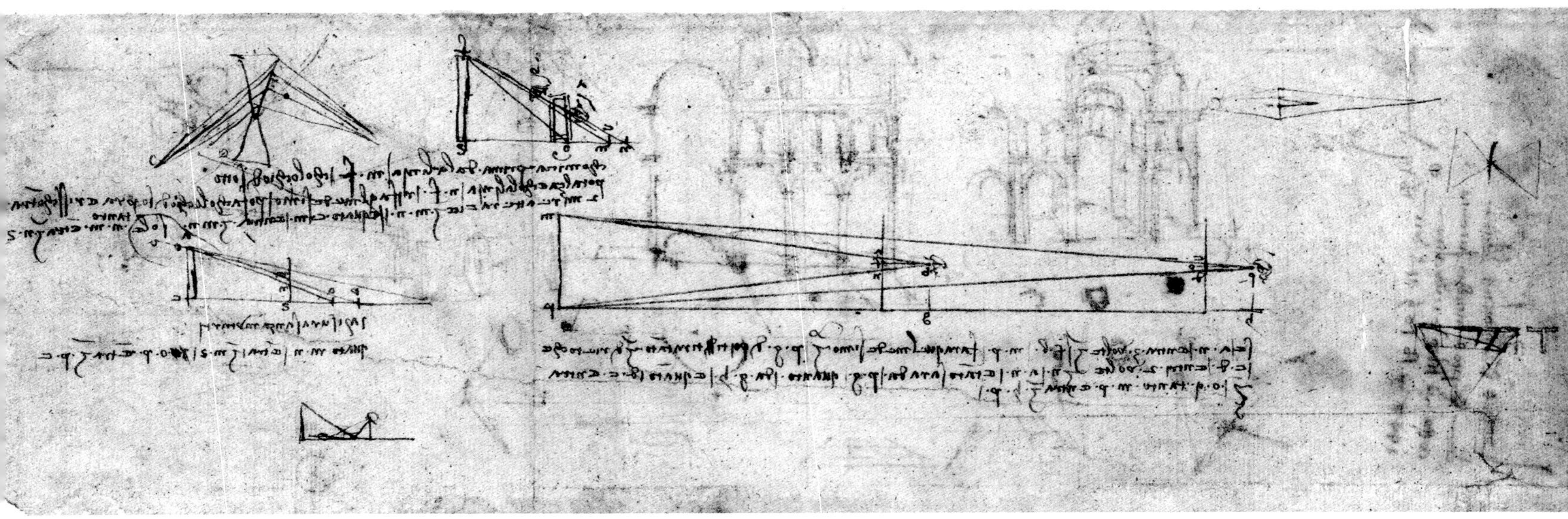

4.4 Leonardo da Vinci, Codex Atlanticus, fol. 119r. © Veneranda Biblioteca Ambrosiana, Milan / De Agostini Picture Library.

'occupying'/displacing each other. . . . So that we conclude that the surface [of objects] is reduced to a point . . . in order to be incorporeal."[76]

The projected image emissions of objects within the eye pass through a nondimensional point. Thus the self-images that nature "paints" proceed through "nothing." The optical process reveals itself as a proof of the point's presence in nature, a majestic, continuous spectacle of loss and rebirth of dimension, form, and beauty, an oscillation between being and nonbeing, a permanent model for artistic reinvention. The optical process provided the sublime paradigm for some of Leonardo's most intimate poetic lines, written at about the same time as his late studies of point and eye. It is no coincidence that these lines paraphrase a passage in Lucretius's long poem on atoms and void: "Look at the light and consider its beauty. / Close your eye and look at it again. / What you see of it, was not before, and what has been, is not anymore. / Who makes it anew, when its cause [the flame] continuously dies?"[77]

The Earth

In addition to optics, the point occupies a central position in other important fields of Leonardo's studies, namely, hydrogeology and physics. In both cases, the power of the point does not lead to the loss and rebirth of dimensions but to a similar regression to "nonbeing" and reemergence of "multitude": geometric irregularity in the case of hydrogeology, and movement in the case of dynamics. Especially after 1505, Leonardo's hydrogeology centers around concerns about the earth's transformations.[78] This is, to be sure, a natural consequence of his more than decade-old studies of erosion—the antagonistic struggle

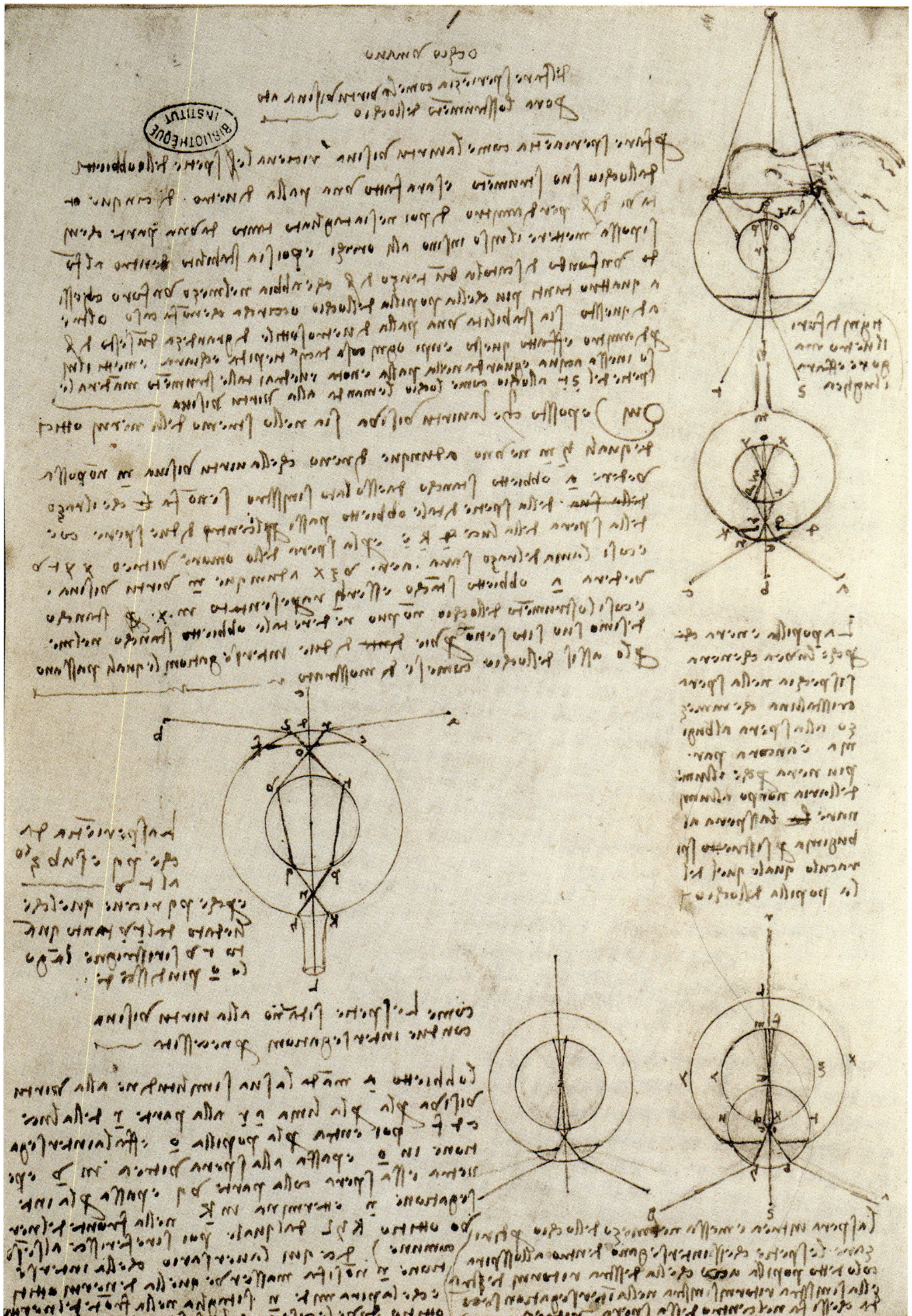

4.5 Leonardo da Vinci, Manuscript D, fol. 3v, Institut de France, Paris. © Bibliothèque de l'Institut de France, Paris.

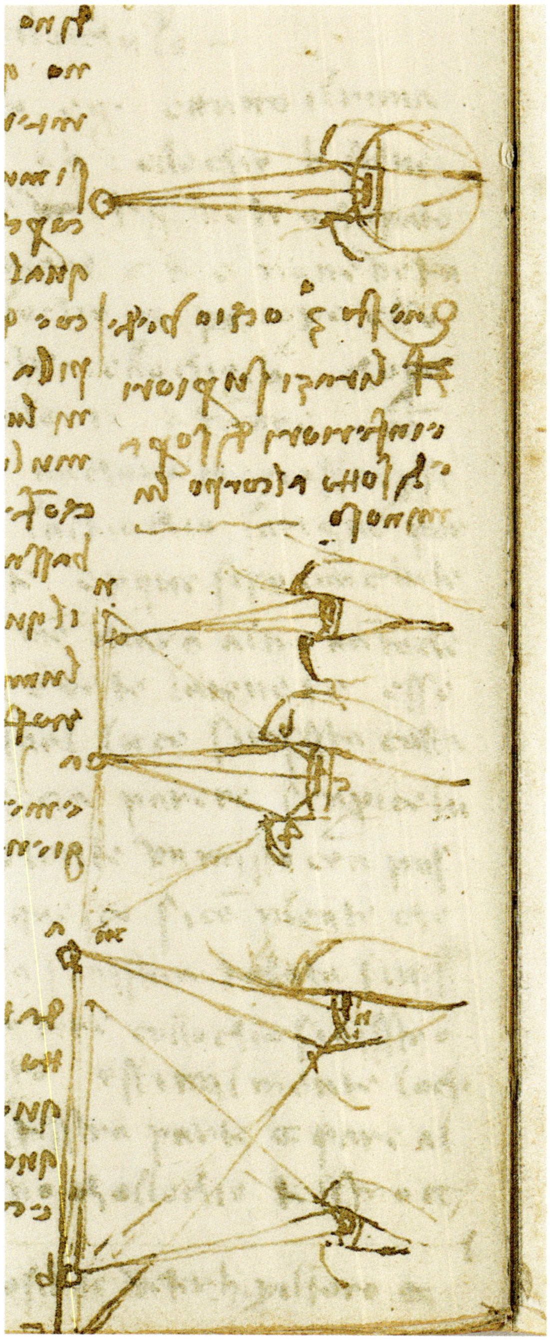

4.6 Leonardo da Vinci, Manuscript D, fol. 9v, Institut de France, Paris. © Bibliothèque de l'Institut de France, Paris.

between water and earth—but also of his long pursuit of the analogy between micro- and macrocosm and the implied "death" of the earthly organism (or machine, for that matter).[79] At about the same time that Leonardo was meditating on the point as a paradoxical entity in nature and abandoning the idea of optical perception at the apex of the visual pyramid, he was also developing potential scenarios for the earth's past and future (especially in Codex Leicester, ca. 1506–8, and MS F, ca. 1508). In both cases, Leonardo conceptualizes an equilibrium of the elements that is related to rest, geometric order, and death.

In this context, Leonardo's conceptual background likewise derives from the influential Aristotelian tradition and, more precisely, from the idea that movement in nature is the result of the mixture of elements in bodies and/or the effect of "accidental," "violent" intervention of external agents (for instance, organisms). While the globe of the earth at some point in the past lost its initial equilibrium (probably because of the movement of the ocean's waves),[80] the movement of the elements, and especially the erosive actions of water, are nothing other than a tireless and long effort to reestablish primordial equilibrium and rest. Following Aristotle's layering of the four elements, rest is achieved only when the lighter elements (water, air, and fire) find their proper place and return to their concentric spheres around the earth's center.[81]

In his earlier manuscripts, Leonardo had already referred to late Scholastic ideas about the earth's equilibrium, elaborated especially by Jean Buridan and Albert of Saxony.[82] According to these scholars, God's providence separated the center of the earth's gravity from the center of its geometrical magnitude, in order to allow for a global disequilibrium that, in turn, perpetuates movement, diversity, and life. Leonardo's doubts about the permanency of this constellation increased when he considered the erosive effects of moving water. Codex Leicester and MS F document another of Leonardo's fascinating intellectual journeys, this time a monologue about the validity of the Scholastic argument. Toward the beginning of MS F, Leonardo translates and quotes (!) a passage in Albert of Saxony's commentary on the pseudo-Aristotelian *De mundo*: "On the world. Every weight tends towards lower places, and high objects will not remain on their height, but by and by all of them will descend; and therefore by and by the world will remain spherical and consequently everything will be covered by water."[83]

For Albert, this remains an improbable hypothesis and serves mainly to praise God's *providentia* (the separation of the two "center points"). But Leonardo takes the assumption with deadly seriousness. His two emendations of Albert's Latin text (two times *col tempo*) emphasize the *time* it takes for the erosive powers of water (and air) to level the earth's surface. Mapping out a future in which all the mountains are worn away, allowing water to cover the earth's surface completely, he concludes that the four elements will be stratified spherically. The point of magnitude will then merge with the point of gravity, and all life on earth will be extinguished. This dreadful consequence of the terrestrial mechanism is outlined in the middle of MS F. Here, Leonardo repeats an abbreviated version of the passage on MS F, fol. 84r, adding laconically, "and it [the earth] will be uninhabitable" (e sarà inabitabile).[84]

Again, the permanent reduction of natural diversity becomes visible—the power of the point. Already in Codex Madrid I, fol. 146v (ca. 1495), Leonardo insists upon the fact that the center of gravity of every single body is *unica e sola*, and that this center is a point without extension. This center point in each object is the punctiform "soul of weights" (anima de' pesi). In projectiles it can be located in their front part, as the "guide of their movements" (guida de' lor moti; Codex Madrid I, fol. 146r). If the elementary spheres on earth would be concentric, the "odd concept"[85] of the earth's twin center would disappear, the two points would merge, perfect geometric order would be established, there would be no motion on earth, and terrestrial life would be extinct. In this process, the "accidental center" (the center of magnitude) of the earth would have to "die" (more) while merging with the "anima" of the earth's weight.[86] The history of the earth would be nothing but a very slow counterpart to the almost instantaneous loss of dimensions in the first part of the optical process, a global *contraction* of diversity, and another triumph of the point (fig. 4.7).

Thomas Leinkauf has convincingly demonstrated that the idea of a spatial and, at the same time, nonspatial (*illud paradoxum*), liminal, punctiform center of the world was crucial to the major geocentric argument of a much later scholar's astonishing work on the subterranean world: Atharasius Kircher and his concept of *centrosophia*.[87] However, as in the paradigm of optics, Leonardo embraced *and* complicated the idea of the point as an immanent (physical) element.

A close reading of Leonardo's manuscripts reveals that even at this moment of fusion of the two points, in the state of perfect geometry, global rest and death will only be transitional. On Codex Leicester 16A:21v (15), Leonardo assumes that the center of gravity of the watery sphere changes continuously as a result of the impact of winds.[88] By introducing the powerful activity of sun and moon, Leonardo acknowledges the dynamics of external

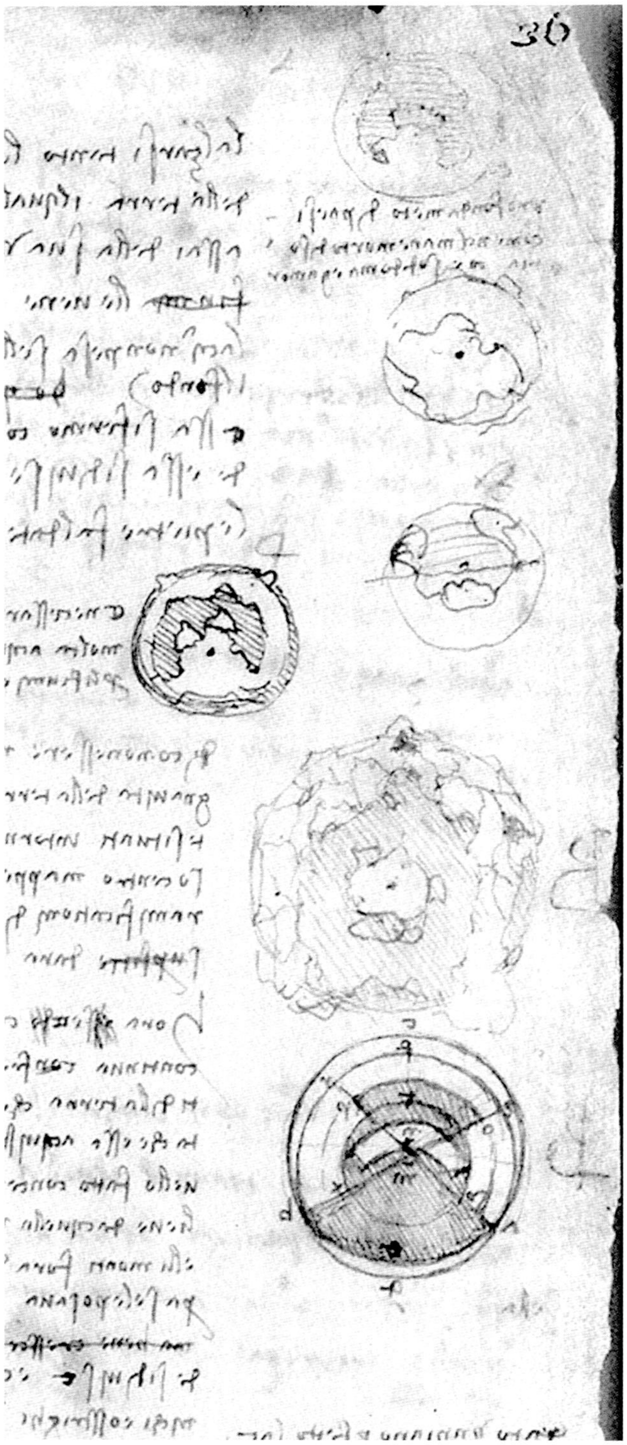

agents that will make sure that disequilibrium on earth continues. This culminates in a monumental *lalde del sole* in MS F (fols. 5r–4v), a hymn that digests countless classical and medieval sources, as Cesare Vasoli has shown. According to this text, not only the light of the stars but also the warmth in the universe and the "souls" of living beings originate in the sun. Equally important, the sun works against the merging of the earth's center points, as other texts of this period claim.[89] Even a spherical, water-covered earth would remain geometrically imperfect, because the sun heats and distends the equatorial seas, while allowing the polar seas to freeze and condense.[90] The equatorial waters are therefore "higher" than the northern and southern areas. Together with the moon's influence on the tides, the sun's heat works against the "perfection of the [watery] sphere" (perfezzione alla sua spera). This means that the merging of the two center points will be challenged, and the unique point *continuously* split in two, an oscillating rather than a restful entity. The consequences of the sun and moon's impact on the earth are dramatic; higher areas of the oceans are continuously set in motion (e piglia moto da ogni parte del colle).[91] Leonardo leaves it to the reader to conclude that disequilibrium will continue and that new agglomerations of soil under water will be most probable, eventually leading to new islands, larger areas of uncovered land, and so on. The telos of the earth—to become centered around a single point—can be conceived only as a moment of *transition*, like the convergence of image emissions, the intersection of rays in the eye, and the point *in atto* that is always neither nothing nor something, "the common border [termine] of naught and line, . . . neither naught nor line." In natural philosophy, optics, and cosmology, the point is *always* about to overcome its own paradoxical being, thus creating the line or collapsing into nothing.

Physics

In physics, Leonardo relied equally on Scholastic predecessors, a common foundation for fifteenth-century ideas about force and "nonnatural" movement, the central agent of which was *impetus*.[92] The concept of impetus as a force transmitted from the moving body to a passive object originated in late antique natural philosophy (John Philoponus, sixth century A.D.) and was reformulated, at the end of the thirteenth century, in the context of commentaries on Peter Lombard's *Sentences*. Two Franciscans were the first to rediscover the idea, Peter John Olivi, the head of the Spirituals, and Franciscus de Marchia, the rector of the order's *studium* in Paris.[93] In contrast to Aristotelian (and Platonic) natural philosophy, both Franciscans negated any decisive role of the medium surrounding an object moved by a "counternatural" force. Instead, they came up with the idea of a force actively transmitted by the mover into the passive mobile, a force that pushed the mobile further even after it lost its contact with the mover. This force was called *vis impressa* by Olivi and *vis derelicta*, the force "left back" in the moved object, by de Marchia. Future debates focused on the question of whether this force—now called *impetus*—diminishes "by itself" or whether it would be perpetual without the counterforces of gravity and friction. Impetus theory was powerfully elaborated by Nicolas Oresme and, again, by Jean Buridan and Albert of Saxony; one of its major exponents in Italy around 1400 was Biagio Pelacani of Parma.

Following this tradition, Leonardo described impetus as something added to each body in a state of nonnatural movement, the paradigmatic example being a stone thrown by an external mover. As a projectile, the stone follows a trajectory that is different from its natural inclination perpendicularly downward, virtually toward the earth's center point (fig. 4.8).[94] However, the force infused into a stone moving "nonnaturally" is not "corporeal"; it neither adds weight to nor changes the form of the moving body. Impetus affects the whole body of the projectile (as the soul in its organism), but its center occupies the *anima del peso*, the center of gravity of the object (in arrows, for instance, in the tip). Impetus adds to its object the resemblance of life, as Nicholas of Cusa had already observed: "The child takes the top which is dead, that is, without motion, and wants to make it alive; . . . the child makes it move with rotational motion as the heavens move. The spirit of motion, evoked by the child, exists invisibly in the top; it stays in the top for a longer or shorter time according to the strength of the impression by which this virtue has been communicated; as soon as the spirit ceases to enliven the top, the top falls" (*Dialogus trilocutorius de possest*).[95] These links between the moving force and life refer to Plato, who defines, in his *Laws*, autonomous movement as the main criterion of life.[96] Plato argues that the elements would be alive if they could move themselves. Consequently, the ubiquity of movement in nature manifests the activity of a living soul of the world. In *Phaedrus* 245c, Plato states that the object that moves, and is moved by others, participates in "life."[97]

For Leonardo, impetus remains a transitional, and, as a point, continuously self-transcending entity.

4.7 Leonardo da Vinci, Codex Leicester, fol. 36r. Courtesy of the Bill and Melinda Gates Foundation, Seattle.

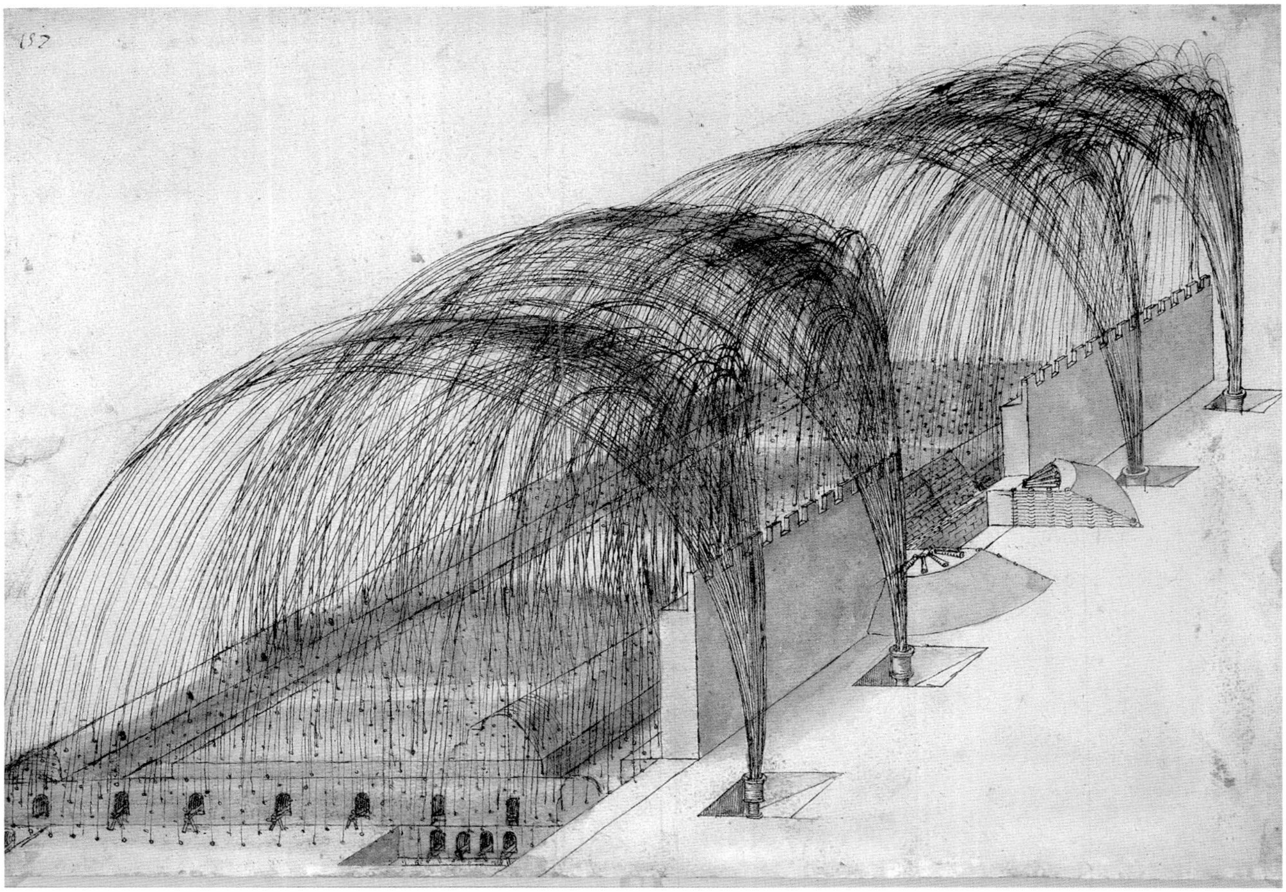

4.8 Leonardo da Vinci, *Military Project*, ca. 1503. Windsor Castle, Royal Library, inv. 12275. Collection Trust / © Her Majesty Queen Elizabeth II 2013.

Significantly, in the majority of his impetus-related texts, he describes impetus as a dynamic entity desiring its own death. Impetus (or force, *forza*) consumes itself, aided by the resistance of the medium (air, water, etc.) and "natural" gravity. "What is force? Force, I say, is a spiritual power, incorporeal and invisible, existing for a short life in such bodies that are, for some accidental violence, outside of their natural being and rest. Spiritual I called it, because in this force there's active life; incorporeal and invisible I say, because the body in which it is born grows neither in weight nor in form; short-living [I say] because [force] always desires to defeat its origin/reason, and, after having defeated it, [this force] kills itself."[98] Or, in a beautiful late observation: "this impetus, evader of the body where it is created, consumes itself and dies together with the movement of this body."[99]

A close parallel to the suicidal qualities of the impetus can be found in Seneca's *Naturales quaestiones* (7.9.3), where the antique philosopher describes a hurricane, concluding that every violent movement desires its own ruin. For Leonardo, suicidal impetus is the origin not only of bodily but also of sensory movement: "Force [forza] is complete and whole throughout itself and in every part of it. Force is a nonmaterial [spirituale] power, an invisible potency which is imparted and infused by accidental violence from animated bodies to inanimate bodies, giving to these *the semblance of life . . . it speeds in fury to its undoing*. . . . Retardation strengthens and speed weakens it. *It lives by violence and dies through liberty* . . . great power gives it great desire for death. . . . *It likes to consume itself* . . . nothing moves without it . . . *no sound or voice is heard without it*."[100]

Its desire to rest characterizes the animating point of impetus as a liminal quality, the force that connects two states of rest through movement. Leonardo's poetics of movement locate the driving force of objects in a pointlike entity "pushing" (spinge) against the body that created this force (dove è nata);[101] the object's movement thus results from the dynamic point's desire "not to be" (disiderio di non essere).[102] To extinguish themselves, these physical forces must squander their energies through movement, leading to the object's rest, which is conceived as a form of "death" of force, or stasis. For the physical world as a whole, this would be a futile desire, of course, because nature is by definition a realm of continuous movement and unrest.[103]

The parallels to Leonardo's meditations on the natural point are obvious. Each point transcends itself, tirelessly connecting, dividing, and overcoming nonbeing and being, a force creating *and* annihilating bodies (dimensions), time, and movement. Every point—the infinitely small entity in which all and nothing coincide—is identical with itself only through nonidentity. The point's desire to overcome its own being, to oscillate continuously to "the other side" (its continuous "suicide," the death of "nothing" being the birth of "something" and vice versa), creates multitude and movement. "Where 'nothing' ends, the thing is born, and where the thing disappears [dove manca la cosa], 'nothing' emerges."[104]

Painting

This chapter could not end without some hints at the importance of the point in Leonardo's art and art theory. It is extremely important to realize that what appears to be a mere mind game on Leonardo's part establishes not only optics but also painting itself as a paradigm and continuation of the most fundamental paradox in nature, a paradox with dynamic implications. Leonardo's style is what first comes to mind. No stylistic trait is more characteristic of Leonardo than the blurred boundaries and transparent substances of his soft bodies and surfaces, his almost weightless draperies and vaporous atmospheres. *Sfumato* can indeed be related to Leonardo's observations regarding the complexity of the visual process, particularly in relation to his optical treatise, the Manuscript D, of about 1508.[105]

Leonardo inherited the standard conviction of traditional optics, namely, that only the centric ray being emitted from or reaching each eye establishes a "sharp" perception of the object's form. However, following his ideas about the paradoxical status of the point and the nonexistence of contour lines, Leonardo went on to challenge this formula. Around 1508, Leonardo came to understand

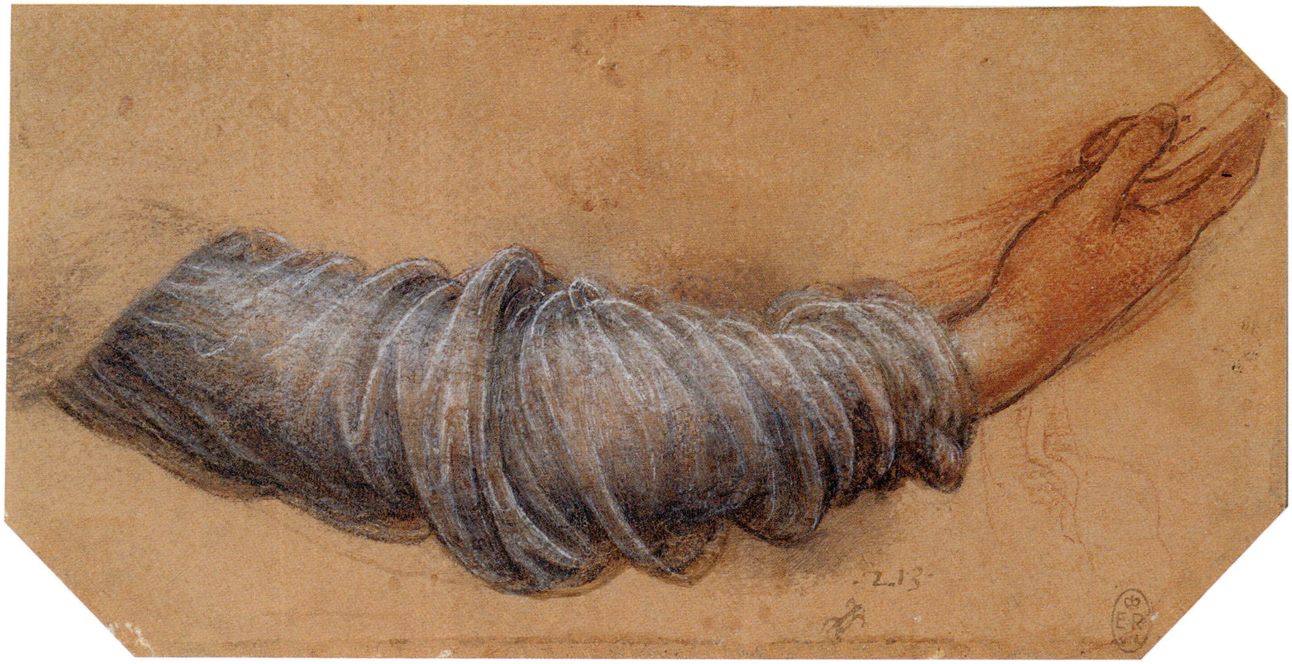

4.9 Leonardo da Vinci, *Study for the Virgin with Child and St. Anne*, ca. 1508. Pen and ink and wash, black and red chalk, heightened with white, 8.6 x 17 cm. Windsor Castle, Royal Library, inv. 12532. Collection Trust / © Her Majesty Queen Elizabeth II 2013.

the line of the centric ray as a liminal entity in a continuity of more or less sharp perceptions. More important, Leonardo observed that this liminal entity oscillates continuously in order to "scan" the object's boundaries and internal differentiations of detail.[106] *Sfumato* exaggerates the fact that physical boundaries are "nothing," paradoxical entities connecting and disconnecting bodies and their surroundings. In this revolutionary view, nothing (the centric ray) meets nothing (the contours of objects), creating a dynamic field of gradual differentiation and a negation of any positive location of forms (fig. 4.9).

Leonardo's oft-repeated formula for the ubiquity of punctiform object emissions in optics (*spezie, similitudini*) derives from one of the most influential Neoplatonic attributes of the soul: "tutto per tutto e tutto in ogni parte."[107] While Leonardo applied, at some point in his early anatomical research, the geometry of the Golden Section to determine the site of the *sensus communis* in the brain, he insisted, following the predominant view on the subject, that the human soul is undivided and omnipresent in its body. The soul is a point, indivisible and everywhere (in the body); it has, as Marsilio Ficino explains, a temporal existence without a spatial (material) extension.[108] In his fascinating study *Leonardo nella Roma di Leone X*, Domenico Laurenza has shown that the late Leonardo in particular would not have agreed with the supposed

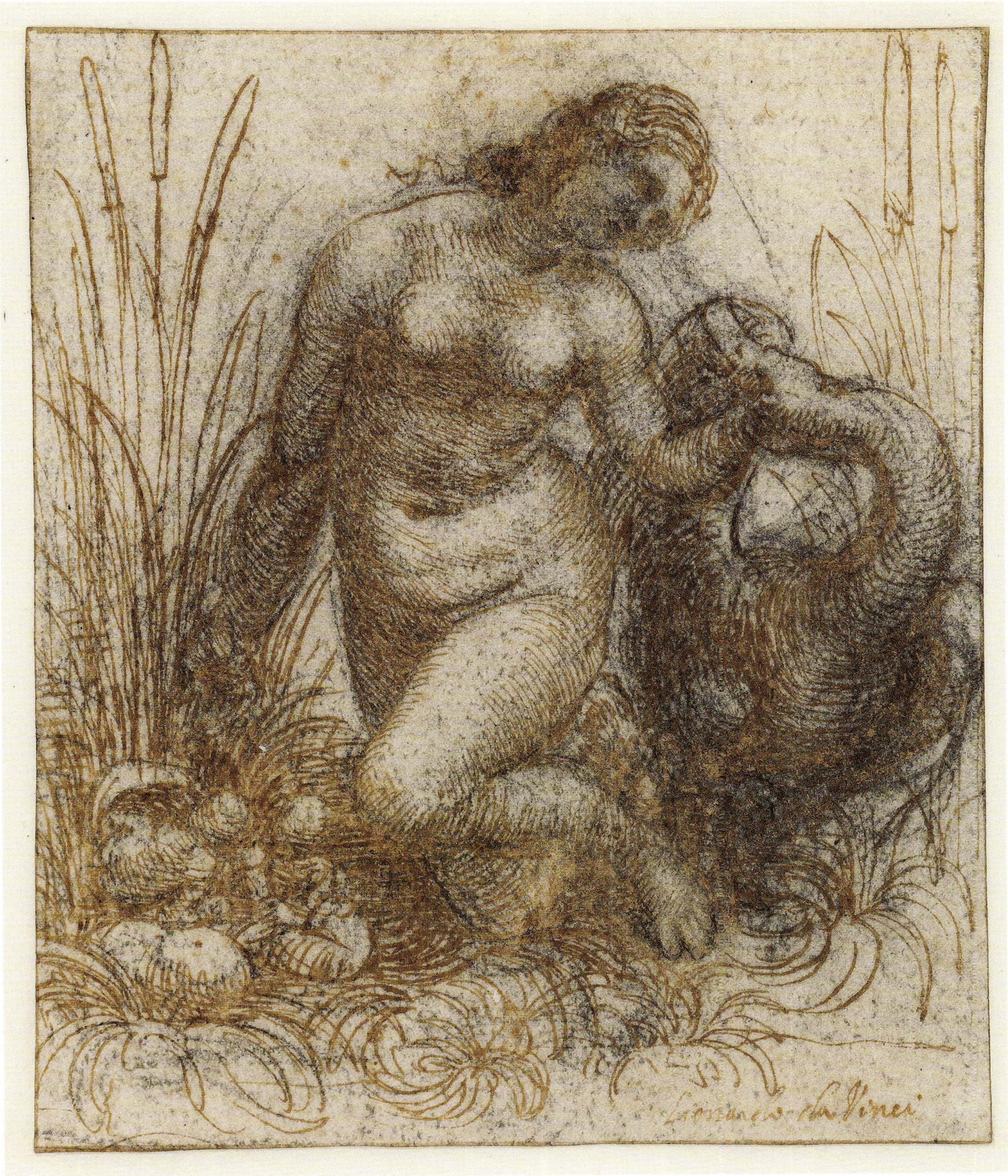
Leonardo da Vinci

nonmateriality of the Neoplatonic soul. Nevertheless, Leonardo continued to conceive the soul—a liminal, spiritual substance—in paradoxical terms, as indivisible and entirely present at every minimal part of the animated body, as a ubiquitous point.[109]

How does the pointlike character of soul relate to art? Brian Rotman compellingly compared the pinhole at the vanishing point of Brunelleschi's first perspective panel to zero. According to this interpretation, the organizing principle of perspectival space is a nonsign and at the same time the condition of every pictorial sign—like zero in postmedieval Western mathematics.[110] But Leonardo goes further. As we have seen, he defines the utmost principle of painting as the infinitely small point that creates the line. *Every* single point of the picture surface is marked by the point, by this paradoxically identical element of transition, just as the soul permeates the entire body. Leonardo discovered one of the most fascinating principles of pictorial representation, namely, that every point of the picture plane is at the same time a positive element on a surface *and* transparent to the fictive appearance of represented objects, figures, atmospheres, and so on. The material element of color is thus added to a material surface in order to transgress material factualness, to become "nothing" (non è della materia di essa superfitie), and vice versa.[111] This becomes even more evident if the depicted surfaces are themselves "nothing," like Leonardo's favorite objects of the *paragone* with sculpture: water, veils, dust, mist, etc., or the transitional states of movement.[112]

4.10 (*overleaf*) Leonardo da Vinci, *Study for a Kneeling Leda*, ca. 1508. Black chalk and pen and ink, 12.5 x 11 cm. Museum Boijmans Van Beuningen, Rotterdam. Loan: Museum Boijmans Van Beuningen Foundation (Collection Koenigs).

The creations of the painter originate in the movement of the point,[113] leading to visual forms on the picture surface (visible points, lines, surfaces) (fig. 4.10). One of Leonardo's texts begins, "If you portray, or if you move some principle of a line," thereby locating the point precisely at the inception of the pictorial act.[114] The transition from naught to being embraces, in this perspective, the paradigm of *creatio ex nihilo* and therefore complements Leonardo's definition of the painter as *signore e dio* of his creations.[115]

The emissions of nature (creating perfect self-portraits) do not simply end in a complete loss of dimensionality—they are transformed, in the eye, into a new image that passed through "nothing." In the eye, the "first" nature becomes a point, negating corporeality, in order to be constituted as a "second" nature.[116] To be sure, this is Leonardo's honorary title for the perfect painter, a painter who transforms himself, like a pure, rational mirror, into the mind of nature.[117] Leonardo's ideal painter overcomes the idiosyncrasies that force him to continually produce *figure come fratelli* and, in the end, nothing but self-portraits.[118] Instead of being a particular "line," one could extrapolate, the painter should become a "point," capable of being the *interprete* of "everything" in nature. At the same time, the perfect painter exploits the category of *possibility* that is so intimately connected to the *tertium ens* of the point: "a Medium between Being and Naught, or the Possibility of Being," as Jacques Gaffarel put it in 1634.[119] The perfect painter would be, to extend Leonardo's analogy further, a "point," capable of creating every *possible* line—everything "per essentia, presentia o' immaginatione."[120] This painter's works would document the triumphs of the point (the permeation of being and naught) in nature. "Necessity forces the mind of the painter to transform himself into the actual mind of nature, and to become an interpreter between nature and art, in order to consider with her the reasons of his demonstrations, forced by her laws, and *in which way the similitudes of the objects that surround the eye converge* [*concorrino*], *with their true simulacra, at the pupil of the eye*."[121]

Notes

1. On "images without viewer," see the essays in Bredekamp and Werner, *Bilder ohne Betrachter*.

2. Given the extent and depth of Leonardo's work in each of these fields, my chapter can be only a rough draft.

3. I use these terms as synonyms, in order to provide a flexible set of nouns for Leonardo's shifting terminology.

4. For this argument, see Kemp, "Crisis of Received Wisdom."

5. See Fehrenbach, *Licht und Wasser*, 229–56; Frosini, "Leonardo da Vinci e il 'nulla,'" 232 ("radicalizzazione dell'attitudine osservativa e—per usare un anacronismo—fenomenologica dinnanzi ad essa").

6. See Pedretti, *On Painting*, 177.

7. I have used Claire Farago's translation, with amendments. See Farago, *Leonardo da Vinci's "Paragone,"* 177–78.

8. "E questo si prova col zero over nulla, cioè la decima figura de la arismetrica, per la quale si figura un o per esso nullo, il quale, posto dopo la unità, il fa dire dieci . . . e così *infinitamente* crescerà." *Libro di pittura*, par. 1 (emphasis added). The implication of the dynamic qualities of zero in a positional numerical system is reflected in a passage in Giovanni da Sacrobosco's widely diffused *Algorismus* (ca. 1240—after zero's distant origins in Babylonian calculations, blurry development in Greek thought, elaboration in Indian-Arabian mathematics, and introduction in the West by the *Liber algorismi*, ca. 1143): "Sciendum quod iuxta 9 limites 9 inveniuntur figure significative 9 digitos representantes qui tales sunt—0 9 8 7 6 5 4 3 2 1—decima dicitur theca vel circulus vel cifra vel figura nihili quoniam nihil significant. Ipsa tamen locum tenens dat aliis significare. Nam sine cifra vel cifris purus non potest scribi articulus." See Menninger, *Number Words and Number Symbols*, 401–3, 410–12, 422–24. For the literature on zero and nothing surrounding the turning of 1999 to the year 2000 (and fueled by the fears of a "Y2K computer bug"), see Kahl-Furthmann, *Problem des Nicht*; Seife, *Zero*; Barrow, *Book of Nothing*; and, especially, Kaplan's most elegantly written *Nothing That Is*. Rotman's *Signifying Nothing* provides a fascinating deconstructivist analysis of the ambiguous character of zero, describing it as "between an empty character . . . and a character for emptiness" (13), a sign for nothing and a sign for the absence of signs. On the history and philosophy of the point, see Federici-Vescovini, *Studi sulla prospettiva medievale*, 213–37; Hagengruber, "Punkt und Mathematik"; and Wohlfart, *Punkt*. For an overview of the mathematics of Leonardo, see Bagni and d'Amore, *Leonardo e la matematica*.

9. *Corpus of the Anatomical Studies . . . at Windsor Castle*, fol. 19151r (Keele and Pedretti 118rB, ca. 1508); cf. *Codice Arundel 263 nella British Library*, fol. 131v (ca. 1505). Marinoni, in "Essere del nulla," underlines the paradoxical properties that connect these geometrical elements, the optical pyramid, nature's imponderable forces, and the soul: "'cose spirituali,' incorporee" (23). Marinoni's reading of Leonardo's oppositions in a Neoplatonic perspective has been convincingly challenged by Frosini, "Leonardo da Vinci e il 'nulla,'" esp. 213 and 231.

10. *Libro di pittura*, par. 31c

11. See Summers, *Judgment of Sense*, passim; Fehrenbach, *Licht und Wasser*, 181–92 ("Moti mentali").

12. This refers to Ibn-al-Haytham/Alhazen's insistence on the physical nature of light rays, as echoed, for instance, in John Pecham's *Tractatus de perspectiva*, chap. 1, 26–27. Cf. Filarete, *Trattato d'Architettura* (Cod. Magliabecchianus), fol. 175v (see also fol. 173v); and Piero della Francesca, *De prospectiva pingendi*, bk. 1, preface. See Field, *Invention of Infinity*, 80–81.

13. "Quae vero intuitum non recipiunt, ea nemo ad pictorem nihil pertinere negabit." Alberti *De pictura* 1.2. Notice the hyperbolic triple negation!

14. Cf., as *pars pro toto*, Nicole Oresme: "linee, puncta, etc. nihil sunt sed solum ymaginantur esse." *Questiones de spera*, qu. 3, quoted in Frosini, "Leonardo da Vinci e il 'nulla,'" 217, with reference to Clagett, "Use of Points in Medieval Natural Philosophy." For Scholastic discussions of *continua*, infinity, and *minima*, and for precursors of Leonardo's immanence of the infinite in the fourteenth century, see Maier, *Vorläufer Galileis*.

15. Euclid, *Thirteen Books of Euclid's Elements*, def. 1.

16. "Ultimi principii, de'quali *in natura* null'altra cosa si puó trovare che sia parte d'essa scientia." *Libro di pittura*, par. 1 (emphasis added). Cf. *Codice Atlantico*, fol. 784ar: The point does not occupy any space, and it exists in nature (si dà in natura); it is mobile and generates the line. See also the *Codices Madrid*, vol. 1, fols. 60v, 109v.

17. "In tali discorsi mentali non accade esperienza, sanza la quale nulla dà di sé certezza." *Libro di pittura*, par. 1.

18. On the early modern history of infinity, see Field, *Invention of Infinity*.

19. Pedretti, "Saggio di una cronologia."

20. *Codice Arundel*, fol. 131r.

21. "Punto non è parte di linia. / L'acqua che tocchi de' fiumi è l'ultima di quella che andò, e la prima di quella che viene; così il tempo presente."

22. "Ogni quantità continua *intellettualmente* è divisibile in infinito." *Codice Arundel*, fol. 131r (emphasis added). This nominalist argument has been elaborated especially by William Ockham and Gregorio da Rimini; see Federici-Vescovini, *Studi sulla prospettiva medievale*, 219–21.

23. "Infra le magnitudine delle cose che sono *infra noi*, l'esse del nulla tiene il principato, e 'l suo ofitio s'astende infra le cose che non hanno l'essere." *Codice Arundel*, fol. 131r (emphasis added).

24. Ibid., fol. 159v.

25. Frosini, "Leonardo da Vinci e il 'nulla,'" 226. The definite article aims at a dynamic inclusion of "nothing" in the realm of an enhanced, more complex "being." The question of whether the *proxima nihilo* of the *prima materia* (a victory of Aristotle over the Bible's *creatio ex nihilo*) should be given the attribute of "unity" (oneness, identity) was answered affirmatively in Ficino *De amore* 6.xv and, quite significantly, denied in Pico della Mirandola *De ente et uno* 3; cf. 7. On the related notion of "chaos" in Leonardo, see Toussaint, "Leonardo filosofo dei contrari." For the complex history of nothingness in religion, see Nishitani, *Religion and Nothingness*. For its history in nineteenth-century science, see Adams, *Nil* (I would like to thank Paul Barolsky for this reference). On the return of the *proxima nihilo* in the modern theory of cosmic vacuum energy, see Barrow, *Book of Nothing*, 8–11. For a modern "nihilist's" critique of the reification of naught, see Lütkehaus, *Nichts*.

26. *Codice Arundel*, fol. 159v. A close parallel to this idea can be found in Jacques Gaffarel's ingenuous *NIHIL, fere NIHIL, minus NIHILO: Seu De Ente, non Ente, et Medio inter Ens, et non Ens POSITIONES XXVI* (Venice, 1634): "[XXXIII] Punctum similiter extremus magnitudinum terminus, ac meta, fere Nihil est, quia magnitudinum licet aliquid sit, ut *eorum initium, et terminus*, non tamen realiter est magnitudo, cum omni careat dimentione; inter ergo magnitudinem, et non magnitudinem situm est, seu inter Ens, et Nihil" (emphasis added). Quoted in Ossola, *Antiche memorie del nulla*, 160.

27. A close parallel to this argument is provided by Martianus Capella's translation of Euclid's definition 1: "Punctum est cuius pars nihil est." See Thomas L. Heath's commentary in Euclid, *Thirteen Books of Euclid's Elements*, 1:155.

28. "Adunque il fine del nulla e 'l principio della linia son in contatto infra loro, ma non congiunti; e in tal contatto è il punto. Del qual punto el nulla è fratello." *Codice Arundel*, fol. 204r. On the genealogical relationship of the "four powers," cf., for instance, MS A, fol. 35v. On music as *sorella* of painting, see *Libro di pittura*, par. 29.

29. "I' modo che si pò dire quantità continua per essere apicate insieme, e discontinue per essere di 2 nature." *Codice Arundel*, fol. 130r.

30. "Tutte le cose che niente ocupano, sono equali infra loro . . . e ciascuna di quelle equale a tutte. Onde in questo caso seguita che la parte sarà equale al tutto e 'l tutto alla parte, el divisibile allo indivisibile, el finito allo infinito. Adunque per quel ch'è detto, la superfizie, la linia, el punto nulla è, perché niente occupa, e tutti e'nulli sono equali a tutti, e tutti a uno, come si prova in aritmetrica. . . . El corpo è vestito di più superfizie, e le superfizie son circundate di linie, e le linie son terminate da punti." *Codice Arundel*, fol. 132r. Cf. a related passage in an anonymous late fourteenth-century commentary on Euclid: "Nec potuit crescere nec decrescere punctus. . . . Omnes igitur puncti visi videntur aequales." Quoted in Federici-Vescovini, *Studi sulla prospettiva medievale*, 227; on apparent echoes of this manuscript in Alberti's and Leonardo's writings, see 224, and for late medieval discussions on the physical or abstract character of point, line, and angle, see 213–37.

31. Frosini, "Leonardo da Vinci e il 'nulla,'" 225–32.

32. Ficino, *Platonic Theology*, 2:6; cf. Plotinus, *Ennéades*, 3:9.4, 5:2.2, 6:8.16, 5:5.9. See Beierwaltes, *Denken des Einen*, 47, 220–22. However, Neoplatonic philosophy emphasized that nature provides an analogy for the triad of the One, multitude, and nothing; on Proclus, for instance, see Beierwaltes, *Denken des Einen*, 234. On the use of analogy in Leonardo's thought, see Economides, "Everything Comes from Everything." On Porphyry of Tyre's theology of God as both "preexistent" (*prousion*) and "nonexistent" (*anusion*), but also as the pure *activity*, efficacy of being (in his commentary on Parmenides), see Beierwaltes, *Denken des Einen*, 198.

33. See Fehrenbach, *Licht und Wasser*, 205–15, 229–56.

34. See Barrow, *Book of Nothing*, 54–64. Aristotle's critique (with reference to continuous movement in fluids) can be found in *Physics* 4.214a31–32.

35. *Codice Arundel*, fol. 173v; cf. fol. 205r (in the context of geological arguments): "Dice l'aversario che il punto è o e' non è, e s'egli è, o egli si pò movere o e' non si pò movere, e se sarà mosso, e' descrive la linia."

36. Augustine *Confessiones* 11; see Flasch, *Was ist Zeit?* On the point of presence as *a-topon* ("nothing," not positively given) and as a "transitional" moving principle in Plotinus, see Beierwaltes, *Denken des Einen*, 196.

37. *Codice Arundel*, fol. 159r. The authoritative discussion of point and line, instant and time, with a focus on Zeno's paradoxes, is in the sixth book of Aristotle's *Physics*: "impossibile est ut aliquod continuum sit compositum ex indivisibilis: et linea sit composita ex punctis" (231a; cf. 235b, 237b); for the analogy in time, see 234a. On the continuum in Aristotle, see Wieland, *Die aristotelische Physik*, 283–85; Böhme, *Zeit und Zahl*, 99–158; and Inwood, "Aristotle on the Reality of Time"; in early modern science, see Mainzer, *Begruendungsproblem des mathematischen Kontinuums*, 125–37. Leonardo, however, seems more closely to follow Euclid, who states in his *Elements*, def. 3, "The extremities of a line are points." For Simon Stevin's later insistence on

"the true and natural beginning" of numbers (and counting) in zero, the *poinct de nombre* (in analogy to the point as origin of the line), see Rotman, *Signifying Nothing*, 28–32. On the line generated by the point in motion, see Aristotle *De anima* 1.4.409a4. A parallel is provided by Proclus, who refers to line as "flux of the point"; see Beierwaltes, *Proklos*, 173. Simplicius follows the same argument; see Heath's commentary in Euclid, *Thirteen Books of Euclid's Elements*, 1:157.

38. "La conversa. Il punto si gienera dove manca la lunghezza della linia; la linia nasce dove finisce la largheza o lungheza della superfizie, e la superfizie è creata nel termine della largheza, lungheza e profondità del corpo." *Codice Arundel*, fol. 159r.

39. Ibid.

40. Ossola, *Antiche memorie del nulla*, 26. Cf. Ficino *Theologia Platonica* 10.2, cited in Marinoni, "Essere del nulla," 25.

41. "Perché il moto è di natura di linia e la linia comincia in punto. Adunque il moto principia ancora lui in punto." *Codices Madrid*, vol. 1, fol. 109v.

42. "Colpo è termine di moto creato in tempo indivisibile, perché è causato nel punto, termine della linia del moto." *Codici Forster*, vol. 3, fol. 32r.

43. See Rotman, *Signifying Nothing*, 57–78.

44. See Beierwaltes, *Denken des Einen*, 343–44.

45. See Kaplan, *Nothing That Is*, 64.

46. This is explained by Petrus of Dacia in 1291; see ibid., 66.

47. Rotman, *Signifying Nothing*, 57.

48. Ibid., 60 (with reference to Colie, *Paradoxia epidemica*).

49. Barrow, *Book of Nothing*, 36–38.

50. Most important on Neoplatonic philosophy of the "One," see Mahnke, *Unendliche Sphäre und Allmittelpunkt*; Beierwaltes, *Denken des Einen*.

51. See Babb, *King Lear*, 235–49; Rotman, *Signifying Nothing*, 78–86.

52. "Est enim illud, quicquid est a se, indivisum et ab aliis divisum quae non sunt ipsum." Pico della Mirandola, *De ente et uno*, chap. 8 (cf. chap. 5).

53. See Kaplan, *Nothing That Is*, 68, 110.

54. See Frosini, "Leonardo da Vinci e il 'nulla,'" 222.

55. It would be tempting to relate Leonardo's paradox to the different Greek terms for the point—*stigme*/puncture (Aristotle) and *semeion*/mark (Euclid, Archimedes; see Heath's commentary in Euclid, *Thirteen Books of Euclid's Elements*, 1:156)—and to read the point, as *semeion*/semen, with Derrida, as *différance*. See Derrida, *Positions*, 44–46.

56. Kemp, "Leonardo's Visual Pyramid"; Ackerman, "Leonardo da Vinci's Eye"; Strong, *Leonardo on the Eye*; Veltman, *Linear Perspective and the Visual Dimensions*; Fehrenbach, *Licht und Wasser*, 115–41, 157–92.

57. See Lindberg, *Roger Bacon's Philosophy of Nature*; Koelsch Loose, "Roger Bacon on Perception."

58. Lindberg, *Theories of Vision from al-Kindi to Kepler*, 58–86; on Alhazen, see Sabra, *Optics of Ibn al-Haytham*.

59. The liminal quality of rays/images is emphasized, for instance, in MS A, fol. 2v: "L'aria è piena d'infinite linie rette e radiose insieme intersegate e intessute sanza occupazione l'una dell'altra; rapresentano a qualunche obbietto la vera forma della lor cagione."

60. See Alberti *De pictura* 1.5.

61. "L'occhio non vede se non per piramide. La prospettiva de'pittori sanza piramide non ha locc." *Codice Arundel*, fol. 232r, ca. 1483–99.

62. See Crombie, *Robert Grosseteste and the Origins of Experimental Science*; Federici-Vescovini, *Studi sulla prospettiva medievale*, 219 ("Punctus est substantia posita"); Lindberg, *Roger Bacon's Philosophy of Nature*, 94–102; Agnoli, *Roberto Grossatesta*.

63. See Frosini, "Pittura come filosofia."

64. "Tutte le potenzie spirituale, quanto piú s'allontanano dalla prima o seconda cagione, piú occupano di sito e piú diminuiscano di lor valitudine." *Codice di Leonardo da Vinci nel Castello Sforzesco*, fol. 11v.

65. On the semiosis of organisms, see Jonas, *Phenomenon of Life*; Maturana and Varela, *Autopoiesis and Cognition*; Weber, *Natur als Bedeutung*.

66. "L'aria pura è capacie di ricievere in sé sanza intervallo di tenpo ogni similitudine di corporea qualità che sia dai sola razi percossa o per la cagione aluminata. . . . Ogni parte d'aria ricieve in sé tutte le spezie de' corpi che la vedano e son veduti da lei 'n un medesimo tenpo." *Codice Arundel*, fol. 185v (1483–99).

67. Plotinus, *Ennéades*, 4:7.6; cf. Summers, *Judgment of Sense*, 93. Augustine *De genesi ad litteram* 4.34.54: "Omnis illa tam ampla immensaque spatia simul uno ictu transiri manifestum est"; Pecham, *Tractatus de perspectiva*, chap. 2; *Codice Atlantico*, fol. 949r: "Chi crederrebbe che sì brevissimo spazio fussi capace delle spezie di tutto l'universo? O magna azione, quale ingegno potrà penetrare tale natura? Qual lingua fia quella che esplicare possa tal maraviglia? Certo nessuna. Questo dirizza l'umano discorso alla contemplazione divina, eccetera."

68. "La virtù visiva è solo un puncto." Piero della Francesca, *De prospectiva pingendi*, bk. 1, chap. 1. "Ogni cosa tira a quel punto; perchè, come ai inteso, quello è il punto centrico, el tuo occhio, col quale ti bisognia fermare ogni cosa, a similitudine come colui, che balestra, che sempre tiene la sua mira a uno dato e fermo punto." Filarete, *Trattato d'Architettura* (Cod. Magliabecchianus), chap. 23, fol. 178v (p. 163 in Oettingen's edition). Leonardo's position can be compared to that of Roger Bacon, who blames the *auctores perspectivae* of his time for neglecting the functions and anatomy of the eye (*Opus Majus*, part 5, bk. 1, distinction 2.1). For a thorough discussion of the relevance of the optical tradition to Quattrocento painting, with an emphasis on the extended area of perception on the surface of the eye or on one of the eye's humors (as a *speculum animatum*), apparently in contrast to Leonardo's arguments, see Summers, *Vision, Reflection, and Desire*, 43–77.

69. Cf. MS C, fol. 19v; MS A, fols. 3r, 10r, 40v, 41r; MS H, fol. 131v; MS I, fol. 97v, all in Leonardo, *Manoscritti dell'Institut de France*; *Codici*

Forster, vol. 3, fol. 36r; *Codice Atlantico*, fols. 85va, 232r, 955r, 1101ar; *Corpus of the Anatomical Studies . . . at Windsor Castle*, fol. 19152r (Keele and Pedretti 118rB). See also Kemp, "Leonardo's Visual Pyramid," 133.

70. See Galluzzi, *Leonardo e i proporzionanti.*

71. Cf. MS D, fol. 4v: "Occhio dell'omo. Che sia vero che ogni parte della popilla abbia virtù visiva e che tal virtù non sia ridotta in punto, come vogliano li prespettivi."

72. Aristotle *De sensu et sensibilibus* 438a.

73. *Codice Arundel*, fol. 71v (ca. 1508); *Codice Atlantico*, fol. 543r.

74. Alhazen *De aspectibus* 2.2.23; Bacon, *Opus Majus*, part 5, bk. 1, distinction 3.1.

75. Cf. MS D, fols. 1v, 2r, 9v (all at the end of the manuscript).

76. "Tucte le parte minime delle spetie penetra l'una l'altra sanza occupatione l'una dell'altre. . . . Come concludiamo noi la superfitie riddursi in punto . . . per essere incorporee." *Corpus of the Anatomical Studies . . . at Windsor Castle*, fols. 19149–19152v (Keele and Pedretti, fol. 118v). Cf. Kemp, "Leonardo's Visual Pyramid," 144. On the difference between mathematical and natural points, see Zubov, "Jean Buridan et les concepts du point"; Federici-Vescovini, *Studi sulla prospettiva medievale*, 213–37; Leinkauf, *Mundus combinatus*, 58–59.

77. "Guarda il lume e considera la sua bellezza. Batti l'occhio e riguardalo. Ciò che di lui tu vedi, prima non era, e ciò che di lui era, più non è. Chi è quel che lo rifà, se 'l fattore al continuo more?" MS F, fol. 49v. Cf. Lucretius *De rerum natura* 5.290–93: "ut noscas splendore novo res semper egere / et primum iactum fulgoris quemque perire / nec ratione alia res posse in sole videri, / perpetuo ni suppeditet lucis caput ipsum."

78. Baratta, *Leonardo ed i problemi della terra*; Kemp, *Leonardo da Vinci*, 114ff.; Fehrenbach, *Licht und Wasser*, 193–215.

79. For the Christian perspective relevant to these pursuits, see Augustine *De civitate Dei* 12.11; Fehrenbach, *Licht und Wasser*, 215–29.

80. Codex Leicester, 1B:36r (1).

81. Cf. ibid., 2A:35v.

82. Cf. MS A, fol. 58v; *Codici Forster*, vol. 3, fol. 7r; *Codices Madrid*, vol. 1, fols. 6r, 114r. See Moody, *John Buridan on the Habitability*; Zubov, *Leonardo da Vinci*, 249ff.

83. "Del mondo. Ogni grave attende al basso e le cose alte non resteran in loro altezza, ma col tempo tutte discenderanno; e così col tempo il mondo resterà sperico e per conseguenza fia tutto coperto dell'acqua." MS F, fol. 84r. For the reference to Albert of Saxony's *Questiones subtilissime in libros Aristotelis*, II, qu. 28, see Solmi, *Fonti di Leonardo*, 55–56.

84. MS F, fol. 52v.

85. Kemp, *Leonardo da Vinci*, 254–55.

86. Cf. Codex Leicester, 2A:35v (3).

87. Kircher's colossal effort to localize God in the *punctum* of the earth's center is both a bold twist in the history of the famous Neoplatonic formula "Deus est sphaera infinita, cuius centrum est ubique, circumferentia nusquam" (*Liber XXIV philosophorum*, propositio 2) *and* a confutation of the traditional rejection of the earth's center as the most material, dark, and shameful place, *fex foeculentissima* (Francesco Patrizi, *Nova de universis philosophia*); see Leinkauf, "'Centrosophia' des Athanasius Kircher," 221, 224–25.

88. "Muovesi il centro dell'acqua e terra insieme g[i]unti quando si move il peso del mare portato da venti."

89. Cf. *Codice Atlantico*, fol. 412v.

90. Codex Leicester, 15A:22r (8).

91. MS F, fol. 70v.

92. See Duhem, *Études sur Léonard de Vinci*, vol. 3, *Les précurseurs parisiens de Galilée*. On Leonardo's dynamics, see Moody's introduction in *John Buridan on the Habitability*, xx; Fehrenbach, *Licht und Wasser*, 239–45. On impetus physics, see Maier, *Zwei Grundprobleme*; Maier, *Metaphysische Hintergründe*; Drake, "Impetus Theory Reappraised"; Wolff, *Geschichte der Impetustheorie*; Wolff, "Mehrwert und Impetus"; Sorabji, *Philoponus and the Rejection of Aristotelian Science*; and (with reservations) Grün, *Vom Unbewegten Beweger*. For a general survey, see Jammer, *Concepts of Force*.

93. See Schneider, *Kosmologie des Franciscus de Marchia*, chap. 3.

94. For the Aristotelian tradition, see Seeck, "Theorie des Wurfs."

95. Quoted in Jammer, *Concepts of Force*, 71. This example is depicted in one of Leonardo's latest manuscripts, MS E, fol. 50v (ca. 1513–14); in the context of his studies of the flight of birds, cf. fol. 50r: "Difinizion dell'impeto. L'impeto è una virtù creata dal moto e trasmutata dal motore al mobile. Il qual mobile ha tanto di moto, quanto l'impeto ha di vita."

96. Plato *Laws* 10.895c–899c.

97. On the tradition of self-motion after Plato, see Gill and Lennox, *Self-Motion.*

98. "Che cosa è forza? Forza dico essere una potenzia spirituale incorporea e invisibile, la quale con breve vita si causa in quelli corpi che per accidentale violenza stanno fori di lor naturale essere e riposo. Spirituale dissi, perché in essa forza è vita attiva; incorporea e invisibile dico, perché il corpo, dove nascie, non crescie in peso ne in forma; di poca vita perché sempre desidera vinciere la sua cagion, e, quella vinta, sé occide." MS B, fol. 63r.

99. "Esso impeto fugatore del corpo dov'è creato si consuma e more insieme col moto d'esso corpo." *Codice Arundel*, fol. 2r (ca. 1508–9).

100. "Forza non è altro che una virtù spirituale, una potenza invisibile, la quale è creata e infusa per accidental violenza, da corpi sensibili nelli insensibili, dando a essi corpi *similitudine di vita . . . corre con furia a sua disfazione. . . .* Tardità la fa grande, e prestezza la fa debole. *Vive per violenza, e more per libertà . . . volentieri consuma se stessa . . .* [ne]ssuna cosa sanza lei si move." *Codice Atlantico*, fol. 826 (ca. 1492), emphasis added.

101. MS C, fol. 6v (ca. 1490).

102. *Codice Atlantico*, fol. 340r (ca. 1497–1500).

103. Aristotle *Physics* 1.185a.

104. *Codice Arundel*, fol. 159r.

105. For a more extended discussion of this argument, see Fehrenbach, "Oszillierende Blick." For different views on *sfumato*, cf. Nagel, "Leonardo and 'sfumato'"; Prater, "Sehnsucht nach dem Chaos"; Bell, "Sfumato, Linien, und Natur"; Wellmann, *Entdeckung der Unschärfe*.

106. "È la maestra dell'altre linie, dalle quali sempre essa è mossa diterminando quel che l'altre vedano e non cognoscano." MS D, fol. 8v.

107. Cf. Plotinus's formula of the indivisible soul, "tota in toto et tota in qualibet parte," a topos echoed in Augustine, Cusanus, and Ficino. Leinkauf, *Mundus combinatus*, 58–60. On Luca Pacioli's use of the formula to mark the ubiquity of a *proportion*, the Golden Section, in any quantity, see Veltman, *Linear Perspective and the Visual Dimensions*, 83; and Frosini, "Leonardo da Vinci e il 'nulla,'" 231.

108. "Opinio autem inordinata et mobilis imaginum multitudo sed substantia puntisque unita, cum anima ipsa in qua est opinio una substantia sit, nullum occupans locum." Ficino *De amore* 7.13. On Ficino's and Pico's notion of the soul as *vinculum naturae totius* and *mundi copula*, see Beierwaltes, *Denken des Einen*, 282ff.

109. Cf. Pascal, *Pensées*: "Car enfin qu'est-ce que l'homme dans la nature? Un néant a l'égard de l'infini, un tout a l'égard du néant, un milieu entre rien et tout" (84).

110. Rotman, *Signifying Nothing*, 14–22. Cf. Luigi Manzini, *Il niente* (Venice, 1634): "La prospettiva ove riguarderebbe, quando non ricorresse al Niente? S'ella non s'industriasse di guidarti con piacevole tradimento de gli occhi alla meta d'un punto, d'un indivisibile imaginato, ch'è pure un niente, come la conteresti tu fra le pompe che, illustrando le maraviglie dell'arte, felicitano il lusso degli occhi? La pittura poi, vana creatrice d'imagini, altro non intende gia' che, fingendoti una figura, dissimularti un niente." Quoted in Ossola, *Antiche memorie del nulla*, 103. See also the objection by Raymond Vidal, *Il niente annientato*, of the same year (1634): "La prospettiva, la pittura, la scultura e somiglievoli arti, non ponno altrimenti consistere nel Niente, perché di loro ragion formale non hanno altro scopo che d'andar pascendo con i lor effetti apparenti e reali la curiosità di quell'occhio che per oggetto ha l'Ente, non il Niente." Quoted in Ossola, *Antiche memorie del nulla*, 144. On the spectator in perspective as "non-empirical punctum," see Bryson, *Vision and Painting*, 110, and the recent critique of the argument in Summers, *Vision, Reflection, and Desire*, 155–65.

111. See the discussion of this aspect of the point in Boehm, "Topos des Anfangs."

112. On transitional objects in painting and their relationship to Leonardo's notion of the point, cf. Pedretti, "Poem to Sculpture."

113. *Libro di pittura*, par. 1.

114. "Fa, che quando ritrai e che tu movi alcun principio di linea . . ." MS A, fol. 109r.

115. *Libro di pittura*, par. 13. Compare this to Aristotle's acceptance of a point in time as marking the *end* of movement, but his categorical denial of an ultimate point at the *beginning* of a process (*Physics* 6.236a). The relationship between Leonardo's theory of the point and artistic creation is considered in Batkin, *Leonardo da Vinci*, 167–82.

116. According to the main line of argument in Summers, *Vision, Reflection, and Desire*, the disembodied *punctum* of perception remains therefore a moment of transition giving way necessarily to an insurmountable corporeality, one that, however, keeps oscillating between being and nonbeing.

117. Cf. *Libro di pittura*, par. 58a.

118. Cf. Kemp, "Crisis of Received Wisdom"; Zöllner, "'Ogni pittore dipinge sé'"; Laurenza, *Leonardo nella Roma di Leone X*.

119. "Hinc inter Ens, et Nihil, Medium datur, scilicet Possibilitas ad Ens." Quoted in Ossola, *Antiche memorie del nulla*, 152. On the realtionship between the philosophy of "nothing" and painting in the seventeeth century, see Stoichita, *Self-Aware Image*, chap. 9.

120. *Libro di pittura*, par. 13.

121. "Necessità costringe la mente del pittore a trasmutarsi nella propria mente di natura e che sia interprete infra essa natura e l'arte, comentando con quella le cause delle sue dimostrationi costrette dalla sua legge, *et in che modo le similitudini delli obietti circostanti all' occhi concorrino colli veri simulacri alla popilla dell'occhio*." Ibid., chap. 40 (emphasis added). Farago's translation—"Praising the causes of nature's demonstrations in this, the painter is compelled by the laws of nature in the way that the similitudes of objects which surround the eye act together with the real simulacri on the pupil of the eye" (*Leonardo da Vinci's "Paragone*," 273)—misses Leonardo's point.

Bibliography

Ackerman, James. "Leonardo da Vinci's Eye." *Journal of the Warburg and Courtauld Institutes* 41 (1978): 108–46.

Adams, Robert Martin. *Nil: Episodes in the Literary Conquest of Void During the Nineteenth Century*. New York: Oxford University Press, 1966.

Agnoli, Francesco. *Roberto Grossatesta: La filosofia della luce*. Bologna: ESD, 2007.

Alberti, Leon Battista. *Über die Malkunst / Della pittura*. Edited by Oskar Bätschmann and Sandra Gianfreda. Darmstadt: Wissenschaftliche Buchgesellschaft, 2002.

Albert of Saxony. *Questiones subtilissime in libros Aristotelis de celo et mundo*. 1492. Reprint, Hildesheim: G. Olms, 1986.

Alhazen. *The Optics of Ibn al-Haytham: Books I–III, On Direct Vision*. Edited by A. I. Sabra. 2 vols. London: Warburg Institute, 1989.

Aristotle. *De sensu and De memoria*. Edited by G. R. T. Ross. New York: Arno Press, 1973.
———. *Physik: Vorlesung über die Natur*. Edited by Hans Günter Zekl. 2 vols. Hamburg: Felix Meiner Verlag, 1987–88.
———. *Über die Seele / De anima*. Edited by Horst Seidl. Hamburg: Felix Meiner Verlag, 1995.
Augustine. *Bekenntnisse / Confessiones*. 4th ed. Edited by Joseph Bernhart. Munich: Kösel-Verlag, 1980.
———. *Der Gottesstaat / De civitate Dei*. Edited by Carl Johann Perl. 2 vols. Paderborn: Ferdinand Schöningh Verlag, 1979.
Babb, Howard S. "King Lear: The Quality of Nothingness." In Babb, *Essays in Stylistic Analysis*, 235–49. New York: Harcourt Brace Jovanovich, 1972.
Bacon, Roger. *The "Opus Majus" of Roger Bacon*. Edited by John H. Bridges. 3 vols. Oxford: Clarendon Press, 1897–1900.
Bagni, Giorgio Tomaso, and Bruno d'Amore. *Leonardo e la matematica*. Florence: Giunti, 2006.
Baratta, Mario. *Leonardo ed i problemi della terra*. Turin: Fratelli Bocca, 1903.
Barrow, John D. *The Book of Nothing: Vacuums, Voids, and the Latest Ideas About the Origins of the Universe*. New York: Pantheon Books, 2002.
Batkin, Leonid. *Leonardo da Vinci*. Bari: Laterza, 1988.
Beierwaltes, Werner. *Denken des Einen: Studien zur neuplatonischen Philosophie und ihrer Wirkungsgeschichte*. Frankfurt am Main: Vittorio Klostermann Verlag, 1985.
———. *Proklos: Grundzüge seiner Metaphysik*. 2nd ed. Frankfurt am Main: Vittorio Klostermann Verlag, 1979.
Bell, Janis. "Sfumato, Linien, und Natur." In *Leonardo da Vinci: Natur im Übergang*, ed. Frank Fehrenbach, 229–56. Munich: Wilhelm Fink Verlag, 2002.
Boehm, Gottfried. "Der Topos des Anfangs: Geometrie und Rhetorik in der Malerei der Renaissance." In *Visuelle Topoi: Erfindung und tradiertes Wissen in den Künsten der italienischen Renaissance*, ed. Ulrich Pfisterer and Max Seidel, 48–59. Munich: Deutscher Kunstverlag, 2003.
Böhme, Gernot. *Zeit und Zahl: Studien zur Zeittheorie bei Platon, Aristoteles, Leibniz, und Kant*. Frankfurt am Main: K. Klostermann, 1974.
Bredekamp, Horst, and Gabriele Werner, eds. *Bilder ohne Betrachter*. Berlin: Akademie Verlag, 2007.
Bryson, Norman. *Vision and Painting: The Logic of the Gaze*. 6th ed. New Haven: Yale University Press, 1995.
Clagett, Marshall. "The Use of Points in Medieval Natural Philosophy and Most Particularly in the 'Questiones de spera' of Nicole Oresme." In Clagett, *Studies in Medieval Physics and Mathematics*, 215–21. London: Ashgate, 1979.
Colie, Rosalie Littell. *Paradoxia epidemica: The Renaissance Tradition of Paradox*. Princeton: Princeton University Press, 1966.
Crombie, Alistair C. *Robert Grosseteste and the Origins of Experimental Science: 1100–1700*. Oxford: Clarendon Press, 1953.
Derrida, Jacques. *Positions*. Translated and annotated by Alan Bass. Chicago: University of Chicago Press, 1981.
Drake, Stillman. "Impetus Theory Reappraised." *Journal of the History of Ideas* 36, no. 1 (1975): 27–46.
Duhem, Pierre M. M. *Études sur Léonard de Vinci*. 3 vols. Paris: A. Hermann, 1906–13.
Economides, Aliki. "Everything Comes from Everything, and Everything Is Made Out of Everything, and Everything Returns into Everything: Leonardo's Analogical (Re)search." Master's thesis, McGill University, 2002.
Euclid. *The Thirteen Books of Euclid's Elements*. 2nd ed. Edited by Thomas L. Heath. 3 vols. New York: Dover, 1956.
Farago, Claire J. *Leonardo da Vinci's "Paragone": A Critical Interpretation with a New Edition of the Text in the "Codex Urbinas."* Leiden: E. J. Brill, 1992.
Federici-Vescovini, Graziella. *Studi sulla prospettiva medievale*. Turin: G. Giappichelli, 1965.
Fehrenbach, Frank. *Licht und Wasser: Zur Dynamik naturphilosophischer Leitbilder im Werk Leonardo da Vincis*. Tubingen: E. Wasmuth, 1997.
———. "Der oszillierende Blick: 'Sfumato' und die Optik des späten Leonardo." *Zeitschrift für Kunstgeschichte* 65, no. 4 (2002): 522–44.
Ficino, Marsilio. *Platonic Theology / Theologia Platonica*. Edited by James Hankins and William Bowen. 6 vols. Cambridge: Harvard University Press, 2001–6.
———. *Über die Liebe, oder Platons Gastmahl / De amore*. 2nd ed. Edited by Paul R. Blum. Hamburg: Felix Meiner, 1984.
Field, Judith V. *The Invention of Infinity: Mathematics and Art in the Renaissance*. Oxford: Oxford University Press, 1997.
Filarete. *Antonio Averlino Filarete's Tractat über die Baukunst*. Edited by Wolfgang von Oettingen. Vienna: C. Graeser, 1890.
Flasch, Kurt. *Was ist Zeit? Augustinus von Hippo, das XI. Buch der Confessiones*. Frankfurt am Main: Vittorio Klostermann Verlag, 1993.
Frosini, Fabio. "Leonardo da Vinci e il 'nulla': Stratificazioni semantiche e complessità concettuale." In *Il volgare come lingua di cultura dal Trecento al Cinquecento: Atti del convegno internazionale, Mantova, 18–20 ottobre 2001*, ed. Arturo Calzona, 209–32. Florence: Leo S. Olschki, 2003.
———. "Pittura come filosofia: Note su 'spirito' e 'spirituale' in Leonardo." *Achademia Leonardi Vinci* 10 (1997): 35–59.
Galluzzi, Paolo. *Leonardo e i proporzionanti, XXVIII Lettura Vinciana*. Florence: Giunti Barbèra, 1989.
Gill, Mary Louise, and James G. Lennox, eds. *Self-Motion: From Aristotle to Newton*. Princeton: Princeton University Press, 1994.
Grün, Klaus-Jürgen. *Vom Unbewegten Beweger zur bewegenden Kraft: Der pantheistische Charakter der Impetustheorie im Mittelalter*. Paderborn: Mentis, 1999.
Hagengruber, Ruth. "Punkt und Mathematik in der Metaphysik des

Tommaso Campanella." In *Verum et Factum: Festschrift für Stephan Otto*, ed. Tamara Albertini, 505–14. Frankfurt am Main: Lang, 1993.

Inwood, Michael. "Aristotle on the Reality of Time." In *Aristotle's Physics: A Collection of Essays*, ed. Lindsay Judson, 151–78. Oxford: Clarendon Press, 1991.

Jammer, Max. *Concepts of Force: A Study in the Foundations of Dynamics*. Cambridge: Harvard University Press, 1957.

Jonas, Hans. *The Phenomenon of Life: Toward a Philosophical Biology*. New York: Dell, 1966.

Kahl-Furthmann, Gertrud. *Das Problem des Nicht: Kritisch-historische und systematische Untersuchungen*. 2nd ed. Meisenheim am Glan: Anton Hain Verlag, 1968.

Kaplan, Robert. *The Nothing That Is: A Natural History of Zero*. Oxford: Oxford University Press, 1999.

Kemp, Martin. "The Crisis of Received Wisdom in Leonardo's Late Thought." In *Leonardo e l'età della ragione*, ed. Enrico Bellone, 27–42. Milan: Scientia, 1982.

———. *Leonardo da Vinci: The Marvellous Works of Nature and Man*. London: Dent, 1981.

———. "Leonardo's Visual Pyramid." *Journal of the Warburg and Courtauld Institutes* 40 (1977): 129–49.

———. "'Ogni dipintore dipinge se': A Neoplatonic Echo in Leonardo's Art Theory?" In *Cultural Aspects of the Italian Renaissance: Essays in Honour of Paul Oskar Kristeller*, ed. Cecil H. Clough, 311–23. Manchester: Manchester University Press, 1976.

Koelsch Loose, Patrice. "Roger Bacon on Perception: A Reconstruction and Critical Analysis of the Theory of Visual Perception Expounded in the 'Opus Majus.'" PhD diss., Ohio State University, 1979.

Laurenza, Domenico. *Leonardo nella Roma di Leone X (c. 1513–16): Gli studi anatomici, la vita, l'arte*. Florence: Giunti, 2004.

Leinkauf, Thomas. "Die 'Centrosophia' des Athanasius Kircher SJ: Geometrisches Paradigma und geozentrisches Interesse." *Berichte zur Wissenschaftsgeschichte* 14, no. 4 (1991): 217–29.

———. *Mundus combinatus: Studien zur Struktur der barocken Universalwissenschaft am Beispiel Athanasius Kirchers SJ (1602–1680)*. Berlin: Akademie Verlag, 1993.

Leonardo da Vinci. *The Codex Hammer of Leonardo da Vinci*. Edited by Carlo Pedretti. Florence: Giunti Barbèra, 1987. (Cited as Codex Leicester.)

———. *Il Codice Arundel 263 nella British Library*. Edited by Carlo Pedretti. Florence: Giunti Barbèra, 1998.

———. *Il Codice Atlantico di Leonardo da Vinci nella Biblioteca Ambrosiana di Milano*. Edited by Augusto Marinoni. 24 vols. Florence: Giunti Barbèra, 1973–80.

———. *Il Codice di Leonardo da Vinci nel Castello Sforzesco*. Edited by Pietro C. Marani and Giovanni M. Piazza. Milan: Electa, 2006.

———. *Codices Madrid*. Edited by Ladislao Reti and Augusto Marinoni. 5 vols. Frankfurt am Main: Fischer Verlag, 1974.

———. *I Codici Forster del Victoria and Albert Museum di Londra*. Edited by Augusto Marinoni. 3 vols. Florence: Giunti Barbèra, 1992.

———. *Corpus of the Anatomical Studies in the Collection of Her Majesty the Queen at Windsor Castle*. Edited by Kenneth D. Keele and Carlo Pedretti. 3 vols. London: Johnson Reprint, 1978–80.

———. *Leonardo da Vinci on Painting: A Lost Book (Libro A)*. Edited by Carlo Pedretti. Berkeley: University of California Press, 1964.

———. *Libro di pittura: Edizione in facsimile del codice urbinate lat. 1270 nella Biblioteca Apostolica Vaticana*. Edited by Carlo Pedretti. 2 vols. Florence: Giunti Barbèra, 1995.

———. *I Manoscritti dell'Institut de France*. Edited by Augusto Marinoni. 10 vols. Florence: Giunti Barbèra, 1987–92.

Lindberg, David C. *Roger Bacon's Philosophy of Nature: A Critical Edition, with English Translation, Introduction, and Notes, of "De multiplicatione specierum" and "De speculis comburentibus."* Oxford: Oxford University Press, 1983.

———. *Theories of Vision from al-Kindi to Kepler*. Chicago: University of Chicago Press, 1976.

Lucretius Carus, Titus. *Welt aus Atomen / De rerum natura*. Stuttgart: Reclam, 1981.

Lütkehaus, Ludger. *Nichts: Abschied vom Sein, Ende der Angst*. 4th ed. Frankfurt am Main: Haffmanns bei Zweitausendeins, 2005.

Mahnke, Dietrich. *Unendliche Sphäre und Allmittelpunkt: Beiträge zur Genealogie der mathematischen Mystik*. Halle: M. Niemeyer, 1937.

Maier, Anneliese. *Metaphysische Hintergründe der spätscholastischen Naturphilosophie*. Rome: Edizioni di Storia e Letteratura, 1955.

———. *Die Vorläufer Galileis im 14. Jahrhundert: Studien zur Naturphilosophie der Spätscholastik*. 2nd ed. Rome: Edizioni di Storia e Letteratura, 1966.

———. *Zwei Grundprobleme der scholastischen Naturphilosophie*. Rome: Edizioni di Storia e Letteratura, 1951.

Mainzer, Klaus. "Das Begruendungsproblem des mathematischen Kontinuums in der neuzeitlichen Entwicklung der Grundlagenforschung." *Philosophia Naturalis* 16 (1976–77): 125–37.

Marinoni, Augusto. "L'essere del nulla." In *Leonardo da Vinci, letto e commentato da Marinoni . . . [et al.]: Letture Vinciane I–XII (1960–1972)*, ed. Paolo Galluzzi, 7–28. Florence: Giunti Barbéra, 1974.

Maturana, Humberto R., and Francisco J. Varela. *Autopoiesis and Cognition: The Realization of the Living*. Dordrecht: Reidel, 1980.

Menninger, Karl A. *Number Words and Number Symbols: A Cultural History of Numbers*. Cambridge: MIT Press, 1970.

Migne, Jacques-Paul, ed. *Patrologiae cursus completus . . . series latina*. 221 vols. Paris: Apud Garniere Fratres, 1844–96.

Moody, Ernest Addison. "Foreword." In Ivor B. Hart, *The Mechanical Investigations*, v–xi. Berkeley: University of California Press, 1963.

———. "John Buridan on the Habitability of the Earth." *Speculum* 16 (1941): 415–25.
Nagel, Alexander. "Leonardo and 'sfumato.'" *RES* 24 (1993): 7–20.
Nishitani, Keiji. *Religion and Nothingness*. Berkeley: University of California Press, 1982.
Ossola, Carlo. *Le antiche memorie del nulla*. 3rd ed. Rome: Edizioni di Storia e Letteratura, 2007.
Pascal, Blaise. *Pensées*. Paris: Librairie Générale Française, 2000.
Pecham, John. *Tractatus de perspectiva*. Edited by David C. Lindberg. New York: Franciscan Institute, 1972.
Pedretti, Carlo. "A Poem to Sculpture." *Achademia Leonardi Vinci* 2 (1989): 11–39.
———. "Saggio di una cronologia dei fogli del Codice Arundel di Leonardo da Vinci." *Bibliothèque d'Humanisme et Renaissance: Travaux et Documents* 22 (1960): 172–77.
Pico della Mirandola, Giovanni Francesco. *L'esprit du Quattrocento: De l'être et de l'un / De ente et uno*. Edited by Stéphane Toussaint. Paris: H. Champion, 1995.
Piero della Francesca. *De prospectiva pingendi*. Edited by G. Nicco-Fasola. Florence: G. C. Sansoni, 1984.
Plato. *The Laws*. Edited by Trevor J. Saunders. London: Penguin Books, 2004.
———. *Phaedrus*. Edited by Christopher Rowe. London: Penguin Books, 2005.
Plotinus. *Les Ennéades de Plotin*. Edited by M.-N. Bouillet. 3 vols. Paris: J. Vrin, 1981.
Prater, Andreas. "Sehnsucht nach dem Chaos: Versuch über das Sfumato der Mona Lisa." In *Ikonologie und Didaktik: Begegnungen zwischen Kunstwissenschaft und Kunstpädagogik; Festschrift für Axel v. Criegern*, ed. Johannes Kirschenmann, 89–105. Weimar: VDG, 1999.
Rotman, Brian. *Signifying Nothing: The Semiotics of Zero*. 2nd ed. Stanford: Stanford University Press, 1993.
Schneider, Notker. *Die Kosmologie des Franciscus de Marchia: Texte, Quellen, und Untersuchungen zur Naturphilosophie des 14. Jahrhunderts*. Leiden: E. J. Brill, 1991.
Seeck, Gustav Adolf. "Die Theorie des Wurfs: Gleichzeitigkeit und kontinuierliche Bewegung." In *Die Naturphilosophie des Aristoteles*, ed. Gustav Adolf Seeck, 384–90. Darmstadt: Wissenschaftliche Buchgesellschaft, 1975.
Seife, Charles. *Zero: The Biography of a Dangerous Idea*. New York: Viking, 2000.
Solmi, Edmondo. *Le fonti dei manoscritti di Leonardo da Vinci*. Turin: Loescher, 1908.
Sorabji, Richard, ed. *Philoponus and the Rejection of Aristotelian Science*. Ithaca: Cornell University Press, 1987.
Stoichita, Victor. The Self-Aware Image: An Insight into Early Modern Meta-painting. Cambridge: Cambridge University Press, 1997.
Strong, Donald. *Leonardo on the Eye: An English Translation and Critical Commentary of MS. D in the Bibliothéque Nationale, Paris, with Studies on Leonardo's Methodology and Theories of Optics*. New York: Garland, 1979.
Summers, David. *The Judgment of Sense: Renaissance Naturalism and the Rise of Aesthetics*. Cambridge: Cambridge University Press, 1990.
———. *Vision, Reflection, and Desire in Western Painting*. Chapel Hill: University of North Carolina Press, 2007.
Toussaint, Stéphane. "Leonardo filosofo dei contrari: Appunti sul 'chaos.'" In *Leonardo e Pico: Analogie, contatti, confronti*, ed. Fabio Frosini, 13–35. Florence: Leo S. Olschki, 2005.
Vasoli, Cesare. *La lalde del sole di Leonardo da Vinci, XII Lettura Vinciana*. Florence: Giunti Barbèra, 1973.
Veltman, Kim H. *Linear Perspective and the Visual Dimensions of Science and Art: Studies on Leonardo da Vinci I*. Munich: Deutscher Kunstverlag, 1986.
Weber, Andreas. *Natur als Bedeutung: Versuch einer semiotischen Theorie des Lebendigen*. Würzburg: Königshausen & Neumann, 2003.
Wellmann, Marc. *Die Entdeckung der Unschärfe in Optik und Malerei: Zum Verhältnis von Kunst und Wissenschaft zwischen dem 15. und dem 19. Jahrhundert*. Frankfurt am Main: Lang, 2005.
Wieland, Wolfgang. *Die aristotelische Physik: Untersuchungen über die Grundlegung der Naturwissenschaft und die sprachlichen Bedingungen der Prinzipienforschung bei Aristoteles*. 2nd ed. Göttingen: Vandenhoeck & Ruprecht, 1970.
Wohlfart, Günter. *Der Punkt: Ästhetische Meditationen*. Freiburg im Breisgau: K. Alber, 1986.
Wolff, Michael. *Geschichte der Impetustheorie: Untersuchungen zum Ursprung der klassischen Mechanik*. Frankfurt am Main: Suhrkamp, 1978.
———. "Mehrwert und Impetus bei Petrus Johannis Olivi: Wissenschaftlicher Paradigmenwechsel im Kontext gesellschaftlicher Veränderungen im späten Mittelalter." In *Sozialer Wandel im Mittelalter: Wahrnehmungsformen, Erklärungsmuster, Regelungsmechanismen*, ed. Jürgen Miethke and Klaus Schreiner, 413–23. Sigmaringen: Jan Thorbecke, 1994.
Zöllner, Frank. "'Ogni pittore dipinge sé': Leonardo da Vinci and 'Automimesis.'" In *Der Künstler über sich in seinem Werk: Internationales Symposium der Bibliotheca Hertziana 1989*, ed. Matthias Winner, 137–60. Weinheim: VCH, Acta Humaniora, 1992.
Zubov, Vasilii Pavlovich. "Jean Buridan et les concepts du point au quatorzième siècle." *Medieval and Renaissance Studies* 5 (1961): 43–95.
———. *Leonardo da Vinci*. Cambridge: Harvard University Press, 1968.

5.

Beyond the Eye

Observing the Unseen in Mathematics and Architecture

ALINA PAYNE

Teofilo Gallaccini (1564–1641) was a Sienese polymath, professor of mathematics and logic at the University of Siena, architecture critic, medic, and philologist, as well as a Neoplatonist and supporter of Galileo, whose activity spanned across the sixteenth and seventeenth centuries. The term "across" is key here, as his interests and work display very clearly the nature of this period of transition, astride two centuries, generally understood as the eve of classical science and the dawn of the Scientific Revolution.[1] And by showcasing a period of transition, Gallaccini's activity is important, I would argue, precisely because it allows us to identify redundancies and shifts in discourses, to see not only what works but what *stops* working and why—and in so doing it turns a more nuanced mirror upon ruptures, caesuras, and revolutions.[2]

Gallaccini's work fanned out over a remarkable number of disciplines and subdisciplines. He wrote treatises on mathematics (a commentary on Tartaglia's *Scienza*

This chapter was submitted for publication in the spring of 2012. References are to publications appearing before that time.

nuova, a treatise on Euclid's book VI, one on the circle, one on angles, and one reputedly on the ellipse); on astronomy (on the astrolabe, a now lost *Cosmografia* on the movement of the planets, after the appearance of three comets in 1618, and an essay on the unity of the skies titled "L'unità del cielo / La monade celeste," in which he leaned equally on John Dee and Galileo); on ballistics and artillery (extensions of his commentary on Niccolò Tartaglia), as well as on waterworks or hydraulics (on the "force" of water).[3] He also made a serious study of architecture, writing an extensive treatise on fortifications and another on port city design, one on perspective, one on ornament (on column capitals), and a travel sketchbook; more important, he wrote a treatise on the errors of architects, on which his fame now chiefly rests, and an exquisitely illustrated treatise on the ideal temple, which is preserved only in part.[4] Finally, to add one other branch to this combination of interests, Gallaccini was trained as a medical doctor (at the Ospedale della Scala in Siena) and not only was active as such in his hometown (as *proto medico* in a supervisory capacity over doctors and pharmacists) but also wrote on anatomy and, according to his biographers, contributed anatomical illustrations to the *Anatomia* of Valverde de Amusco.[5]

A native of Siena, Gallaccini interacted with painters there (Francesco Vanni in particular), spent time in Rome and Naples, and reported on artworks, old and new, that he encountered there. In Siena he entered the intellectual milieu of the Accademia dei Filomati, and eventually became a professor of mathematics and logic at the University of Siena, a position he held for almost thirty years, until his death. He wrote a significant amount—some of his texts were clearly meant for teaching—but published only one work (his doctoral dissertation) in his lifetime. He was rediscovered in the mid-eighteenth century in the ambit of Venetian neoclassicists Father Lodoli, Antonio Visentini, Consul John Smith, and Alessandro Pompei as an early prophet shoring up the emerging antibaroque position gathering strength among them.[6]

What we can take from this short review of Gallaccini's activities is that he was a scientist working mostly in three domains: mathematics and its applications, medicine, and the arts. In his oeuvre the three were tightly interwoven. For example, his treatise on angles is mostly on their use in the arts of *disegno*, though with links as far distant from the arts as agriculture and astronomy. Architecture seems to have been another natural extension of his mathematical interests, as it too belongs (through geometry) to the world of numbers, as does perspective, on which he wrote a treatise clearly focused on describing a mathematical space and its architectural props (fig. 5.1). But what is particularly striking is that throughout his work Gallaccini is less focused on perfect forms (in architecture) or theorems as such (in mathematics) than on forces and movement. Indeed, not only is he working on artillery, ballistics, and fortifications as well as machines, on the movement of planets and of comets, and on rivers and moving water, but when he comments on Euclid and looks at angles and circles, he does so through the prism of bodies in movement (fig. 5.2).

In short, much of Gallaccini's oeuvre seems to have been an effort to understand motion—of projectiles and planets, of earthly bodies and celestial ones—or, put another way, a lifelong effort to understand forces (for which there was not as yet a clear concept, though they were manifested visibly as movement) and to chart movement

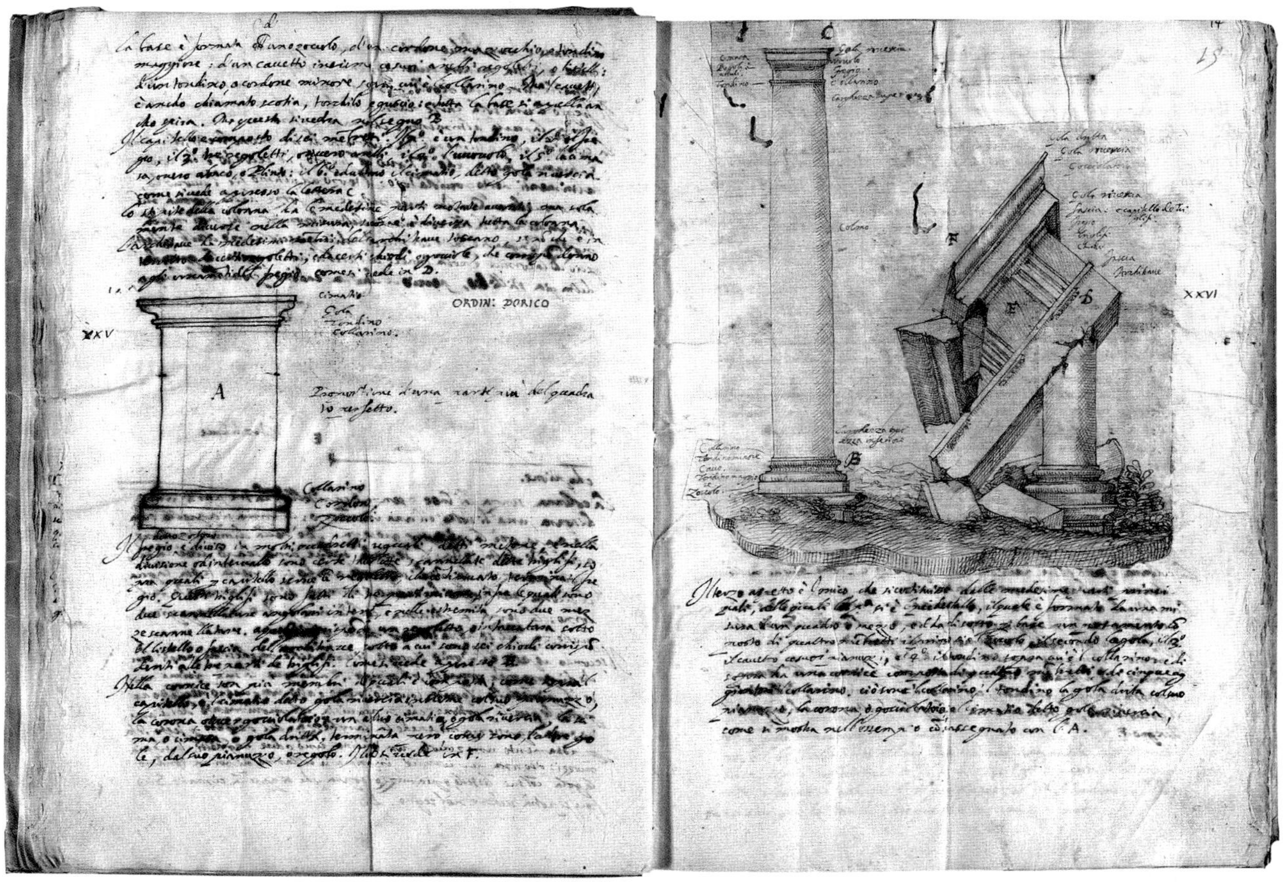

5.1 Teofilo Gallaccini, "Teoriche e pratiche di prospettiva scenografica." Biblioteca comunale degli Intronati, Siena, MS L.iv.4. © Biblioteca comunale degli Intronati, Siena (29.8.2013).

as trajectories. How such abstractions could be seen and made visible to others, and how they were understood to be the visible truth of empirical phenomena, is an obstacle that Gallaccini encountered and tried to overcome. In the end, in attempting to connect mathematical abstractions with empirical observation—rather than illustrate it as abstract diagrams—Gallaccini attests to the presence of a moment of caesura in scientific thinking, or of a confrontation with the fact that demonstration images need not and cannot always be of things seen. To be sure, the geometrical constructions carried the memory of the hand's and its tool's choreography, a visual trace of movements across the page. And to some degree such constructions may be deemed to be representational. But the analyses of mechanics are not, for the world of forces cannot be seen. And it is this clash between the desire to

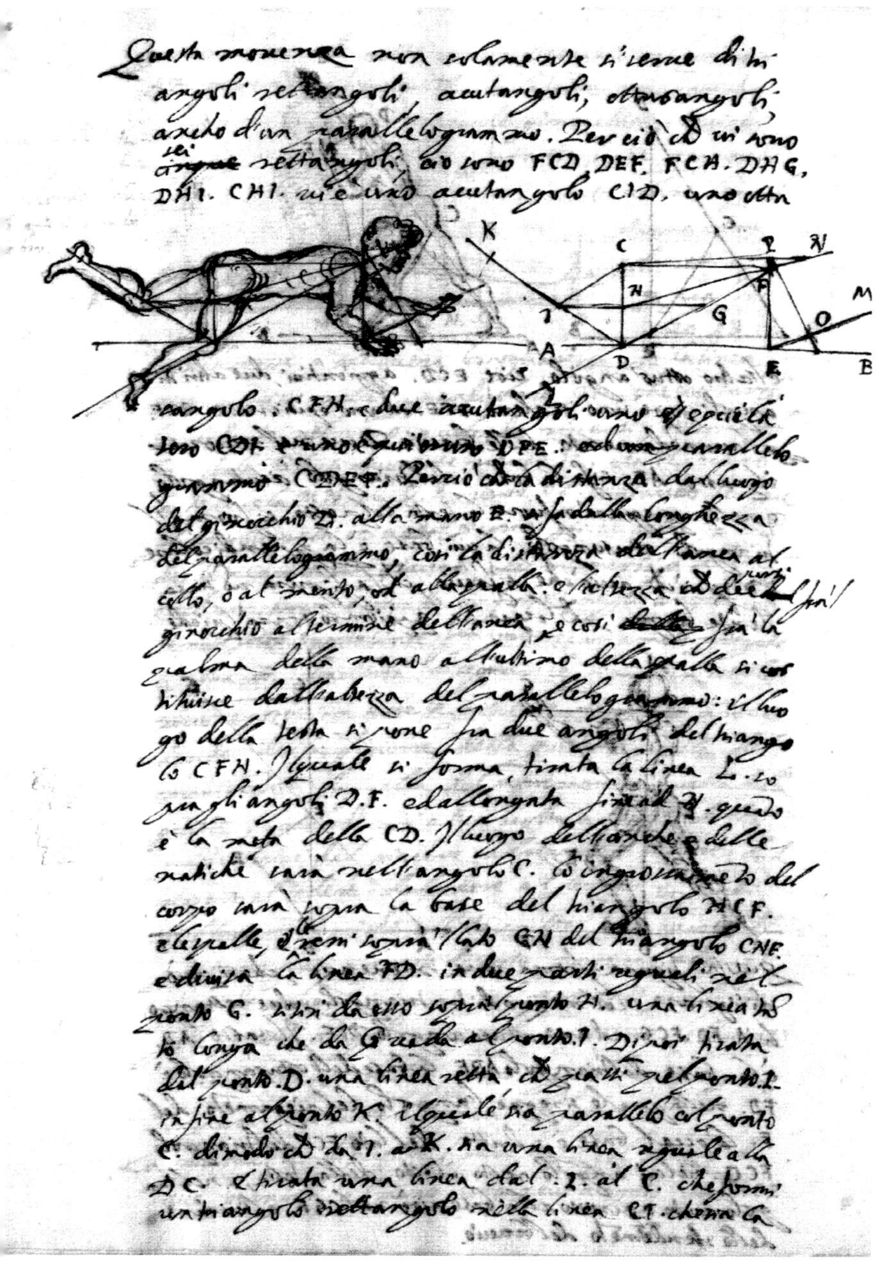

5.2 Teofilo Gallaccini, "Perigonia, overo degli angoli." Biblioteca comunale degli Intronati, Siena, MS L.iv.5, fol. 81v. © Biblioteca comunale degli Intronati, Siena (29.8.2013).

see and illustrate and the impossibility of doing so that Gallaccini confronts.

Matematica Media

Gallaccini's interests certainly followed the trend of the times, for the three new mathematical sciences that developed in the seventeenth century and owed much to Galileo were the science of motion, the science of waters, and the science of resistance of materials, all of which were fundamental components of mechanics.[7] However, since mechanics was in its most exciting phase in the 1640s, when new connections were also being made between mathematics and the physical world, Gallaccini's career, which started in the 1590s and ended with his death in 1641, unfolded in the period preparatory to this moment.[8] His own expertise had offered him points of entry into this buoyant field; most important, it is also here that the third area of his interest—anatomy and medicine—converged, and it is to this three-way dialogue among mathematics, mechanics, and anatomy that his synthesizing mind and thirst for visual explication were particularly drawn.

Indeed, Gallaccini connected anatomy and mathematics within the new developing field of mechanics and applied the insights that this connection provided him to most of the work his inquisitive mind turned to. Neither mathematics nor astronomy was devoid of a mechanics perspective, and as this connection intensified at the time, the anatomical references he associated with these fields also grew in significance in his work. On the face of it, the use of the human body as a means of illustrating and understanding various phenomena was part of such a long-standing tradition of anthropomorphism, especially in architecture—one of Gallaccini's principal interests—that its absence would have been more conspicuous in his oeuvre than its presence. Yet the way in which Gallaccini applied the analogy was both subtly and radically different from that of his predecessors. His interest was not in appearances—in proportions, or numbers, or symmetry, or even form—but in the joints and networks that make up organisms and that he illustrated and described in all their anatomical detail. In fact, his reading of the human body was a mechanics-driven one. Moreover, the connections he made were not between anatomy and fields whose objects were tangible and graspable materially, like architecture, linking physical bodies with other physical bodies. Instead, he drew parallels with mathematics and astronomy, whose objects are either fully abstract or so immense and distant that they become virtually so, thereby suggesting a profound need to make them empirically accessible.

Thus, like his attempt to argue for a single universe and a dense lunar body through his direct observation of eclipses, his anatomical *comparanda* in the "Sfera" (the second part of the "Mondo celeste" manuscript on astronomy) were used to understand and justify the motions of the planets through the analysis of the motions and articulations of the human body and its organs, that is, through another set of empirically accessible data. Most interesting are his analyses of the movements of the earth and the location of the magnetic poles in direct analogy with the movements of the human head and eyeballs (fig. 5.3). Like much of his thinking, his conclusion on the earth's movement is based on an analogy rather than on complex astronomical observations of parallax between the earth and *novae* (new stars), which Galileo and others had tried (unsuccessfully) to use in this context.[9]

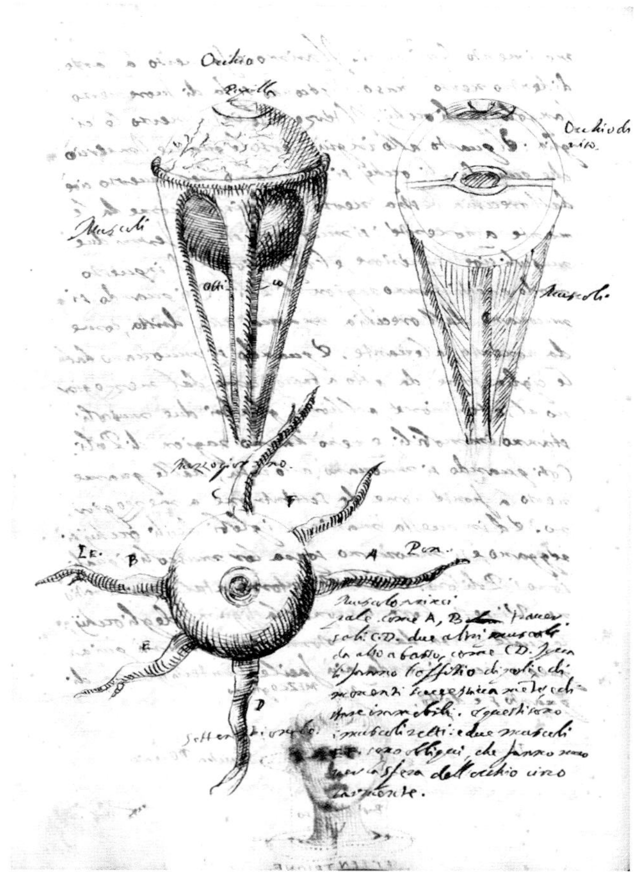

5.3 Teofilo Gallaccini, "La sfera [La monade celeste]." Biblioteca comunale degli Intronati, Siena, MS L.vi.31, fol. 127v. © Biblioteca comunale degli Intronati, Siena (29.8.2013).

This organicist view of the world was a mainstay in Gallaccini's conception of the universe and its parts, and he described the earth lyrically—"corpo di questo grande animale del mondo è un globo composto di più parti di varia natura" (the body of this great animal, the world, is a globe composed of parts of many natures)—in one of his academic discourses.[10] The consistent Neoplatonic orientation in his work is evident here, too: the relationship

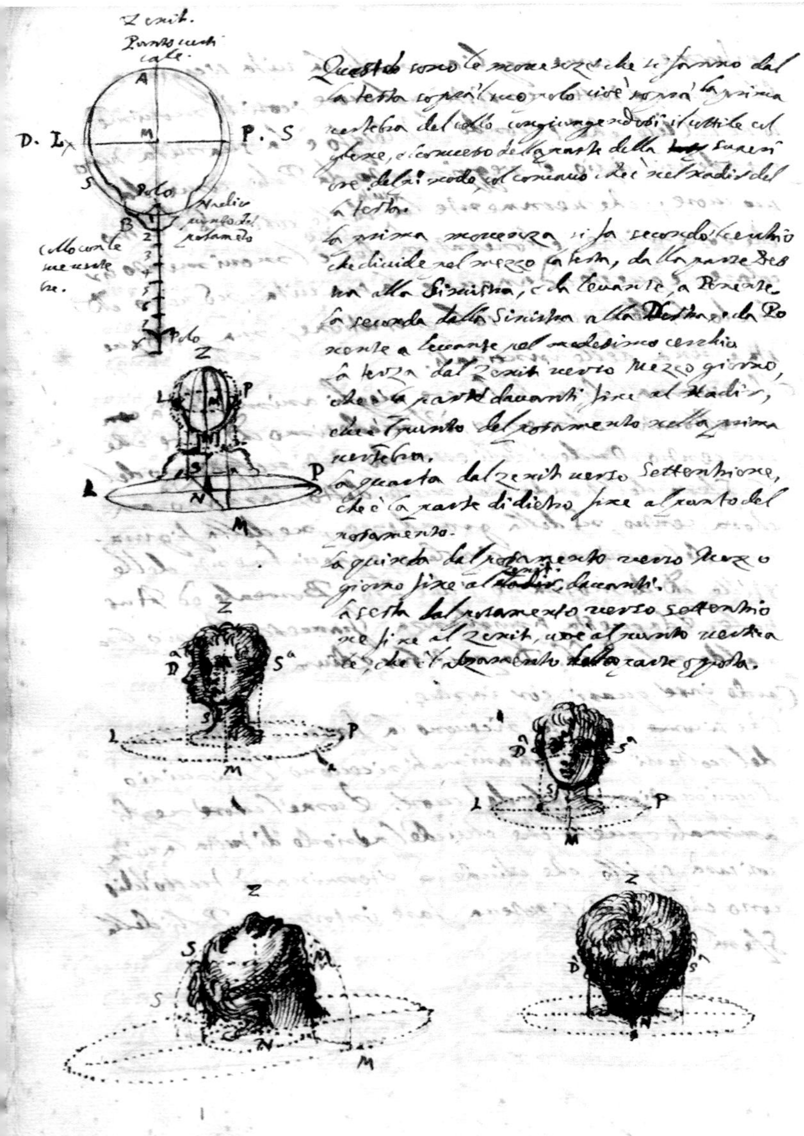

5.4 Teofilo Gallaccini, "La sfera [La monade celeste]." Biblioteca comunale degli Intronati, Siena, MS L.vi.31, fol. 129v. © Biblioteca comunale degli Intronati, Siena (29.8.2013).

of microcosm-macrocosm (the *piccolo mondo* versus the *grande mondo*, in his words) receives a very physical demonstration. The rotation of the eyeball in east-west and north-south directions based on pairs of muscles in tension and repose, and his understanding of joints as fulcrums for movement, provided a means of explicating the terrestrial poles and thus confirming the validity of the astronomers' claims.[11]

But insight worked both ways. On the one hand, Gallaccini's professional knowledge of anatomy allowed him to go one step beyond visual analogies and intuit balances of forces that the human body presented; on the other, his knowledge of astronomy also enhanced his conception of human anatomy. More important, this analogy—an example of circular motion in three dimensions around an axis, complete with very accomplished drawings illustrating cranial motion—brings a measure of visual punch or eloquence to his argument, an argument, moreover, that could not yet be taken for granted (fig. 5.4). That the earth moved around its axis was not what the ancients had held, and this investigation in itself was in line with controversial innovations in the field.[12]

The same interest in the place of mathematics in the physical universe and especially in the geometrical forms of the body's motions also occurs in Gallaccini's "straight" mathematical treatises-cum-lecture notes, the "Cerchio" and "Perigonia." Although on the face of it these are commentaries on Euclid, both works nevertheless branch out into the mathematical analysis of movement. However, if Gallaccini starts by describing circles (as he does angles) in terms of the folding or reaching movement of the human body, the next step of the analogy, which he develops in his notes, logically takes him to use the body's movement to understand basic forces. Thus he refers to Galen (a frequent

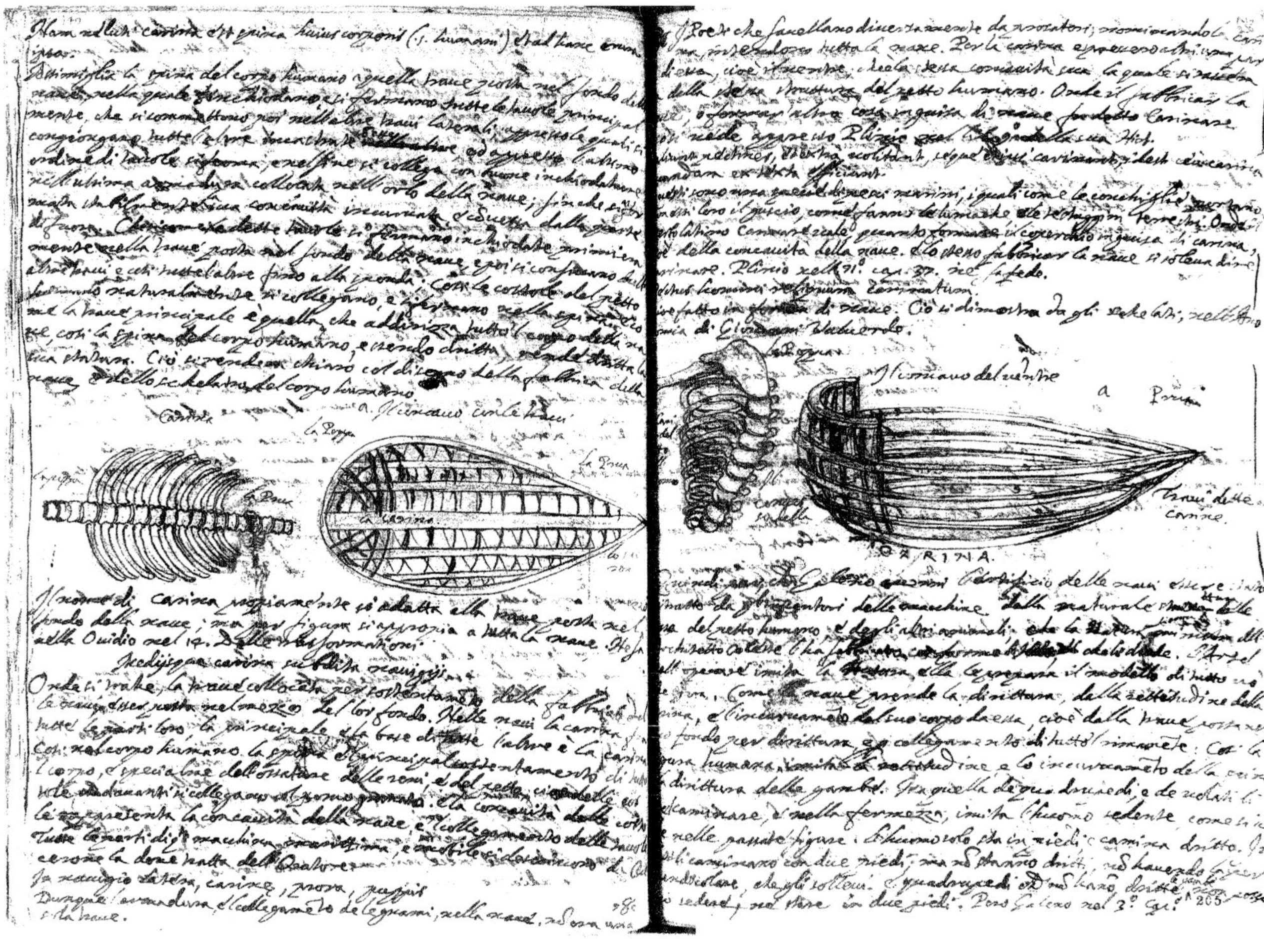

5.5 Teofilo Gallaccini, “Miscellanea.” Biblioteca Apostolica Vaticana, Vatic. Chigi, MS I.vi.206, pp. 138–39.

reference for him) and to the *Anatomia* of Valverde de Amusco, to the principles behind levers (from “Perigonia”), to skeletal construction in wood, and especially to the interweaving of members in shipbuilding (fig. 5.5).

> The human spine is similar to the beam that is located at the bottom of the ship, in which are nailed and held together all the planks, which are then laid tight against the secondary beams; next to which are laid all the others which are not keyed in, up to the last row of planks which are nailed securely to the edge beam of the ship; until such point when its curved concavity, convex from outside, is securely formed. As those said planks are held together primarily

5.6 Teofio Gallaccini, "Miscellanea." Biblioteca Apostolica Vaticana, Vatic. Chigi, MS I.vi.206, p. 206.

through nailing to the beam located at the bottom of the ship, and then are reinforced by the other beams and all the rest up to the edge; thus the ribs of the human chest are naturally connected and are held firm by the spine. And as the main beam is that which organizes all the body of the ship, so the spine of the human body, being straight, causes its stature to be straight likewise.[13]

However, as in his astronomy work, buried among the references to traditional source material are also innovations, and these stem precisely from the connections he makes between anatomy and mathematics. Where in his "Perigonia" and in the "Cerchio" he analyzes the movements of the human body as a means of visualizing geometry, in his later notes Gallaccini also includes birds and raises the interesting issue of folding forms (fig. 5.6).[14] The presence of birds as *exampla* is not insignificant, as they provide the best vehicle for investigations into the folding of bodies, especially evident in their wings as a form of living origami. This is a dimension that contemporary mathematical treatises did not normally include as an explicitly formulated issue. Folding forms tie into aerodynamics as well as into the analysis of simple geometric volumes that both contain and fan out into complex structures and as such suggest other directions that his thinking and teaching might have taken. That folding was a significant issue for Gallaccini that also crossed into architecture is also demonstrated in his proposal for a simplified way to construct the Ionic volute as a system of folding planes. However, most revealing of his mindset is the deconstruction of the folding human form into a sequence of concave and convex angles that resolve themselves into a series of forces and corresponding buttresses that ensure the figure's equilibrium.[15] Gallaccini's analysis starts with the figure in repose (as a stable body in equilibrium), from which he seeks to understand the causes of its motion, and proceeds with a long section on how the body folds during movement. Moving logically from one primary cause to its effects and down the string of consequences, he concludes that folding produces angles and that the nature of these angles and their relationship to one another ensures equilibrium during movement.

The illustrations and in particular the abstract diagram that he derives from his observations, which recall a roof truss on its side, may also be reminiscent of Dürer's *Vier Bücher über menschlicher Proportion* (3.528)—a text he knew well and annotated at length—and also of Leon Battista Alberti's analogies between human anatomy and architectural structures, with which he was also very familiar.[16] But the folding sequence as a prerequisite of structural soundness in both birds and humans, albeit in different ways, signals a more complex and original line of thinking. Gallaccini sought to understand the abstractions of dynamics through the most basic system available for study: the human body. And his citations are not taken from artists (though they were evidently useful in promoting the human analogy). Instead, as a practicing medical doctor and mathematician, he turns to different sources than do his contemporaries, and he blends Aristotle's *Mechanics* with Galen's *Dell'uso delle parti*, which is interesting in itself. Gallaccini evidently knew both texts particularly well, as he was said to have written a commentary on the *Mechanics*, and his notes on Galen in this manuscript almost amount to a commentary. Thus he uses the concept of weight (*gravezza*) as a starting point for an understanding of motion: because all bodies in equilibrium are vertical to the center of the earth, he argues, any change in this position is the primary cause of movement. This interest in a mechanics of organic bodies is already detectable in his early work on anatomy—recently found among the manuscripts collected by Targioni Tozzetti in the eighteenth century—where one of the few illustrations he includes is that of the shoulder articulation shown beside a pulley, evidently included to explain the human mechanism at work (fig. 5.7). The image is not far from the argument developed in the "Sfera" on articulations

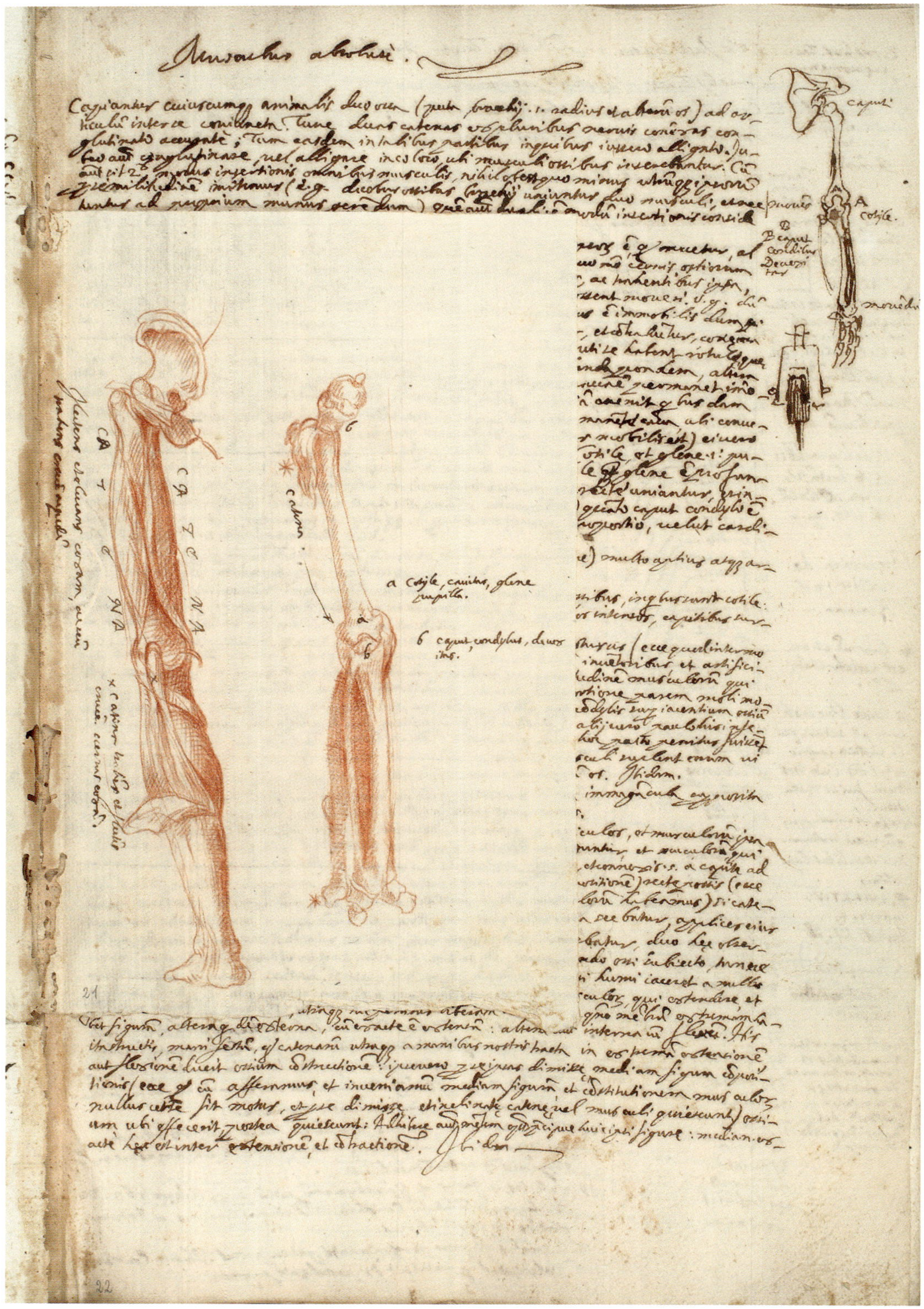

5.7 Teofilo Gallaccini, "Anatomia." Biblioteca nazionale centrale, Florence, G Targioni Tozzetti 68, fol. 22r.

as human poles. Indeed, one can almost see his thought process evolve from one work to another.

Gallaccini's focus on birds and their folding suggests that he observed them closely, as he did shells and other organic bodies, and that however dependent his work may have been on written sources, it rested on empirical observation as well. This is, of course, not so far from Leonardo's incisive and almost hyperrealistic painted and drawn observations that sought to explain phenomena as seen by the eye, visually penetrating them step by step as if by borehole, with the aid of logic rather than microscopes. And Gallaccini's bodies in movement as a form of mathematical demonstration recall the Codex Huygens illustrations, themselves derived from Leonardo (figs. 5.8 and 5.9).[17] Yet there is an important distinction to be made here. The Codex Huygens is a text and series of images preparatory to a treatise for artists on the human figure—its proportion, its movement, and its representation in perspective. It shares an interest in analyzing the geometry of the body's movement, but it is not intended to be a treatise on mathematics, anatomy, or mechanics, and certainly not of all three rolled into one. And while there are undoubtedly similarities—the references to microcosm and macrocosm and to points of rotation, and the images themselves (though Gallaccini's are far more diagrammatic)—they remain just that, superficial similarities deriving from a shared culture. There are none of Gallaccini's insightful applications of body-based mechanics to the understanding of forces (of which there is no mention), the movement of planets, or the functioning of the magnetic poles. Where the Huygens images are ultimately part of an analysis of sight and perspective construction (how bodies in movement can be measured and plotted in a geometrized perspective space, especially in foreshortened

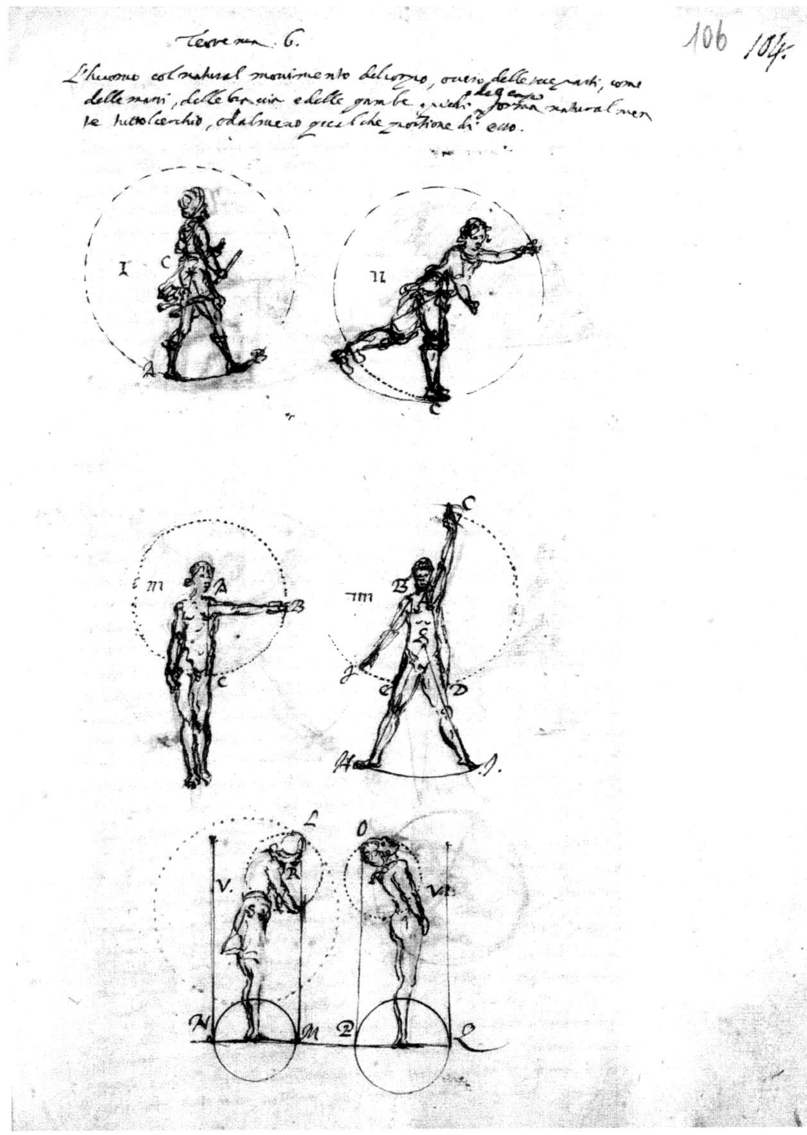

5.8 Teofilo Gallaccini, "Della natura del cerchio e del'operazione del compasso." Biblioteca comunale degli Intronati, Siena, MS L.iv.5, fol. 104r. © Biblioteca comunale degli ntronati, Siena (29.8.2013).

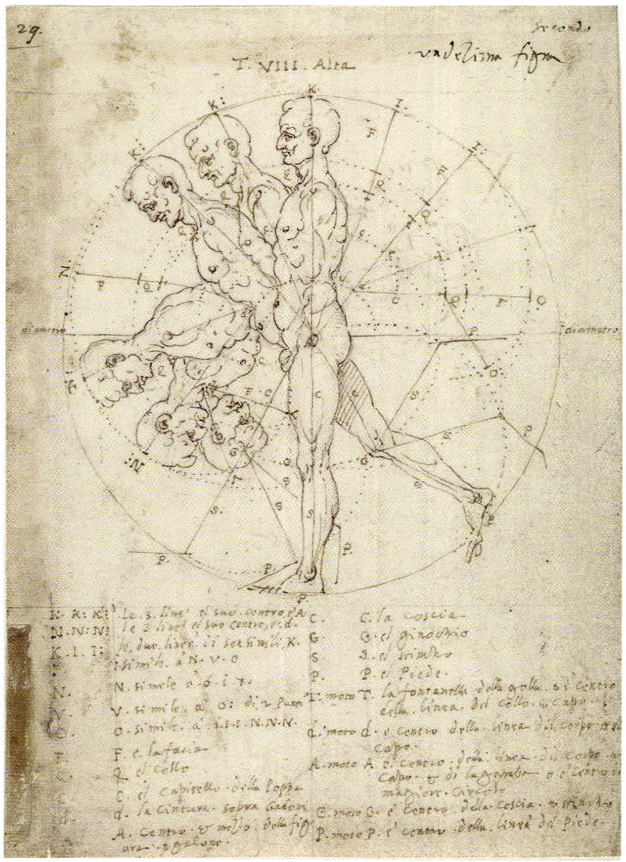

5.9 Codex Huygens, "Undezima figura," fol. 29. The Pierpont Morgan Library, New York, 2006.14:29. Purchased 1938.

form), Gallaccini's are about mathematics itself and about the importance of visualizing abstract geometrical diagrams, of embodying what cannot be seen. His is a work on enlivened mechanics, not on the body itself.

Gallaccini calls the domain that brings together these various forms of embodied mathematics "matematica media." This was a term already employed by mathematician and astronomer Egnatio Danti with reference to geometry, though one that Gallaccini explicates at greater length.[18] Where traditionally the intersection between so-called pure and practical mathematics included optics, cosmography, the sphere, and various aspects of drawing as a means toward machine construction, Gallaccini adds anatomy and medicine: "[After pure mathematics] follows the intermediate [applied] mathematics, which participates in the science of the sphere, of cosmography, from it derives medicine, it participates in anatomy, and natural philosophy. This discourse is partly natural, partly spherical and cosmographic, partly medical and anatomical. Principally it cannot be well understood without prior knowledge of the terrestrial globe, the common habitat of the living."[19] This is a departure from earlier divisions, even when these were as expansive as Danti's or, earlier, as Luca Pacioli's. As the latter describes it in the dedication of his *De divina proportione* (1498), the mathematical disciplines include arithmetic and geometry, astrology, music, perspective, architecture, and cosmography—but not anatomy.[20] Mathematician Bernardino Baldi had used human staffage to explain his work on mechanics, but he did not go so far as to embrace anatomy proper. Gallaccini's introduction of the human body both visually and as an argument is apparently unique, and unique to his combination of skills—among them anatomy as well as drawing—which allowed him to make such links, the better to persuade his students, academic listeners, and would-be readers.

Of course, as ever, Gallaccini's exploration of organic analogies and his insights into mechanics included an apparently heterogeneous group of readings and references. Thus, in his manuscript notes, in a turn that brings together John Dee's esotericism (his hieroglyphs as visual representations of meaning and form) with

Gallaccini's own version of *matematica anatomica* and mechanics, he attends to the "rationis motus brevis humanitas." Once again, these expansive connections occur in his miscellaneous notes (where, alongside Dee's hieroglyphs, he includes references to Galen, Pliny, and Valverde de Amusco with the same enthusiasm with which he quotes Hermes Trismegistus and Marsilio Ficino), while in his treatises and lectures his statements are less wide-ranging.[21] This combination of sources reveals the Neoplatonic basis that binds together various aspects of Gallaccini's thought and that, as an authoritative philosophical corpus, by the 1640s permits him to engage in this form of comparative science.

The Eloquence of the Visible

For Gallaccini, the human body functioned not only as a way to describe or understand but also as a form of visual eloquence. His description of the circle is a case in point: "For if we look at the form of the members, at the connections between the bones, and at the movement of each member, we will see in them the circular form either whole or at least a portion of the larger or smaller circle. . . . Thus man, gathering in himself all the movements, and mutations, together with the circular movement, is worthy to be described as a microcosm."[22] An eloquent device that could be illustrated and that resonated with his audiences, unlike the mathematical abstractions that could not, the body's anatomy facilitated the explication of abstract phenomena in a concrete manner. Perhaps the didactic function of his texts encouraged Gallaccini to seek such parallels, for he needed to help his students visualize complex phenomena, as well as persuade them of their validity. Another incentive for the use of visual eloquence was the courtly audience to whom such mathematical works were also often addressed in this period, and Gallaccini, with his strong ties to the Piccolomini family in Siena, whose palace he frequented, as well as with his active participation in the academic debates of the Filomati, was no exception in engaging such groups.[23]

One lens through which to view the body's appearance as a device of visual eloquence is the Neoplatonic analogy between microcosm and macrocosm, which was a consistent trope in Gallaccini's work since his youthful dissertation on *De verum amore* (1595) and represented his participation in the contemporary reorientation in science away from the traditional Aristotelian corpus.[24] But there was more at stake than this. In his scientific work—in the Tartaglia commentary and the treatise on ballistics, for example—the need for visual eloquence is unspoken yet palpably present. For example, in his discussion of projectile motion, Gallaccini clearly recognizes two distinct components: "violent motion" (or impetus) and "natural fall" (this has a horizontal component and is thus not quite free fall, which is vertical only, but is related to it, as it is affected by gravitational forces). The impetus trajectory is the straight line that merges into a quarter circle to which it is tangent (so he argues in the Tartaglia commentary)—the quarter circle is the beginning of free fall (AB is impetus; CD is natural fall, as he explains in the text). What he shows in his treatise on artillery (fig. 5.10) takes the thought further: not only is he dividing the motion in two, but he tries to model it even more accurately by dividing it into smaller segments. As the long and plodding text indicates, it is difficult to put all this into words; images—that is, visual representations—were essential, particularly in a mathematical culture where notations were far from

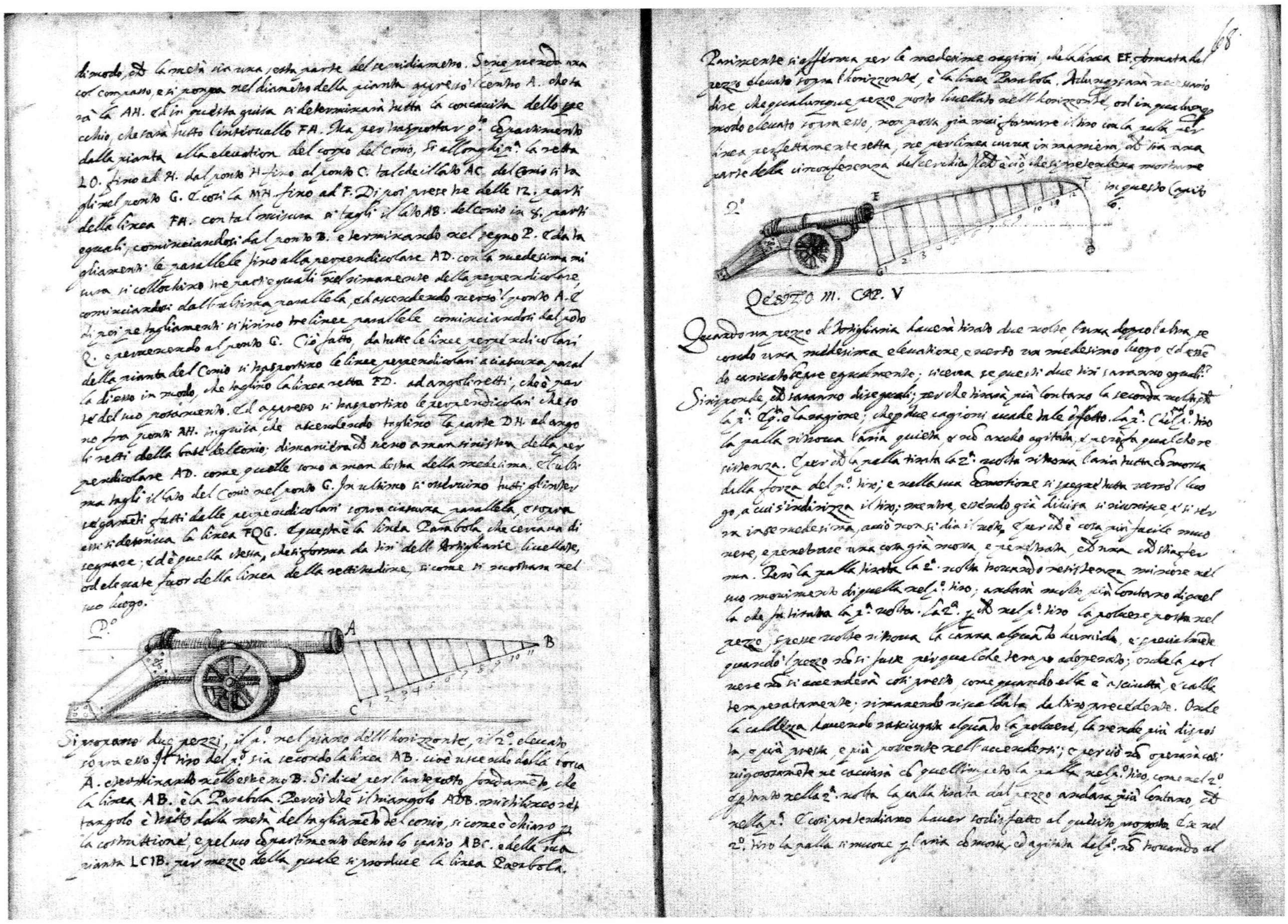

5.10 Teofilo Gallaccini, "Delle ragioni dei tiri: Dell'artiglieria." Biblioteca comunale degli Intronati, Siena, MS L.iv.2, fols. 67v–68r. © Biblioteca comunale degli Intronati, Siena (29.8.2013).

standardized and the means of communication were difficult. A discursive explanation adds complication and laboriousness that a drawing immediately relieves. In short, the image is crucial.

Gallaccini's effort to use the human body to illustrate scientific abstraction responded to this need. Indeed, beyond the analogies and insights he derived from anatomy in his scientific writings, the images of human bodies with which Gallaccini filled his texts was an unusual feature of his work. Yet images were not simply a heuristic device but constituted a tool in themselves.[25] When Gallaccini describes the illustrations of the human body for his geometric understanding of motion and equilibrium, he refers to them as *sensata dimostrazione*.[26] We find

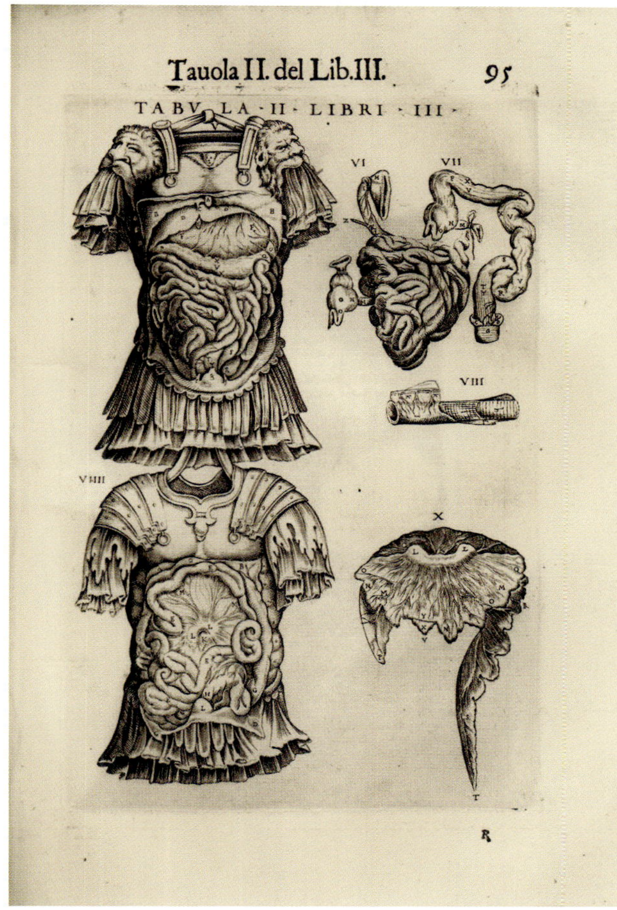

5.11 Juan Valverde de Amusco, "Tavola II. del Lib. III." In *Anatomia del corpo umano* (Rome: Salamanca e Lafreri, 1560), 205.

this insistence that a "sensate" (or physical) component is necessary to mediate the abstraction of mathematical demonstrations again and again in his work. Perhaps Gallaccini was indebted for this insight to Galileo, who had argued that "l'esperienza sensata e la maestra della certezza filosofica" (physical experience is the master of philosophical certainty).[27] But there may have been sources more specific to his own practice that prompted such insight. One reason why Gallaccini was so conscious of the value of the physical dimension was undoubtedly his medical education and career. And his medical studies may also have kindled his interest in the representation of the human body as a tool for recording observation.[28] Moreover, since painting and dissection often intersected, anatomy may well have ignited Gallaccini's interest in the visual arts and in the importance of visual eloquence in the first place. Indeed, the illustrations in Vesalius's *De corporis fabricai*, like those in Valverde's *Anatomia*, were executed by artists of the stature of a Francesco Salviati and indicate that the discipline itself demanded close attention to the visible and a corresponding ability to reproduce it so as to effectively communicate information (fig. 5.11).

At least since Leonardo, observation by way of drawing had worked as an epistemological tool.[29] The drawing that arrests moving phenomena, be they organic bodies, clouds, or waters, and thus allows close and in-depth observation, was not only a means of depicting an *istoria* more persuasively (because closer to nature) but also a means of understanding the natural phenomena themselves. In his *Trattato della pittura* Leonardo had made this quite clear: "No human invention may be called true science, if it does not pass through mathematical demonstrations. And if you were to say that the sciences, which begin and end in the mind, have any truth, that is not accepted, but rejected for a number of reasons; first, because such mental discourses do not draw from experience, *without which nothing can be known with certitude*" (emphasis added).[30] Indeed, Leonardo seems to be speaking here in the voice of Galileo, who would later argue for the *esperienza sensata*. But even if he does not talk about drawing in this passage, Leonardo does focus

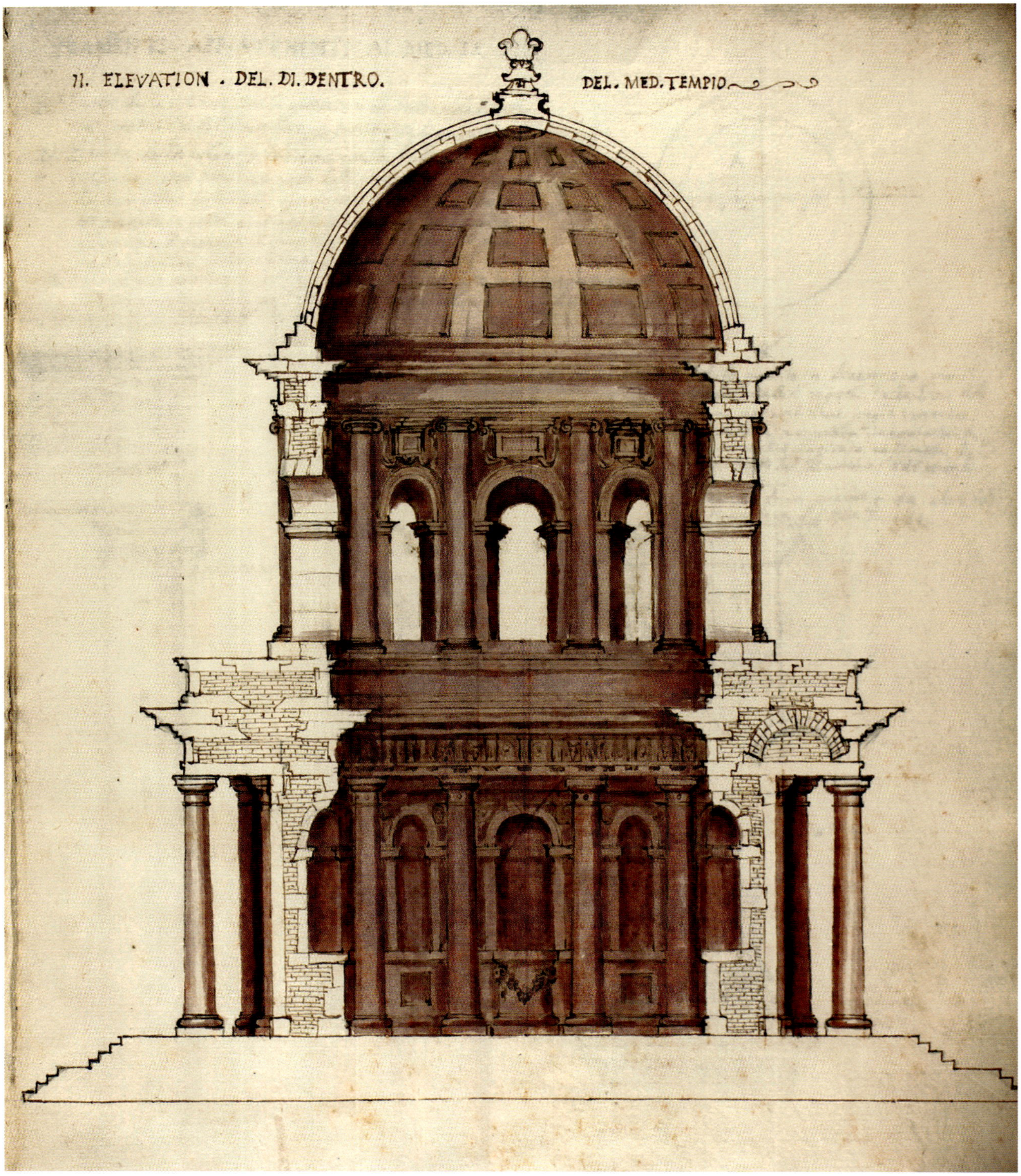
11. ELEVATION . DEL. DI. DENTRO.
DEL. MED. TEMPIO

on the need to ground mathematical abstraction in experience, in what is physically apprehensible. Gallaccini calls it *sensatamente*, and his argument for the utility of *disegno* in mathematics makes the point abundantly clear: the drawing is an intermediary between experience and (philosophical) abstraction.

> *Disegno* is also useful to the understanding of mathematics, for not everyone's intelligence/mind [*ingegno*] can see lines, figures, or bodies, and their formation. On the contrary, the *disegno* alone, and no other part can form all the objects of the mathematician without committing any error. Indeed, the mathematician uses *disegno* as an instrument to devise figures, which he uses to explain himself, and every demonstration is solved by his need for a *physical* [*perceptible*] example, that is easy to understand.[31]

Though, for Gallaccini, drawing is not necessarily equated with observation, it is nonetheless an essential tool for translating abstraction into something that the mind/body relates to: "Since it is through the eye that the rules and demonstrations are best learned (as for all mathematics); the simple verbal description is insufficient, but the *physical/sensuous* example is necessary . . . in such a way that without seeing, this science, like all the other mathematics, would be most difficult to learn."[32] His academic discourse "Dello scintillar delle Stelle," delivered at the Accademia dei Filomati in Siena, is a perfect example of his approach. Although he cannot adduce images, he uses the next-best thing: gesture. The pulsating light of the stars, which he is at pains to describe, he illustrates by the rhythmic opening and closing of the hand.[33]

Architecture, Science, and the Unseen

It is clear that in both his astronomical and mechanics-based work, Gallaccini is concerned with visualizing abstractions: celestial motion, the forces that make things move (or remain in equilibrium). The human body is his paradigm, for it gives physical substance to these abstractions and makes it possible to analyze them empirically. Even in architecture, he looks at movement, stationary and solid though it is: his images of collapsed assemblages (fig. 5.1) of columns and entablatures, precariously balanced cornices and moldings, even irrationally heaped columns, imply the movement caused through breakage, whether by decay, earthquake, or construction faults (which, after all, is one of his interests). These images may seem conceits, but even if they are, they show an underlying and pervasive interest in equilibrium and disequilibrium, movement and rest.

More interesting still is Gallaccini's treatise on the ideal Christian and pagan temples.[34] In his two designs—which he describes exhaustively, from plan through elevation and details of the orders—he includes a remarkable set of sections (fig. 5.12). His approach is unusual in a design drawing, that is, in a drawing focused on presenting an ideal building. Indeed, this is a hybrid drawing: a presentation/design drawing, on the one hand, and a construction drawing, on the other, in which he shows the bricks and masonry infill and the stone slabs tying the structure together. If we compare this with sections illustrated by Sebastiano Serlio or Andrea

5.12 Teofilo Gallaccini, section of temple, in "Il tempio." Warburg Institute, London.

Palladio in their treatises—the latter traditionally seen as the most profession-oriented of the great generation of Renaissance architectural treatises—it becomes clear how unusual Gallaccini's simultaneous presentation of the building appearance and its internal makeup is. Here, too, the human body must have played a role, especially for a medical doctor and author of a treatise on anatomy. The building itself is not presented as analogous to the human body, as had been the case since the recovery of Vitruvius in the Renaissance; but the slice through it is closely connected to a human anatomy, permitting a simultaneous view of how inside and outside relate to each other without privileging one or the other. Gallaccini's anatomical experience must have precipitated an understanding of the need to see both what can and what cannot be seen in one image. This analytical effort on his part is not, strictly speaking, about movement, but it is about the mechanics of preventing it: it is about bracing the wall and the structure against movement, against sheer, against differential settlement. The stone slabs and keying elements tie the walls together and become a solid supporting skeleton, rigid and stable precisely because it is monolithic, made out of one material that is extremely resistant to compression.

When Gallaccini gives the architectural "not seen" as much prominence as the seen, and makes such a prominent feature of his building section, he opens a new chapter in thinking about architecture. Like movement's trajectory, forces at work inside a structure are also invisible to the eye except through their consequences. To be sure, he does not speak of forces—this will happen much later, in the context of engineering design, that is, an approach to design based not so much on geometry as on the mathematical calculation of forces and the resistance of materials. And yet once Gallaccini focuses on materials and their interaction with one another, and their effects on the structure as a whole, he signals an act of attention that is a sea change away from Palladio's or Serlio's interests.[35] His treatise on architectural errors ("Degli errori degli architetti," ca. 1625) is about the effects of forces on architecture, even if he does not call them so by name. Once he shows the wall to be a system of interlocking stones and filler (masonry), the traditional gestural analogy of human body to structure is essentially irrelevant. Swelling bases and flexing architraves, like compressed brackets and columns with entasis, are simply theater or metaphors—the reality of the wall behind it has no connection to it.[36] The thin rhetorical veneer of architecture is shown to be just that. When structures made of stone and brick or rubble fail, as Gallaccini illustrates, they crack, they fissure, they break—they do not swell like human muscles. The analogous model is simply wrong.

Of course, Gallaccini cannot model what actually happens inside the structure. Not even in 1748 does Giovanni Poleni (who holds Niccolò Bernoulli's chair at Padua) yet have the tools to describe them when he analyzes the cracks in the cupola and walls of St. Peter's—even though his explanation of the catenary curve (in the wake of Galileo) and its application to arches is a far cry from what Gallaccini could do one hundred years earlier (fig. 5.13).[37] So Gallaccini has to content himself with repeating again and again the importance of bracing the structure in a three-dimensional unit ("congiugnimento" and "legamento reale" are his terms).[38] And he adamantly insists that the architectural caprices of broken members are like the work of "silversmiths, woodworkers, engravers, *stuccatori* and painters"—in other words, of artisans who mold and glue, solder and paint, paste and varnish, but do not build self-supporting structures.[39] In the end, both words and mathematics fail him.

So what does this quirky scientist-cum-architecture critic add to the topic of vision and its instruments? On the eve of the Scientific Revolution, this is a man struggling with ideas and intuitions he has no names for. Although he does not seem to have the concept of force, he is profoundly focused on movement. In itself, this is in line with the preoccupations of Gallaccini's generation: treatises with titles like *De motu* were current among the luminaries in the field (among them Girolamo Borro, Filippo Fantoni, Francesco Piccolomini, and others who were also fascinated by the movement of water, like Benedetto Castelli and Luca Valerio, and who stand on the shoulders of the ancients—Aristotle, Papus, Hero et al.). What is interesting here, however, is that Gallaccini looks at the movement of terrestrial and celestial bodies as *one problem* (in the sense that he uses the human body to explain the rotation of the earth—and this is not only an analogical example but also intended as a way to understand its actual workings), and for him, clues from one realm move to the other without any apparent obstacles. He is no Newton, but he is in deadly earnest, trying to grope his way through some problems that worry him, and his overly wordy demonstrations testify to the overwhelming weight of tradition. But despite all of that, he is also prepared to envision messy solutions—of trajectories, as of architectural drawings—that emphasize the simultaneous presence of an inside and outside, of a seen and not seen.

Most of all, Gallaccini is pushing against the limits of knowledge, and he seeks, without saying it in so many words, to model what is not there, what can't be seen—trajectories (of cannonballs as of stars, of body movements as of planets), the traces of forces, the geometry of motion. In order to develop the mathematics to understand these things, he needs images, and yet to obtain the images, he needs the mathematics. He needs abstractions to be enlivened, and in the absence of other means, he must fall back upon the human body. Gallaccini's travails, then, represent an example of the symbiotic relationship between mathematics, physics, and images in which architecture, straddling these

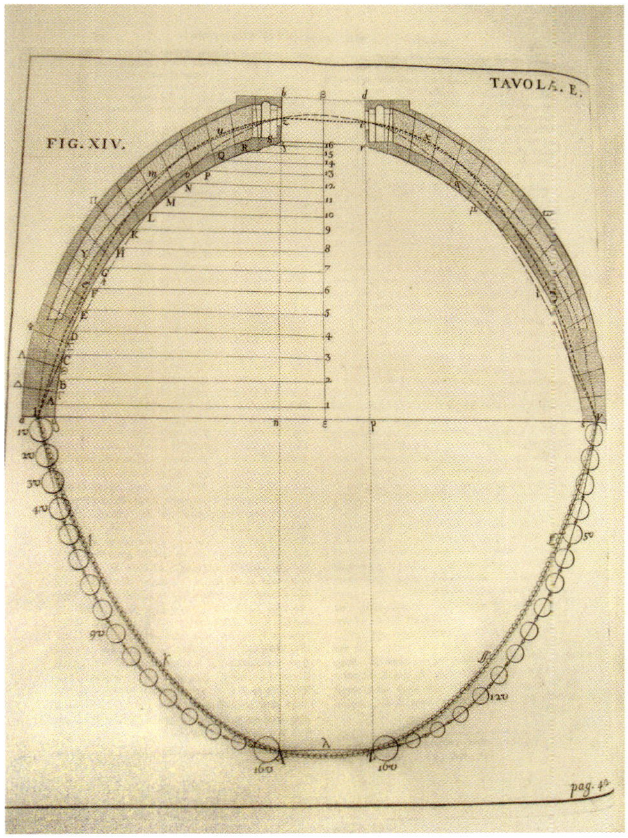

5.13 Giovanni Poleni, *Memorie istoriche della Gran Cupola del Tempio Vaticano* (Padua: Stamperia del Seminario, 1748), 42, fig 14.

domains, has its own, perhaps modest place, as a domain of inquiry and the origin of eloquent representational strategies. Trapped inside a vicious circle, Gallaccini reaches back to the perennial analogy of the human body. But even in movement, the human body does not answer his questions, and ultimately he does not see any better what cannot be seen.

Notes

1. Although useful to describe in a form of shorthand the changes that the sciences experienced in the early modern period, the concept of revolution (first used in the 1930s by Gaston Bachelard and then by Alexandre Koyré) has been challenged and much qualified. Steven Shapin expressed the paradox most concisely in his *Scientific Revolution*: "There was no such thing as the Scientific Revolution, and this is a book about it" (1). Already in the 1950s, Thomas Kuhn (who was otherwise a supporter of the revolution concept) presented the "Scientific Revolution," in his book *The Copernican Revolution*, as a phenomenon that took a good two centuries to play itself out. Conversely, for support of the idea of the Scientific Revolution, see Henry, *Scientific Revolution*. All translations are my own unless otherwise noted.

2. For the bibliography on Gallaccini, see Partsch, "Teofilo Gallaccini"; and, for more up-to-date information, Payne and Fara, "Teofilo Gallaccini e la critica architettonica." Most recently, see Payne, *Telescope and the Compass*.

3. Gallaccini, "Della Nuova Scienza"; "Delle ragioni dei tiri"; "Perigonia, overo degli angoli"; "Della natura del cerchio"; "Esposizione del Sesto Libro di Euclide"; "Trattato di gnomonica"; "Idea della fortificazione"; and "Della natura e della forza dell'Acqua corrente."

4. Gallaccini, "Trattato sopra gli errori degli architetti"; "Degli errori degli architetti"; "Tempio"; "Trattato de capitelli delle colonne"; "Idea della fortificazione"; "Dei porti di mare"; "Commentary to Sebastiano Serlio's Books"; and "Teoriche e pratiche di prospettiva scenografica."

5. Gallaccini was *proto medico* as well as *medico fisico* (as he was described in his marriage act). It is in this capacity that he attended his friend and Filomato colleague Celso Cittadini in 1615. Elsewhere in his writings, Gallaccini describes himself as "Medico, Filosofo, e pubblico lettor di Matematica nello Studio di Siena, e fra gl'Accademici chiamato il Difettuoso." Title page of "Discorsi accademici." See Payne, *Telescope and the Compass*, 7–24. On his writings, see his "Anatomia." The second edition (in Italian) of Valverde de Amusco's *La anatomia del corpo umano* came out in Venice in 1586, too early for Gallaccini, who was still studying at the time, to have contributed to the work, so Gallaccini's biographers must be mistaken.

6. For Gallaccini's biography, including the latest documentary discoveries, see Payne, *Telescope and the Compass*, chap. 1.

7. Meli, "Mechanics."

8. Blair, "Natural Philosophy."

9. Bucciantini, *Galilée et Kepler*, 151.

10. Gallaccini, "Discorsi accademici," fol. 2v.

11. "I Poli già considerati si trovano nel Mondo grande, cioè nella Sfera celeste, e nella Terrena, e nel globo Magnetico, tratto dalle viscere del globo della Terra. Non sarebbe fuor di proposito hora dimostrare i Poli del piccol Mondo. Imperciò che se si considerano le membra del corpo humano, vedremo espressamente, che Poli delle braccia sono le spalle: i Poli delle mani sono le congiunture nel luogo de polsi che si appella il carpo: i Poli delle gambe e delle coscie sono gli assi de fianchi, e 'l polo del busto è l'osso sacro: e 'l Polo del collo è la 8ª Vertebra, cioè 'l 8º nodo della spina: e 'l Polo del capo è la prima Vertebra posta nella sommità della spina. Nella qual Vertebra è impernata e bilicata la testa; di maniera che sopra essa si muove la sfera della testa con movimento mezzo circolare, voltando da man destra a man sinistra, o per contrario, od abbassandosi ed alzandosi davanti e di dietro; di maniera che il primo mezzo cerchio, formato dal movimento laterale, è tagliato dal secondo ad angoli retti Sferali, come si vede in questa dimostrazione. Ma meglio si vedrebbe considerando l'ossatura della testa, dove si vedrebbe il Polo del capo esser fermo nella prima Vertebra, sì come apparisce in questa figura. Oltre acciò appariscono anchora evidentemente i Poli ne gli occhi nostri, i quali son di forma di piccoli globi, e non son fatti per stare otiosi, ed immobili; essendo posti nella testa, come sentinelle, e come guardie di tutta la rocca del corpo humano, e di tutte le sue parti; onde sono stati creati mobili, e non hanno ricevuto un movimento solo, ma quattro regolati movimenti. Imperciòche gli occhii nostri non posson veder tutti gli oggetti visibili da qualsivoglia positione, o da alto, o da basso, o da man destra, o da man sinistra; ma solo da quella parte, che è volta per linea retta verso la cosa visibile in qualunque positione si trovi, e che corrisponde alla rettitudine del centro dell'humor cristallino, e della pupilla dell'occhio, secondo la ragione della Prospet-

tiva ottica. Di modo che quante sono le positioni delle cose visibili, tante debbano esser le positure degli occhii, e tanti i movimenti; che se gli occhi fussero immobili, vedrebbero solamente quelle cose, che gli sono opposte direttamente, e le cose, che si vedessero sarebbero di pochissimo numero, sì come si avvertisce da Galeno nel 10 lib. Dell'uso delle parti del corpo humano, nel cap. 8." Gallaccini, "La sfera," in "Mondo celeste," fols. 125v–126v.

12. Kuhn, *Copernican Revolution*, 148–49.

13. "Assimiglia la spina del corpo humano a quella trave posta nel fondo della nave, nella quale *s'inchiodano* e si fermano tutte le tavole principalmente, che si commettono poi nell'altre travi laterali; appresso le quali si compongano tutte l'altre incastrate non essere, ed appresso l'ultimo ordine di tavole si ferma e nel fine si collega con buone inchiodature nell'ultima armadura collocata nell'orlo della nave; finchè si è formata stabilmente la sua concavità incurvata e convessa dalla parte di fuore. Che siccome le dette tavole si fermano inchiodate primieramente nella trave posta nel fondo della nave, e poi si confermano delle altre travi e così tutte l'altre fino alla sponda; così le costole del petto humano naturalmente si collegano e si fermano nella spina. E come la trave principale è quella che addirizza tutto 'l corpo della nave, così la spina del corpo humano, essendo dritta, rende dritta la sua statura." Gallaccini, "Miscellanea," fol. 204v.

14. Ibid. In the notes, he refers to the work on "Perigonia" as being in the past. See the contemporary Accademia della Crusca definition of the term *piega:* "Quel raddoppiamento di panni, drappi, carta, o simili, che si ripiegano."

15. "Sempre gli angoli sono sostenuti da concavi, e questi prendono forza da convessi. la debolezza di quelli si fortifica dal vigore di questi; mentre tutta la figura si rega dalla perpendicolare detta [la figura mentre camina]." Gallaccini, "Miscellanea," fol. 200v. "Di modo che la rettitudine dell'angolo nella statura humana è la cagione della fermezza, e della sicurezza dello stare, e del sedere, come si conoscerà per questa *sensata* dimostrazione." Ibid., fol. 205v (emphasis added).

16. Gallaccini owned and annotated both books (Dürer in Latin translation by Joachim Camerarius and Alberti in Italian translation by Cosimo Bartoli). See Dürer, *Institutiones geometricae*, BCS, XXI.A.14; Alberti, *Architettura*, KHI, H. 758.

17. On the Codex Huygens, see Panofsky, *Codex Huygens;* and, more recently, Berra, "Storia dei canoni proporzionali."

18. "La Geometria è una scienza di bene, et rettamente misurare, et si divide nella: PRATICA, la quale và misurando le cose secondo ciascuna delle tre misure per longo, largo, et profondo, applicandovi per tutto i numeri. Dimostra la quantità delle linee, delle superficie, et de' corpi. SPECULATIVA, che considera i principii, cioè i punti, linee, superficie, et corpi, comparando le linee alle linee, le superficie alle superficie, et i corpi à i corpi secondo la uguale, et ineguale ragione, che hanno fra di loro, senza l'applicatione de' numeri, et di essa i primi principi sono: Diffinitioni . . . Petitioni . . . Axiomati. . . . MISTA (per dir così), la quale considera i principii come fa la speculativa, ma ci aggionge i numeri, et le misure nella materia sensata come fa l'Architettura, et altre simili mecaniche." Danti, *Scienze matematiche*, 8. This division also shows how geometry/mathematics could be merged with philosophy/*speculativa*.

19. "Segue appresso la matematica media, la quale partecipa della scienza della sfera, della cosmografia, e parte da quella Medicina, e dell'Anatomia, e della Filosofia naturale. Questa ragione parte e naturale, parte sferica & cosmografica, parte medica ed anatomica. Primieramente non si puo bene intendere senza preporre la notizia del globo terrestro, commune habitatione de viventi." Gallaccini, "Miscellanea," fol. 200v. For a related classification of mathematics, see also fol. 198r.

20. Camerota, *Linear Perspective in the Age of Galileo*, 105–6.

21. Gallaccini, "Miscellanea," fol. 197.

22. "Perciò che se riguardiamo alla figura delle membra, à congiognimenti de l'ossa, ed al movimento di ciascun membro, inessi vedremo la figura circolare od intera, od almeno qualche portione di cerchio maggiore, o minore. . . . Onde l'huomo abbracciando in sé tutti i movimenti, e mutationi, insieme col movimento circolare, meritamente si rende degno della preprogativa di microcosmo." Gallaccini, "Della natura del cerchio," fol. 90r–v.

23. See Dupré, "Visualization in Renaissance Optics."

24. See, for example, Gallaccini, "La sfera," in "Mondo celeste," fol. 126v, where he refers to "maggior e minor mondo."

25. On the topic of the epistemological rather than heuristic role of images as it was reflected in Enlightenment science, see Shea, *Science and the Visual Image*; for a cross-historical reading, see, for example, the essays in Baigrie, *Picturing Knowledge*.

26. "Di modo che la rettitudine dell'angolo nella statura humana è la cagione della fermezza, e della sicurezza dello stare, e del sedere, come si conoscerà per questa sensata dimostrazione." Gallaccini, "Miscellanea," fol. 205v.

27. Quoted in Redondi, *Galileo eretico*, 337.

28. On the history of observation and its relation to objectivity as a modern prerequisite of scientific work, see Daston and Galison, *Objectivity*.

29. On the history of making observation visible in science, and especially on the distinction between what is made visible and proof, see Breidbach, *Bilder des Wissens*, 78–79.

30. "Nessuna umana inventigatione si può dimandare vera scientia, se essa non passa per le matematiche dimostrationi. E se tu dirai che le scienzie, che principiano e finiscano nella mente, abbiano verità, questo non si concede, ma si nega per molte ragioni; e prima, che in tali discorsi mentali non accade esperienzia, senza la quale nulla da di se certezza." Quoted in Ackerman, "Early Renaissance Naturalism and Scientific Illustration"; see also Camerota, *Linear Perspective in the Age of Galileo*, 106.

31. "È buono il Disegno ancho alla intelligenza delle Matematiche, perciò che non rende nuovo l'ingegno di ciascuno al veder linee,

figure, o corpi, ed alla formation di loro. Anzi il disegno solo, e non altra parte può formar tutte le cose del Matematico senza commettere alcuno errore. Anzi il Matematico si serve del disegno come di stromento per fabbricar le figure, per le quali si spiega, e si sciogle ogni dimostratione havendo esso bisogno *d'esempio sensato*, e facile ad essere inteso." Gallaccini, "Discorsi accademici," fol. 41v (emphasis added). See Payne, *Teofilo Gallaccini*, 85.

32. "Imperciò che per l'occhio (sì come tutte le Matematiche) meglio si apprendono le regole, e le dimostrationi; che non basta la semplice vocale esplicatione, ma vi bisogna *l'essempio sensate* . . . di modo che senza 'l vedere questa scienza, come ancho l'altre Matematiche, difficilmente si potrebbe imparare." Gallaccini, "Teoriche e pratiche di prospettiva scenografica," fol. 29v (emphasis added).

33. "E fanno in quella guisa, che si fa serrate tutte e cinque le dita d'una mano a modo di piramide, o di pina, cioè accozzate insieme tutte le punte delle dita del pari, e di poi tutte in un tempo aperte, e disciolte, sì che si veda aperta tutta la mano, e subbito ritornate nello stesso modo, e quindi aperte con prestezza." Gallaccini, "Discorsi accademici," fol. 11r.

34. See Gallaccini, "Tempio."

35. "Che quando la pietra esteriore del muro non si collega bene coll'interiore, facilmente l'una si stacca dall'altra, come, quando si fa la muraglia di mattoni a faccia netta, alcune volte, o per diacciato, o per forzamento di peso, o per altro accidente, tutta la parte di fuori del muro si scrosta, o si scompone, e cade; onde il muro rimane scortecciato, rozzo e indebo-lito, non rimanendovi altro, che il ripieno. Il medesimo avviene, quando nelle mura si pongono pietre lavorate sottili, e quasi nella superficie, le quali perché non s'incastrano nelle muraglie, facilmente cadono, se non si tengono legate con grappe di ferro. Nè però questo basta, perché non si legano tutte, ma poche, e quelle si pongono nei filari più bassi; ma quelle, che vi si posan sopra, si fidano nella calcina, la quale tanto tiene, quanto dura di resistere al peso, ed a qualunque accidente." Gallaccini, "Trattato sopra gli errori degli architetti," 37–38.

36. See Payne, *Architectural Treatise in the Renaissance*, 163–65, 178–81; and Payne, "*Ut poesis architectura*," 147–48.

37. Poleni, *Memorie istoriche.*

38. See especially Gallaccini, "Trattato sopra gli errori degli architetti," 50, 70.

39. "Nè si può inventare, se non si prende troppa licenza, e se altri non si vuole accostare al costume barbaro, a grottescamenti, a ghiribizzi, ed alle fantasie degli Orefici, e degli Argentieri, dei Maestri di legname, degl'Intagliatori, degli Stuccatori, e dei Pittori. . . . E per dirla chiaramente, quando tutto il corpo dell'ornamento è perfetto, senza d'esse; come, quando ai pilastri s'aggiungono, o termini, o risaltamenti di cornici, o nuove membra posticce, e riportate, che rendono il lavoro troppo secco, troppo trito, ed ignobile, e non corrispondente alla sodezza, e alla magnificenza del rimanente, come si vede in Roma nel second'Ordine della faccia, e de' fianchi di S. Pietro, e negli ornamenti fra le colonne. Dove si mostrano gli ornamenti, più d'opera di legname, e di stucco, che di pietra; poichè non rappresentano la sodezza della pietra, come fanno le cornici, le colonne, e i pilastri. Che la maniera dell'ornamento, che è proprio del legno, e dello stucco, non si conviene alla pietra: conciossiachè nello stucco, e nel legno non si disdice usare qualche licenza, e l'aggiunger qualche capriccio di propria invenzione; perciocchè in cotali lavori gli ornamenti aggiunti son tutte cose posticce, e non hanno legamento reale col tutto, e non nascono insieme con esso; ma tutte si legano con ferramenti, con chiodi, e con colle. Conciossiachè l'invenzione delle cose d'Architettura si può soltanto esercitare nel ritrovamento dei disegni, e dei modelli delle fabbriche, di qualche parte loro principale, o d'altre cose appartenenti ad esse, ma non mai negli ornamenti particolari del tutto, e delle parti." Ibid., 46.

Bibliography

Manuscript Sources by Teofilo Gallaccini

"Anatomia." Biblioteca nazionale centrale, Florence, MS G Targioni Tozzetti 68.

"Commentary to Sebastiano Serlio's books IV and III." Biblioteca comunale degli Intronati, Siena, MS S.ii.4.

"Degli errori degli architetti." Biblioteca Apostolica Vaticana, Vatic. Chigi, MS G.i.12, 171–273.

"Dei porti di mare." Biblioteca comunale degli Intronati, Siena, MS L.iv.3.

"Della natura del cerchio e del'operazione del compasso." Biblioteca comunale degli Intronati, Siena, MS L.iv.5.

"Della natura e della forza dell'Acqua corrente e dei fiumi dello Stato di Siena, e delle Fonti di Siena." Biblioteca comunale degli Intronati, Siena, MS L.vi.32.

"Della Nuova Scienza di Niccolò Tartaglia Matematico Bresciano." Biblioteca comunale degli Intronati, Siena, MS L.iv.2.

"Delle ragioni dei tiri: Dell'artiglieria." Biblioteca comunale degli Intronati, Siena, MS L.iv.2.

"Discorsi accademici." Biblioteca comunale degli Intronati, Siena, MS L.iv.1.

"Esposizione del Sesto Libro di Euclide." Biblioteca comunale degli Intronati, Siena, MS L.vi.34.

"L'idea della fortificazione ad uso dell'architettura militare e dell'arte della guerra." Biblioteca comunale degli Intronati, Siena, MS S.iv.2.

"Miscellanea." Biblioteca Apostolica Vaticana, Vatic. Chigi, MS I.vi.206.
"Il mondo celeste." Biblioteca comunale degli Intronati, Siena, MS L.iv.31.
"Perigonia, overo degli angoli." Biblioteca comunale degli Intronati, Siena, MS L.iv.5.
"Il tempio." Private collection, on permanent loan to the Warburg Institute, London.
"Teoriche e pratiche di prospettiva scenografica." Biblioteca comunale degli Intronati, Siena, MS L.iv.4.
"Trattato de capitelli delle colonne." Biblioteca comunale degli Intronati, Siena, MS S.iv.3.
"Trattato di gnomonica." Biblioteca comunale degli Intronati, Siena, MS L.vi.35.
"Trattato sopra gli errori degli architetti." British Museum, King's Library, MS 281.

Printed Sources

Ackerman, James. "Early Renaissance Naturalism and Scientific Illustration." In *The Natural Sciences and the Arts: Aspects of Interaction from the Renaissance to the Twentieth Century*, ed. Allan Ellenius, 1–17. Uppsala: Almqvist & Wiksell International, 1985.
Alberti, Leon Battista. *L'architettura*. Edited and translated by Cosimo Bartoli. Florence: Lorenzo Torrentino, 1550.
Baigrie, Brian S. *Picturing Knowledge: Historical and Philosophical Problems Concerning the Use of Art in Science*. Toronto: University of Toronto Press, 1996.
Berra, Giacomo. "La storia dei canoni proporzionali del corpo umano e gli sviluppi in area Lombarda alla fine del cinquecento." *Raccolta Vinciana* 35 (1993): 159–310.
Blair, Ann. "Natural Philosophy." In *The Cambridge History of Science*, vol. 3, *Early Modern Science*, ed. Lorraine Daston and Katharine Park, 399–403. New York: Cambridge University Press, 2006.
Breidbach, Olaf. *Bilder des Wissens: Zur Kulturgeschichte der wissenschaftlichen Wahrnehmung*. Munich: Wilhelm Fink Verlag, 2005.
Bucciantini, Mario. *Galilée et Kepler: Philosophie, cosmologie et theologie a l'époque de la Contre-reforme*. Paris: Les Belles Lettres, 2000.
Camerota, Filippo. *Linear Perspective in the Age of Galileo: Lodovico Cigoli's "Prospettiva pratica."* Florence: Leo S. Olschki, 2010.
Danti, Egnatio. *Le scienze matematiche ridotte in tavole*. Bologna: Compagnia della Stampa, 1577.
Daston, Lorraine, and Peter Galison. *Objectivity*. New York: Zone Books, 2007.
Dupré, Sven. "Visualization in Renaissance Optics: The Function of Geometrical Diagrams and Pictures in the Transmission of Practical Knowledge." In *Transmitting Knowledge: Words, Images, and Instruments in Early Modern Europe*, ed. Sachiko Kusukawa and Ian Maclean, 1–39. Oxford: Oxford University Press, 2006.
Dürer, Albrecht. *Institutiones geometricae*. Paris: Christian Wechel, 1535.
Gallaccini, Teofilo. *Trattato di Teofilo Gallaccini sopra gli errori degli architetti*. Venezia: Per G. Pasquali, 1767.
Henry, John. *The Scientific Revolution and the Origins of Modern Science*. London: Palgrave Macmillan, 1997.
Kuhn, Thomas S. *The Copernican Revolution*. Cambridge: Harvard University Press, 1957.
Meli, Domenico Bartolomeo. "Mechanics." In *The Cambridge History of Science*, vol. 3, *Early Modern Science*, ed. Lorraine Daston and Katharine Park, 632–33. New York: Cambridge University Press, 2006.
Panofsky, Erwin. *The Codex Huygens and Leonardo da Vinci's Art Theory*. London: Warburg Institute, 1940.
Partsch, Susanne. "Teofilo Gallaccini." *Allgemeines Künstlerlexicon* 47 (2005): 517–18.
Payne, Alina. *The Architectural Treatise in the Renaissance*. New York: Cambridge University Press, 1999.
———. *The Telescope and the Compass: Teofilo Gallaccini and the Dialogue Between Architecture and Science in the Age of Galileo*. Florence: Leo S. Olschki, 2012.
———. "*Ut poesis architectura*: Tectonics and Poetics in Architectural Criticism Circa 1570." In *Antiquity and Its Interpreters*, ed. Alina Payne, Ann Kuttner, and Rebekah Smick, 145–58. New York: Cambridge University Press, 2000.
Payne, Alina, and Giovanni Maria Fara. "Teofilo Gallaccini e la critica architettonica a Siena fra XVI e XVII secolo." In *Architetti a Siena: Testimonianze della Biblioteca comunale tra XV e XVIII secolo*, ed. D. Danesi, M. Pagni, and A. Pezzo, 141–90. Cinisello Balsamo: Silvana, 2009.
Poleni, Giovanni. *Memorie istoriche della Gran Cupola del Tempio Vaticano*. Padua: Stamperia del Seminario, 1748.
Redondi, Pietro. *Galileo eretico*. Microstorie 7. Turin: Einaudi, 1987.
Shapin, Steven. *The Scientific Revolution*. Chicago: University of Chicago Press, 1996.
Shea, William R., ed. *Science and the Visual Image in the Enlightenment*. Canton, Mass.: Science History Publications, 2000.

6.

Dante's Eyes and the Abysses of Seeing

Poetical Optics and Concepts of Images in the *Divine Comedy*

GERHARD WOLF

Around the year 1300, optics became a major paradigm in Western literature and art. While literature used optical devices on a metaphorical level (often but not always informed by the latest scientific achievements),[1] art was concerned with the geometry of seeing in the new pictorial spaces created by Giotto and his contemporaries.[2] In the same period there was also a new interest in the spirituality of seeing, especially in the context of debates about so-called individual judgment—the anticipation of the personal face-to-face encounter with God, not only at the end of time but immediately after death.[3] Dante's three-part *Divine Comedy*, the focus of this chapter, demonstrates an awareness of all of these dimensions of seeing and is thus an ideal text for the study of premodern shifts in the concepts of seeing and its instruments. In fact, even if the experience of the traveler "Dante" is multisensorial, the dominant sense in the *Comedy* is sight, and it is no coincidence that *occhi* is the noun that occurs most frequently in

I would like to thank Alina Payne for her patience and support, Theresa Holler for discussion and help with the editing of the text, Rebecca Milner for correcting my English, and the anonymous reader for her/his valuable comments.

the poem, for eyes serve as the active instrument and screen of "Dante's" exploration of hell, purgatory, and paradise.[4]

The principal concern of this chapter is a preliminary analysis of the relationship between optics (i.e., the nature of seeing and light) and theories of images. In particular, I am interested in exploring the relationship between what was thought to be visible and what could be represented visually in the fourteenth century. This includes the question of—or the relationship between—what was considered invisible and what was unrepresentable. The four terms—visible, invisible, representable, unrepresentable—can in fact be crossed in every possible combination. For visibility and visual representability are not the same thing: there are invisible things (for example, sound and inner motion) that fourteenth-century artists nevertheless attempted to represent by drawing in a more or less codified visual language or by pursuing new representational possibilities (for example, in attempts to simulate speech in a visual form).

This brings us to the famous *ekphrasis* of sculpted reliefs in Canto X, lines 25–54, of *Purgatory*.[5]

> . . . and as far as my eye could make its flight, now on the left and now on the right, such this terrace seemed to me. Not yet had we moved our feet on it when I perceived that the encircling bank (which, being vertical, lacked means of ascent) was of pure white marble, and was adorned with such carvings that not only Polycletus but Nature herself would there be put to shame.
>
> The angel who came to earth with the decree of peace, wept for since many a year, which opened Heaven from its long ban, before us there appeared so vividly graven in gentle mien that it seemed not a silent image: one would have sworn that he was saying "Ave," for there she was imaged who turned the key to open the supreme love, and these words were imprinted in her attitude: "Ecce ancilla Dei," as expressly as a figure is stamped on wax.
>
> "Do not keep your mind on one part only," said the sweet master, who had me on the side where people have their heart; wherefore I moved my eyes and saw beyond Mary on the same side as was he who prompted me, another story set in the rock; wherefore I went past Virgil and drew near to it, that it might be displayed before my eyes.[6]

In these verses we encounter a complex poetic fiction describing the evocation of speech by means of a visual work of art (to be called *visibile parlare* in *Purgatory* X, line 95), but this in turn is done through words, in an artistic act of "painting with words" (creating stone reliefs, in this case).[7] The reduction to one medium, exalting one sense through another, is an early form of *paragone*, understood as a competition between artistic media with an explicit or implicit reference to sense physiology.[8] In Dante's *Comedy* this means an exaltation of sight (including its possible obscuration), which is in fact the predominant theme of the pilgrim's experience (that of Dante the pilgrim, as he is usually called to distinguish him from Dante the poet). The *Comedy* is the drama of Dante's eyes and Dante's gaze, permanently attracted by a specific point (a person, an optical effect, etc.), and permanently driven, either by force or by his own curiosity, to go beyond that point: the verbs *volgere gli occhi* and *vagare* recur throughout the poem. This process functions like a motor keeping the narration running, but its dynamic differs in the three parts, as Dante employs different models of looking and changing models of light.

In the *Inferno*, problems of seeing are created by meteorological conditions: obscurity, smoky dark air, fire, and so on.[9] Traveling in the abyss, Dante-pilgrim confuses the giants for towers and is blinded; only the sound of the wind tells him that he is traveling downward. As Suzanne Akbari has shown, the *Inferno* is dominated by a model of seeing that follows the extramission theory of vision.[10] Virgil and Dante project their gazes into the obscurity; in a certain way, they illuminate the *Inferno* by looking; the dominant element is fire.

In the *Purgatory*, the most "human" space in the cosmos of the *Comedy*, the realm of art and sensual experience corresponds more closely to the earthly "equipment" of the senses and perceptive faculties,[11] even if the dynamics and limits of perceptibility are constantly worked out from canto to canto: clear views are altered by optical deception, distortions, and blindings—for example, by the arrival of the shining figure of an angel and finally by Beatrice's arrival.

The ascent to the mountain in *Purgatory* leads from the first *girone* (that of the reliefs referred to above) to the visionary experience of and in the Earthly Paradise (*Purgatory* XXVII–XXXIII), which shifts from a dream space to that of "real appearances," finally revealing Beatrice in an optical mise-en-scène that transcends the faculties of the earthly eye and thus prepares the pilgrim for the transhuman visual dynamics of paradise itself.

Paradise again starts with deception, but in the inversion of a classical myth: the pilgrim shifts reality by way of a reflection (Canto IV). He believes that he is confronted with the mirror of Narcissus, whereas he actually encounters the true images of souls. This leads to complex optics of light in the following cantos, in which the pilgrim's visual capacities are transfigured step by step to enable him finally to emerge into an instantaneous vision of the Trinity.

Despite its obvious interest, the *Inferno* is not the focus of this chapter; instead, *Purgatory* and *Paradise*, and their changing opticality in relation to the role that images and (fictitious) works of art play in the *Comedy*, concern us here. We will proceed in chronological order, following the pilgrim's journey. In the first quotation from *Purgatory* we briefly observed the description of works of art created by God himself and described by the poet in "pictorial" terms. The cycle dealing with visual arts and showing the punishment of the proud (*superbi*) comes to an end in Cantos XIII–XIV, when the pilgrim and his guide have passed into the realm of the invidious (a word that refers to seeing, and more precisely to looking for someone else's fortune). At the beginning of *Purgatory* XV (lines 1–36), the two travelers encounter the fusion of a natural event, the sunrise, and the appearance of an angel, in anticipation of the streams of light that will dominate the Earthly Paradise and finally paradise itself. Dante's eyes are not prepared to perceive the intensity of the angelic radiance, and he uses his hands to shield himself from it:

> As much as between the end of the third hour and the beginning of the day appears of the sphere that is always playing like a child, so much now appeared to be left of the sun's course toward nightfall: it was evening there and here it was midnight. And the beams were striking us full in the face, for the mountain was now so far circled by us that we were going straight

> toward the sunset, when I felt my brow weighed down by the splendor far more than before, and the things not known were a wonder to me. Wherefore I lifted my hands above my eyebrows and made for me the shade that lessens excess of light.
>
> As when the beam leaps from the water or the mirror to the opposite quarter, rising at the same angle as it descends, and at equal distance departs as much from the line of the falling stone, even as experiment and science show; so it seemed to me that I was struck by light reflected there in front of me, from which my sight was quick to flee.
>
> "What is that, gentle father," I said, "from which I cannot screen my eyes so that it may avail me, and which seems to be moving towards us?"
>
> "Do not marvel if the family of Heaven still dazzles you," he answered me; "this is a messenger that comes to invite to the ascent. Soon will it be that the seeing of these will not be hard for you, but as great a delight as nature has fitted you to feel."[12]

In this passage, optical effects are described on two levels: the first is that of the narrative itself, for the pilgrim is blinded and protects himself from the excess of light. The second is that of a metaphor introduced to explain the nature of the angelic light. The poem refers here to an optical phenomenon widely discussed in Arabic and Latin optical treatises on experience and science, or, contracted into one, on an optical experiment.[13] The metaphor is rather conventional: it is not about the refraction of light but about the reflection of a ray on a specular surface (in fact, Dante constructs a universe of mirrors and reflections).[14] The meaning of the metaphor is clear: the angel is not himself a source of light but a mirror of the divine. The *Comedy* thus presents an early appearance of a ray model of light according to the perspectivist optics of the time,[15] and it corresponds to the intromission theory of seeing. Its element is no longer fire but water, and this will become predominant in the later cantos, even if in *Paradise* the sea of light is water and fire in one.

Remaining in purgatory but approaching the top of the mountain where the Earthly Paradise is located and where the pilgrim encounters his new guide, Beatrice, we are confronted with direct and metaphorical descriptions of luminous effects. "And I saw the flames advance, leaving the air behind them painted, and they looked like moving paint brushes, so that overhead it remained streaked with seven bands in all those colors whereof the sun makes his bow, and Delia her girdle" (Canto XXIX, lines 73–78).[16] The angelic rainbows recall less the vapor trail of airplanes than Giotto's flying angels in the Arena Chapel, with their bodies ending in multicolored paint strokes. In this passage, with its explicit reference to the practice of painting, we can see homage to the famous artist-colleague already mentioned in *Purgatory* XII. Dante's optics are indebted to a geometrical theory of sight concerned with the relation of objects to the observer's eyes, whereas his various descriptions of the rainbow are based on Aristotle. As I have argued above, Dante's universe is created and becomes visible by a continuous emanation of light from the divine source; light functions here as an *informing* power. The "bodies" seen by the traveler are in reality "shadows"—that is, somatomorphic souls. In fact, the souls themselves have an "informing power" and build aerial bodies that reassemble their terrestrial precedents, possessing sensorial capacities as well as the ability to suffer. The dynamics of the "informing" process are explained at length in *Purgatory* XXV and are compared to a ray of light that is transformed into colors in a thicker medium.[17]

An overview of the optical effects described in the *Comedy* and the optical metaphors used would allow us to extract some basic elements of Dante's theory of light and vision. Simon A. Gilson has analyzed Dante's position in the context of contemporary scientific developments. According to Gilson, Dante does not follow the most advanced scientific models;[18] one could add, however, that he uses older ones in a new and ingenious way. Akbari follows a different path: rather than look at Dante's optics in general, she discusses the use and transformation of optical models within his work and in relation to its poetic dimension. Regarding the *Comedy*, she shows convincingly how Dante changes "optics" from the *Inferno* to *Purgatory* to *Paradise*, or, rather, how he constructs his universe and "his" own position within it.[19] However, neither author is interested in the connection between optics and visual aesthetics in the *Comedy*, or, to be more precise, between seeing and the concept of images.

Before examining this relationship more closely, let us return for a moment to the pilgrim-viewer while he ascends to the Earthly Paradise and finally arrives in paradise, where the perspectivist model of light dominates. Of the three ways in which light rays "travel," according to thirteenth-century science (namely, directly, reflected, and refracted), Dante seems to know (or at least primarily consider) only the first two. Optics in the *Comedy* are not conceived of as an isolated geometry of sight or the expansion of light; rather, the poem intertwines elements of physics, medical science, physiology, perception theory, and geometry. Furthermore, light is not only conceived of as consisting of rays; there is also a kind of wave model, or, better still, a concept of the sea of light in *Paradise*, to which we will turn later. In medieval optics, light radiation is not the exclusive principle of visibility; instead, Roger Bacon privileges the multiplication of species, small, nearly insubstantial *peliculae* (films) of the bodies to be seen. In any case, Dante, and to some extent the optical theories themselves, combines a psychology of gaze and sight with physics and geometry. Dante also draws on another tradition that involces, so to speak, the chemical processes inherent in seeing. I refer to the role of the gaze and light in Sicilian and French love poetry from the first half of the thirteenth century.[20] Here, we find poetic elaborations of the classical topoi of sense physiology and optics: the eyes are like a burning glass; the glances of the beloved lady hurt the heart of the lover. It is a combination of emission and immission through eye-to-eye contact, while the heart "evaporates." On the one hand, the lady is actively burning the viewer with her radiance (her *occhi micidali*), while on the other she is entering into the eyes to such a degree that the poems question the dynamics of vision: "How can such a great lady enter the small hole of my eye?" asks the lover in a poem by Giacomo da Lentini, a poet at the court of Frederick II and one of the earliest authors writing in *volgare*.[21] The lady can also appear to be statuelike, as a *donna petrosa* who ignores the lover—themes also elaborated in Dante's *Vita nuova* and *Rime petrose*. Once more, the point is that we are concerned with the poetics of a scientific discourse and a play with scientific discourses in fictional texts—the two illuminate each other.

Returning to our reading of the *Comedy*, the model of the burning eyes we encounter in love poetry has either been overcome here or transfigured, for the ultimate source of light is God himself, as we have seen. It is divine love itself that lets Beatrice's eyes shine, and finally those of Dante the pilgrim, too.[22] Until he reaches this ultimate brilliance, Dante's eyes are troubled. When he is first confronted with Beatrice in *Purgatory* XXIX, they are not yet prepared to look into her face, and here we turn to the passage cited above about the candle flames with rainbow colors. This canto is one of the most painterly of the whole *Comedy*, demonstrating a real shift in its visual dimension, which is announced by these iridescent "banners" and the reference to the artist's brush. What follows is the well-known description of the allegorical triumph of the church, slowly approaching the banks of the river, accompanied by the twenty-four elders and four animals with their shining feathers full of eyes. The *currus ecclesiae* is drawn by a griffin (a combination of lion and eagle) signifying Christ himself in his double nature. The *ekphrasis* of the triumph is subtly combined with the pilgrim's purification and the gradual unveiling of Beatrice, who is placed on the triumphal vehicle. Dante-pilgrim is guided by Matelda on the opposite bank of the Lethe, where he first sees the procession and its reflection in the waters. He confesses his sins and looks toward Beatrice with still insecure eyes ("luci, ancor poco sicure") (Canto XXXI, line 79). While crossing the river, Matelda pushes him into it and he is submerged, forced to drink of its waters. Finally, he approaches the griffin, led by the four virtues, who advise him not to spare his eyes ("le viste non risparmi," line 115) when he looks into the emeralds of Beatrice's eyes, "from which Love once had aimed his shafts at you" (ond' Amor già ti trasse le sue armi) (line 117). This is a clear allusion to the optics of love according to the early poetry, and the verse continues in this way: "A thousand desires hotter than flame held my eyes on the shining eyes that remained ever fixed on the griffin" (lines 118–20).[23]

Beatrice's eyes, however, become a mirror of the sun (the son of God): in her eyes, the monstrous double animal is separated into two beings that appear to oscillate from one moment to the next. This optical miracle, which reminds us of the famous rabbit/duck *Kippfigur*, is the point of transformation of the rays of mutual love into a joint recognition of the divine in a perspectivist dimension.[24] In fact, the double animal "raggiava" (line 122), a word referring to rays that transmit, in Beatrice's eyes also splits in two the "cosa" in its "idolo" (lines 125–26). In this scene, her eyes are not yet directed toward Dante, thus opening a triangle of perception that allows her "emeralds" to serve as a mirror for the pilgrim. Finally, in line 135, the three theological virtues ask Beatrice to turn her eyes toward the pilgrim, "who has moved so many steps to see you" (che per vederti, ha mossi passi tanti!), and show him her second beauty: her mouth. This could be the end of his journey, but obviously it is not, even if it is a moment of fulfillment of the ten years' thirst (such that all the other senses are annihilated). Only by a somewhat violent gesture ("per forza") is the pilgrim's face turned around, accompanied by the exclamation of the divine women: "Troppo fiso!" (Canto XXXII, line 9). Looking around, the pilgrim's eyes are momentarily dazzled, as if he had looked into the sun for a long time. As the procession approaches the miraculous withered tree that comes to life, Dante-pilgrim hears heavenly music that he can hardly bear, and he falls asleep. In a subsequent, wonderful passage, the description of the experience of hearing and falling asleep is translated into a pictorial problem: "I did not understand the hymn, and it is not sung here, which that company then sang, nor did I bear to hear the music to the end. If I could portray how the pitiless eyes sank to slumber, hearing of Syrinx, the eyes whose long vigil cost so dear, like a painter who paints from a model, I would picture how I fell asleep: but whoso would, let him be one that can depict slumber well. I pass on, therefore, to when I awoke, and tell that a splendor rent the veil of my sleep, and a call, 'Arise, what are you doing?'" (*Canto* XXXII, lines 61–72).[25]

These lines are particularly interesting not only for the way in which the transcendent and indescribable auditory experience is replaced by a metaphor that in turn results in an ontologically inevitable failure in the process of representation. The attention shifts from Dante-pilgrim to Dante-author, who is introduced on a metalevel of the literary narrative. The split between the two "Dantes" is the consequence of the paradox of a person describing himself falling asleep. What is striking is the double structure of the metaphor: on the one hand, the pilgrim is compared to Argos and the involuntary closing of his eyes by the play of the Syrinx of Hermes; on the other, the author confronts the reader with an aporia, comparing himself to an artist trying to draw from a model (suggesting an outside view of himself, bridging the time of the narrative and the time of the actual process of writing). The use of the terminology of drawing and the visual arts is quite interesting, as is the motif of the final breaking of the veil by means of waking up, with references to the transfiguration. The cross-sensory dimension in the *Comedy*, especially the relationship between hearing and seeing, is certainly fundamental for Dante's work and has not yet been examined in detail in scholarly literature. In the passage cited above, it opens a space for the "author" as he appears in the text, again offering a reference to the visual arts. As we saw in the ascent to Mount Purgatory, the dominant visual arts are sculpture and relief, culminating in the *ekphrasis* of *Purgatory* X–XII. Here, Dante introduces a theory of artistic mimesis and rhetoric, and defines the artist as belonging to the "superbi," not excluding himself from this sin, a "necessary vice" of the artist, so to speak. It is no coincidence that in the circle of the works created by the God-artist, the somatomorphic souls are forced to look down owing to the heavy weights they are carrying (a strange encounter between materiality and the shadowlike bearers of the burden). And, finally, let us not forget that it is Dante's art that invents the God-artist and allows the pilgrim's eyes to shift restlessly between motion ("vagare") and fixation.

From the arrival in the Earthly Paradise (and the change in guidance from Virgil to Beatrice), we enter the realm of *pictorial* metaphors, allegories, and sceneries, where the images appear in the substance of light according to the art of *painting*. In the last canto of *Purgatory*, Beatrice concludes her explanation of the allegorical tree and her admonition of Dante with the words "'I would also have you bear it away within you—and if not written, at least depicted—for the reason that the pilgrim's staff is brought back wreathed with palm.' And I: 'Even as wax under the seal, that does not change the imprinted figure, my brain is now stamped by you; but why do your longed-for words soar so far beyond my sight, that the more it strains the more it loses them?'" (Canto XXXIII, lines 76–84).[26]

To understand this passage, we must consider the parallel appearance of the pilgrim's front and the marble reliefs in *Purgatory* X–XI. The pilgrim's front is sometimes referred to as marble or stone, and is inscribed, according to Canto IX, by seven P's (for the deadly sins), which will be canceled one by one in his ascension to the Earthly Paradise. In *Purgatory* XXXIII, the two levels finally converge, and Dante's brain (whose *intelletto* is made of stone, according to Beatrice in line 73) is now compared to wax. This is imprinted by a seal, just as in Canto X, line

45, where, in the relief of the Annunciation, the angel's words to the Virgin are expressed in his gesture ("atto") "come figura in cera si suggella" (like a figure in wax). In this way, Dante elaborates the analogies between the concepts of artistic mimesis, the incarnation, and the psychology of perception by means of the imprint. In *Purgatory* XXXIII, this simile is immediately followed by Dante's admission that Beatrice's words surpass (literally: fly beyond) his ability to see (and he therefore cannot yet follow her), indicating that this still needs to be achieved; Dante's eyes need to be prepared for the divine vision. In fact, only in *Paradise* XXXIII, lines 112–14, is this fulfilled (even if only momentarily): "But through my sight, which was growing strong in me as I looked, one sole appearance, even as I changed, was altering itself to me."[27] It is not that the divine light itself has substantially changed, but its intensity increases in the upward journey, and the pilgrim's growing visual strength (so beautifully evoked in *Purgatory* XXXIII, line 78: "il bordon di palma cinto") allows him to see the same *appearance* in a new way, for he himself has changed during his journey. The new quality of seeing is linked to the perfection of Dante-pilgrim.

At this point, let us briefly examine the optics of *Paradise*. The most elaborate and famous optical experiment of the *Comedy* is found in *Paradise* II, on the sphere of the moon. Again concerned with reflection, it shows the relation between the undiminished intensity of light and the measurements of reflected figures in mirrors. Giovanni di Paolo has illustrated this passage in a brilliant visual interpretation with a subtle play between a literal rendering and its transformation into a pictorial argument (fig. 6.1). It is important to note that in Dante's *Paradise* the space changes (after all, there is no firm ground on which the pilgrim could leave his footprints, and there is a notable absence of the somatomorphic souls that are reduced or contracted to flames), whereas Beatrice seems to remain constant in the way she appears in the Earthly Paradise of the last cantos of the *Purgatory*. In Botticelli's illustration, paradise becomes the stage for the dancelike encounter between Beatrice and the pilgrim in flight (fig. 6.2), whereas the text elaborates the "methexis" in the divine light of Beatrice's eyes and the Aristotelian concept of seeing as a form of touch:

> *ché dentro a li occhi suoi ardeva un riso*
> *tal, ch'io pensai co' miei toccar lo fondo*
> *de la mia gloria e del mio paradiso*[28]
>
> (Paradise *XV, lines 34–36*)

However, at a certain point this smile could become dangerous for the pilgrim. Once they have reached the sphere of Saturn, where Dante's eyes and mind are again fixed on the face of Beatrice, he notices that she does not smile, and she immediately explains that if she smiled he would burn to ashes like Semele (who did so when she was exposed to the view of her lover Zeus, whom she had desired to see face to face). Her beauty, Beatrice continues, becomes more intense while they are ascending the heavenly palace, so bright that his human eyes could not bear it unless that beauty is tempered. She invites him to direct his eyes toward an image ("figura") and let his mind follow: "Ficca di retro a li occhi tuoi la mente / e fa di quelli specchi a la figura / ch 'n questo specchio ti sarà parvente" (*Canto* XXI, lines 16–18). His eyes should serve as a mirror, to make the image ("figura") appear in the mirror of his mind: the author again operates with a conventional model of visual perception. What follows in Canto XXI is the staging of a light theater where sound is

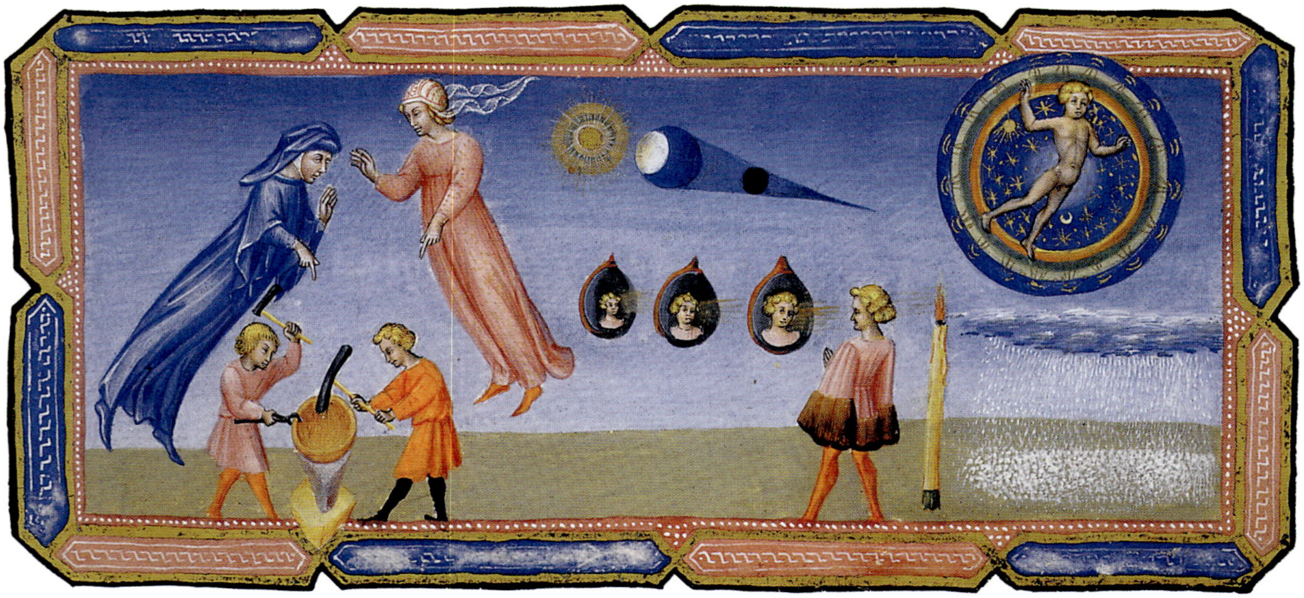

6.1 Giovanni di Paolo, *Paradiso II*, 1444–50. The British Library, London, Yates Thompson MS 36, fol. 132r.

absent, so that Dante asks for the reason for the absence of the "dolce sinfonia del paradiso" that he enjoyed in the other spheres. The reply—"Tu hai l'udir mortal si come il viso" (line 61)—means that his human condition necessitates a splitting and tempering of sound and vision; there is no singing for the same reason that there is no smile (lines 62–63). The optical spectacle in the sphere of Saturn is one of the densest descriptions of light in the *Comedy*: within the translucent crystal sphere of gold in which the sunlight gleams ("di color d'oro in che raggio traluce"), Dante-pilgrim sees a ladder rising up so high that he cannot follow it with his own sight ("luce"). From this ladder, lights ("tanti splendor") descend in such a way that it seems that every light ("lume") in heaven is diffused on the ladder's rungs. The metaphor that follows, of the morning and evening flight of jackdaws (to warm their feathers), explains the multidirectionality of light, again combining the model of ray with a flux of light pouring down the ladder. However, by means of the black birds, the simile deliberately denies an optical evocation of the changing colors of feathers. Among the lights, Dante encounters the *fiamella* of Peter Damian, who explains to him the dynamics of ascent and participation in the divine light. Peter is enveloped in a mantle of light but is also reached directly by a divine ray that penetrates his luminous wrap; he seems to be a light machine, rotating like a millstone, put in motion by the central ray. And light in all its manifestations is the essence of love, the energy that keeps the whole world in motion, according to the last words of the poem. The speech of Peter Damian

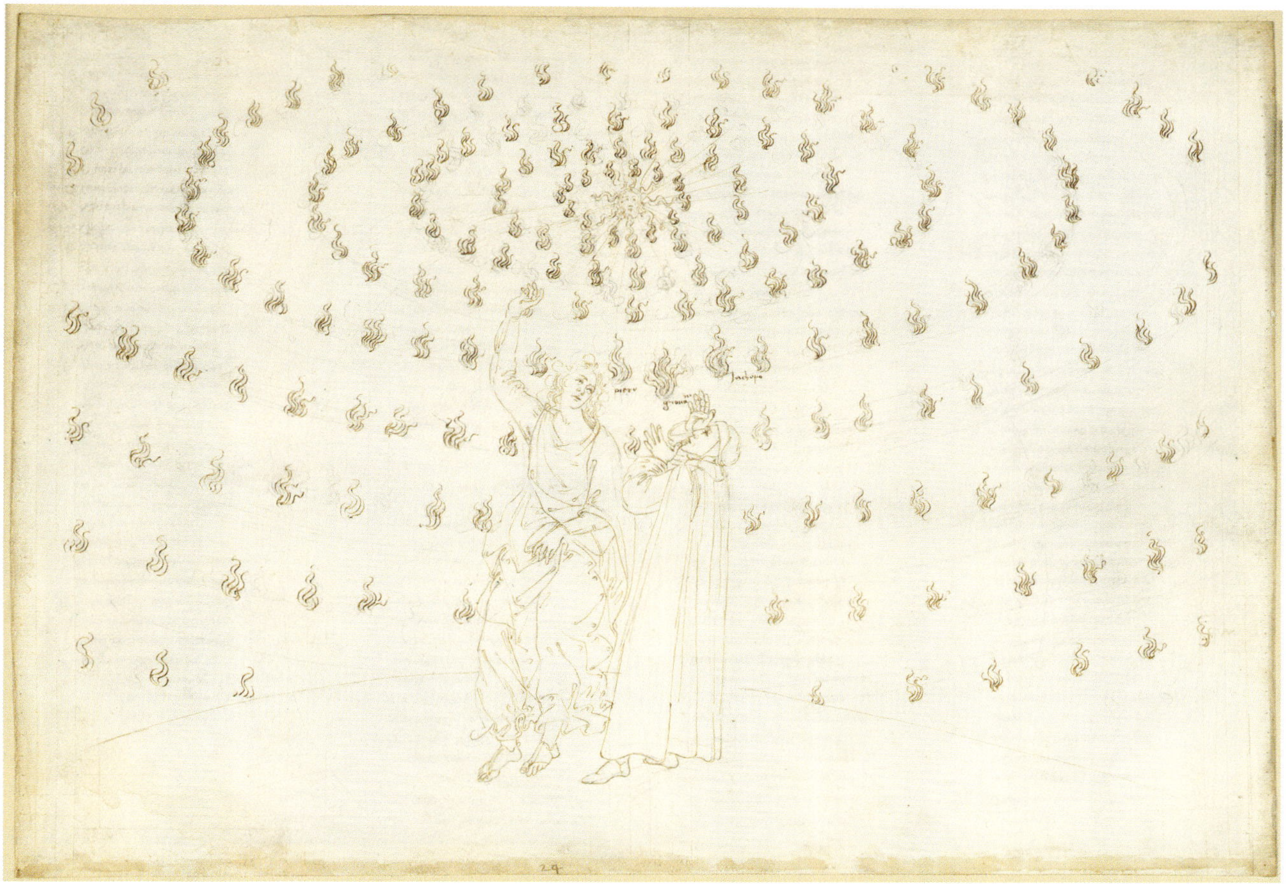

Sandro Botticelli, *Paradiso XXV*, 1444–50. Drawing, 365 x 258 mm. Berlin Kupferstichkabinett, Codex Hamilton 201. © bpk, Berlin / Kupferstichkabinett, SMB / Philipp Allard. **6.2**

in Canto XXI marks a critical point in the ascent of Dante-pilgrim, who is reaching the ultimate threshold, which the human eye and ear, in principle, cannot surpass. Peter opposes a model of ascent, in his self-description as involved in circular light and as brought up by the central ray, with the impossibility of human perception to have access to the divine vision—a kind of negative theology with respect to the men living on earth. For them, God remains in the abyss of the eternal ordinance, cut off from all created beings' vision ("ne lo abisso / de l'etterno statuto quel che chiedi, / da ogne creata vista è scissa") (Canto XXI, lines 94–95). And Peter's words include an admonition to Dante for his future work. Canto XXI of *Paradise* ends in a festive rotation of the *fiamelle* descending the heavenly ladder (the *scala santa*, as it is called), becoming more beautiful with each orbit. They stop at a certain point, breaking out in a cry to which nothing on earth is comparable and which remains incomprehensible to Dante, a thunder in his ears. Canto XXI is again concerned with the dynamics of vision and audition; the pilgrim perceives a kind of tempest, starting with the evocation of the lightning of Beatrice's smile, whereas the thunder is suspended until the end of the canto, where it finally breaks out, arresting the rotation of the light balls. The tempest is revealed to be a didactic performance, as Beatrice explains to him protectively (pale as he is) at the beginning of Canto XXII.

The sphere of Saturn is an apparently dramatic interlude in the pilgrim's experience in paradise, which is a stage for cosmic projections. Even if it is true that the later cantos of *Purgatory* and *Paradise* are suffused with rays of heavenly light, with the dazzling radiance of an infinite number of rays, the perspectivist model is not sufficient to define the optics of *Paradise*. The always "thirsty" and insatiable pilgrim is exposed to a sea of light, and has been brought to the limits of his natural perceptive "instruments," as we have seen (and heard) in Canto XXI. A final blinding (as a rite of passage) will prepare the pilgrim for the last step of vision. This blinding and return of sight is described in Canto XXVI, immediately before the pilgrim encounters Adam and then transcends the dimension of human sight. This *plus ultra* equates him with Adam and Adam's "trapassar del segno" (overpassing of the bound [line 117], although "segno" has a broader meaning than that given in Singleton's translation, referring to the divine system of signs). It also links him with Ulysses, Adam's counterpart in the *Inferno*, Canto XXVI, who trespasses the columns of Hercules, installing a horizontal order of terrestrial exploration that crashes on the rocks, on the mountain of *Purgatory*, which represents the vertical axis of the world.[29] The course of universal history shown in the pure mirror of Adam is counterbalanced by the drama of the pilgrim's eyes as a personal experience. In fact, the pilgrim's seeing remains a bodily, sensual experience, while he becomes able to perceive cosmic spatiality in a new way that transcends the earthly *perspectiva naturalis*. A striking example is the pilgrim's last glance at Beatrice in *Paradise* XXXI, lines 72–78. This canto is crucial, as it is concerned with the preparation for the final revelation, on the penultimate stage. It strongly reintroduces the subject of pilgrimage and confronts the pilgrim with Beatrice's farewell and his new guide, Saint Bernard. The eyes of Dante-pilgrim circle around the heavenly rose, looking up and down, and his gaze takes in the whole of paradise—its form in general—but without looking closely at any one part (Canto XXXI, lines 52–54).[30]

At this point he encounters Saint Bernard and, looking into his face, compares himself to a pilgrim from Croatia who comes to see the Veil of Veronica in St. Peter's. Before doing so, Dante asks the venerable "ov' è ella," directing his gaze to the far distance, where Beatrice is sitting on a throne, forming a crown in reflection of the eternal rays (note that once again the light emanating from God is described as rays). But, and this is the surprising effect, even if she is apparently far away, "from the region which thunders most high no mortal eye is so far distant, were it plunged most deep within the sea, as there from Beatrice was my sight. But to me it made no difference, for her image came down to me unblurred by aught between" (Canto XXXI, lines 73–78).[31]

The pilgrim speaks to Beatrice but gets no verbal answer; rather, he has the impression that she looks at him and smiles. The passage offers a model of vision that is not subject to a medium that obscures the view (much like the optical experiment in *Paradise* II). This is the first time that Dante uses the word "effige," linking it directly to "distance." The term appears just once more, to denote the human picture in the circle of the son in *Paradise* XXXIII, line 131. Turning his gaze to his immediate surroundings, the pilgrim sees Saint Bernard, who—in anticipation of that final vision—is compared to the miraculous image of Christ of Veronica, and the pilgrim understands that Bernard will lead him to the view of the Virgin Mary in the center of the "candida rosa."

Dante's eyes are finally prepared for the view of the Trinity, albeit in a contraction of time, much like a flash of light. This final *momentum* is prepared by a complex metaphor that again compares the time span between the pilgrim's vision and the author's attempt to write it down with the whole course of human history: "A single moment makes for me greater oblivion than five and twenty centuries have wrought upon the enterprise that made Neptune wonder at the shadow of the Argo. Thus my mind, all rapt, was gazing, fixed, motionless and intent, ever enkindled by its gazing" (Canto XXXIII, lines 93–99).[32] This metaphor merits a thorough study.[33] It is the final reference to Dante's beloved myth of Jason and the Argonauts, exalting the role of the exploratory sea travels in the *Comedy*. Jason's adventurous journey to Kolchis is the most challenging human enterprise, at least before Dante's cosmic journey. In *Paradise* II, lines 1–18, Dante's adaptation of the myth of Jason reveals its poetological dimension when he compares his poem with the boat ("following behind my ship that singing makes her way") (Canto II, line 3),[34] and his public to the Argonauts, promising them even greater things to admire than Jason plowing the fields with the fire-breathing oxen. The metaphor is represented in Giovanni di Paolo's first illustration of *Paradise* II, showing Jason plowing a field in the form of an island and the boat with two spectators beside him (fig. 6.3). The Argo metaphor in Canto XXXIII shows the fulfillment of the promise: there is nothing that creates more "letargo" than this single point, meaning the instantaneous vision of the Trinitarian God. It is clear that the passage refers to the final split between Dante the author and Dante the pilgrim, and the impossibility of directly transmitting the visionary experience in the poem. Here, we could speak about the optics of memory, or the "perspectival" relation of an author permanently commenting on his own process of writing and remembering the description of the journey itself, in a kind of double optics. From this point of view, the poem is nothing but a shadow, in accordance with *Paradise* I, lines 22–24: "O divine Power, if you do so lend yourself to me that I may show forth the image of the blessed realm which is imprinted in my mind."[35] This poetological invocation to Apollo displays the passage between the optics of light and the concepts of images in the *Comedy*. In Canto XXXIII, the shadow returns in the Argo metaphor. If we accept the reading that Neptune is thought to be in the abyss of the waters and sees the (shadow of the) ship passing over him, we would find an analogy with the position of Dante-pilgrim, in Canto XXXI, looking upon Beatrice as if from the bottom of the sea. It is obvious that the first level of the Argo metaphor equates Dante-author with Neptune, and subsequently reintroduces the analogy between Dante-pilgrim and Jason, in order to show the fissure between them. This is the final point of departure for the description of the ultimate immersion in the vision of God, permanently interrupted by the invocating voice of the author: here, as in *Purgatory* XXXI, the pilgrim is

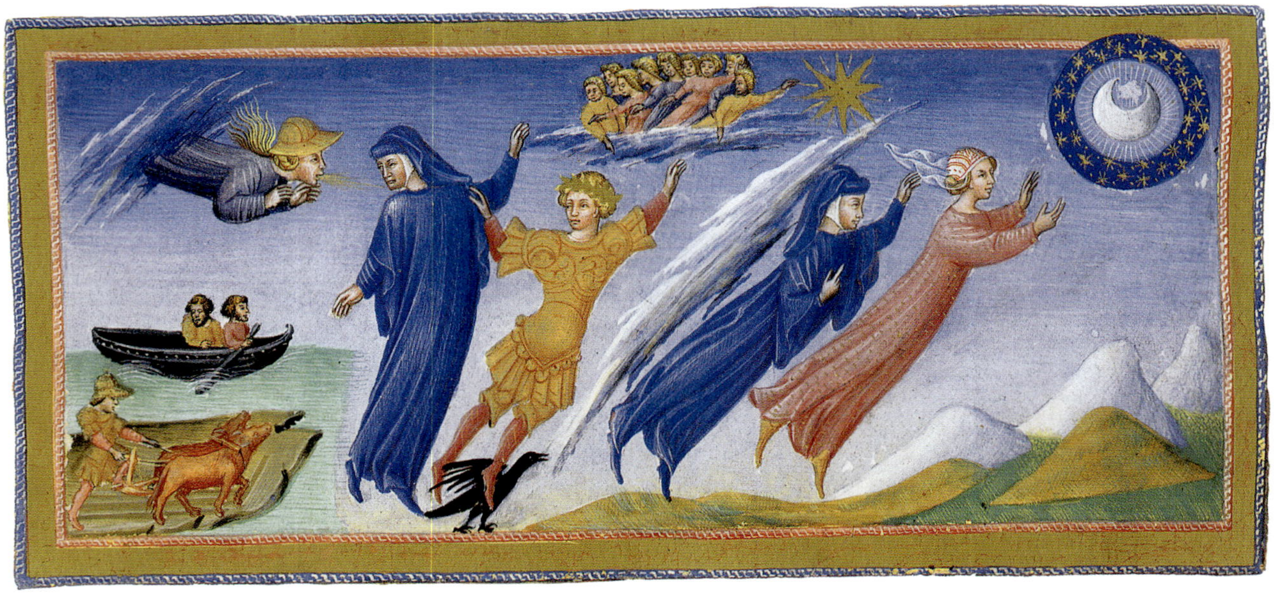

6.3 Giovanni di Paolo, *Paradiso II*, 1444–50. The British Library, London, Yates Thompson MS 36, fol. 131r.

exposed to the danger that he is lost in his vision, becoming an angel with his eyes fixed on God. So he needs to wake up, as in *Purgatory* XXXII, and the veil must be torn, not so as to see the divine face but so as to find himself back on earth, whereas the description of the vision reveals itself to be poetic fiction.

Once again we are confronted with the realm of painting: the vision of the Trinity is pictorial. It consists of three circles of different colors and one periphery, two like rainbows and one (that of the Holy Spirit) like fire. The show culminates in the circle of the son: "That circling which, thus begotten, appeared in Thee as reflected light, when my eyes had dwelt on it for a time, seemed to me depicted with our image within itself and in its own color, wherefore my sight was entirely set upon it" (*Paradise* XXXIII, lines 127–32).[36] Dante's eyes perform another circle to discover it as "pinta de la nostra effige" (painted with our effigy). This encircled face of Christ obviously follows the model of the *vera icona*, the imprint of the true image of Christ in St. Peter's introduced in *Paradise* XXXI on a metaphorical level. "Pinta de la nostra effige" sets into motion a complex chain of references leading back to Genesis 1:26, where God says that he has created men in "our image and likeness." In the exegesis of this passage (by Gregory of Nyssa and others), God is seen as painter.[37] While we have already been confronted with God the sculptor (in the sense of an active artist), God produces a reflection of himself by means of his own light, the painterly substance of paradise, in divine anticipation of the earthly incarnation and the appearance of

God in the human face. There is a doubling not only of Dante by means of his inspiration in author and pilgrim, but also of God, who is painting himself in his own creation (by means of the Holy Spirit), and this allows Dante to project—within the inner circle of the Trinity—a final and collective self-portrait. And after this circumspection,[38] Dante's eyes have finally fully emerged, but in this moment the vision comes to an end and it appears to him as if he has had a momentary insight into the squaring of the circle, the irresolvable basic mystery and enigma of geometry. With this simile, the *Comedy* gives a final affirmation of the geometrical nature of its universe and optics, but also of the fundamental role of images/pictures in it, for at the end, the pilgrim tries to measure how the "imago" (in Latin, and thus a clear reference to the Vulgate text) conforms to the circle.

Notes

1. For a general overview and extensive bibliography, see the most recent study by Simone Tarud Bettini, *Luce, Amore, Visione.*

2. See Traska, *Gesellschaft der Räume;* and Camille, "Before the Gaze," which investigates the relationship between theoretical models of optics in terms of senses and processes of image making in the Middle Ages. The most recent study on this subject is Büttner, *Giotto und die Ursprünge.*

3. Regarding the interplay between optics, theology, and religious life, see Denery, *Seeing and Being Seen.*

4. Cf. Tollemache, "Occhio."

5. For the English version, I follow Charles S. Singleton's translation, though in the case of *Paradise*, Canto XXXI, my paraphrases are inspired by Allen Mandelbaum's translation. The Italian text is taken from Giorgio Petrocchi's 1966–67 edition.

6. ". . . e quanto l'occhio mio potea trar d'ale, / or dal sinistro e or dal destro fianco, / questa cornice mi parea cotale. // Là sù non eran mossi i piè nostri anco, / quand'io conobbi quella ripa intorno / che dritto di salita aveva manco, // esser di marmo candido e addorno / d'intagli sí, che non pur Policleto, / ma la natura lì avrebbe scorno. // L'angel che venne in terra col decreto / de la molt'anni lagrimata pace, / ch'aperse il ciel del suo lungo divieto, // dinanzi a noi pareva sí verace / quivi intagliato in un atto soave, / che non sembiava imagine che tace. // Giurato si saria ch'el dicesse 'Ave!'; / perché iv'era imaginata quella / ch'ad aprir l'alto amor volse la chiave; // e avea in atto impressa esta favella / 'Ecce ancilla Deï,' propriamente / come figura in cera si suggella. // 'Non tener pur ad un loco la mente,' / disse 'l dolce maestro, che m'avea / da quella parte onde 'l cuore ha la gente. // Per ch'i' mi mossi col viso, e vedea / di retro da Maria, da quella costa / onde m'era colui che mi movea, // un'altra storia ne la roccia imposta; / per ch'io varcai Virgilio, e fe' mi presso, / acciò che fosse a li occhi miei disposta."

7. See Collareta, "Visibile parlare"; Kablitz, "Jenseitige Kunst oder Gott als Bildhauer"; and Stierle, "Das System der schönen Künste im *Purgatorio*," in *Große Meer des Sinns*, 178–205.

8. See Pfisterer, *Donatello und die Entdeckung der Stile*, 247–54, and Pfisterer, "Phidias und Polyklet von Dante bis Vasari," which presumes an inverse *paragone* between Donatello's *Annunciation* for Santa Croce in Florence and the Dantean text.

9. Concerning the natural world of Dante's *Comedy*, see Boyde, *Dante Philomythes and Philosopher;* and, more recently, Ledda, *Poesia della natura.*

10. Akbari, *Seeing Through the Veil*, 147–53. See also the fundamental studies Parronchi, "Perspettiva dantesca"; Gilson, "Dante and the Science of 'Perspective'"; and Gilson, *Medieval Optics and Theories of Light.*

11. Regarding the "human" landscape and the sensual world of purgatory, see Pegoretti, *Dal "lito diserto" al giardino.*

12. "Quanto tra l'ultimar de l'ora terza / e 'l principio del dí par de la spera / che sempre a guisa di fanciullo scherza, // tanto pareva già inver' la sera / essere al sol del suo corso rimaso; / vespero là, e qui mezza notte era. // E i raggi ne ferien per mezzo 'l naso, / perché per noi girato era sí 'l monte, / che già dritti andavamo inver' l'occaso, // quand' io senti' a me gravar la fronte / a lo splendore assai piú che di prima, / e stupor m'eran le cose non conte; // ond' io levai le mani inver' la cima / de le mie ciglia, e fecimi 'l solecchio, / che del soverchio visibile lima. // Come quando da l'acqua o da lo specchio / salta lo raggio a l'opposita parte, / salendo su per lo modo parecchio // a quel che scende, e tanto si diparte / dal cader de la pietra in igual tratta, / sí come mostra esperïenza e arte; // cosí mi parve da luce rifratta / quivi dinanzi a me esser percosso; / per che a fuggir la mia vista fu ratta. // 'Che è quel, dolce padre, a che non posso / schermar lo viso tanto che mi vaglia,' / diss' io, 'e pare inver' noi esser mosso?' // 'Non ti maravigliar s'ancor t'abbaglia / la famiglia del cielo,' a me rispuose: / 'messo è che viene ad invitar ch'om saglia. // Tosto sarà ch'a veder queste cose /

non ti fia grave, ma fieti diletto / quanto natura a sentir ti dispuose.'"

13. See Gilson, *Medieval Optics and Theories of Light*, 112.

14. Apart from the sources quoted above, see also Mocan, *Trasparenza e il riflesso.*

15. See Akbari, *Seeing Through the Veil*, 158–60.

16. "E vidi le fiammelle andar davante, / lasciando dietro a sé l'aere dipinto, / e di tratti pennelli avean sembiante; // sí che lí sopra rimanea distinto / di sette liste, tutte in quei colori / onde fa l'arco il Sole e Delia il cinto."

17. See Belting, *Bild-Anthropologie*, 189–211; Klemm, *Bildphysiologie*, 9–11.

18. See Gilson, *Medieval Optics and Theories of Light*, chap. 3.

19. See Akbari, *Seeing Through the Veil*, chap. 6.

20. See Stewart, *Arrow of Love*; Tarud Bettini, *Luce, Amore, Visione.*

21. "Or come pote sí gran donna intrare / per gli ochi mei, che sí piccioli sone?" Giacomo da Lentini, sonnet 22.

22. On Beatrice's eyes, see Bertolucci Pizzorusso, "Gli smeraldi di Beatice."

23. "Mille disiri piú che fiamma caldi / strinsermi li occhi a li occhi rilucenti / che pur sopra 'l grifone stavan saldi."

24. See Akbari, *Seeing Through the Veil*, 165.

25. "Io non lo 'ntesi, né qui non si canta / l'inno che quella gente allor cantaro, / né la nota soffersi tutta quanta. // S'io potessi ritrar come assonnaro / li occhi spietati udendo di Siringa, / li occhi a cui pur vegghiar costò sí caro; // come pintor che con essempro pinga, / disegnerei com'io m'addormentai; / ma qual vuol sia che l'assonnar ben finga. // Però trascorro a quando mi svegliai, / e dico ch'un splendor mi squarciò 'l velo / del sonno, e un chiamar: 'Surgi: che fai?'"

26. "'Voglio anco, e se non scritto, almen dipinto, / che 'l te ne porti dentro a te per quello / che si reca il bordon di palma cinto' // E io: 'Sí come cera da suggello / che la figura impressa non trasmuta, / segnato è or da voi lo mio cervello. // Ma perché tanto sovra mia veduta / vostra parola disïata vola, / che piú la perde quanto piú s'aiuta?'"

27. "Ma per la vista che s'avvalorava / in me guardando, una sola parvenza, / mutandom'io, a me si travagliava."

28. "For in her eyes was blazing such a smile that I thought with mine I had touched the limit both of my beatitude and of my paradise."

29. See Stierle, *Große Meer des Sinns*, 293–94.

30. "La forma general di paradiso / già tutta mïo sguardo avea compresa, / in nulla parte ancor fermato fiso."

31. "Da quella regïon che piú sú tona / occhio mortale alcun tanto non dista, / qualunque in mare piú giú s'abbandona, // quanto lí da Beatrice la mia vista; / ma nulla mi facea, ché süa effige / non discendëa a me per mezzo mista."

32. "Un punto solo m'è maggior letargo / che venticinque secoli a la 'mpresa / che fé Nettuno ammirar l'ombra d'Argo. // Cosí la mente mia, tutta sospesa, / mirava fissa, immobile e attenta, / e sempre di mirar faceasi accesa."

33. See Curtius, "Argo."

34. "Al mio legno che cantando varca."

35. "O divina virtú, se mi ti presti / tanto che l'ombra del beato regno / segnata nel mio capo io manifesti."

36. "Quella circulazion che sí concetta / pareva in te come lume reflesso, / da li occhi miei alquanto circunspetta, // dentro da sé, del suo colore stesso, / mi parve pinta de la nostra effige: / per che 'l mio viso in lei tutto era messo."

37. See Preimesberger, "Jahwist."

38. See Frojmovic, "Giotto's Circumspection." Frojmovic studies the female figure with "evaporating" eyes over the north door of the Arena Chapel and convincingly identifies her as an allegory of "Circumspection," as can be found in Barberino's *Documenti d'amore*. In this context, it refers to the divine "virtus" to look around the whole cosmos, whereas in the case of Dante (not mentioned in the article), "circunspetta" means precisely the instantaneous "transhuman" ability of the pilgrim's eyes to "look around" the cosmic circles of the Trinity.

Bibliography

Akbari, Suzanne Conklin. *Seeing Through the Veil: Optical Theory and Medieval Allegory*. Toronto: University of Toronto Press, 2004.

Alighieri, Dante. *La Commedia secondo l'antica vulgate*. Edited by Giorgio Petrocchi. 4 vols. Milan: Mondadori, 1966–67.

———. *The Divine Comedy*. Translated by Charles S. Singleton. 6 vols. Princeton: Princeton University Press, 1970–75.

———. *The Divine Comedy: A Verse Translation*. Translated by Allen Mandelbaum. 3 vols. New York: Bantam Books, 1982–84.

Belting, Hans. *Bild-Anthropologie: Entwürfe für eine Bildwissenschaft*. Munich: Fink, 2001.

Bertolucci Pizzorusso, Valeria. "Gli smeraldi di Beatice." *Studi Mediolatini e Volgari* 17 (1969): 7–16.

Boyde, Patrick. *Dante Philomythes and Philosopher: Man in the Cosmos*. Cambridge: Cambridge University Press, 1981.

Büttner, Frank. *Giotto und die Ursprünge der neuzeitlichen Bildauffassung: Die Malerei und die Wissenschaft vom Sehen in Italien um 1300*. Darmstadt: Wiss. Buchgesellschaft, 2013.

Camille, Michael. "Before the Gaze: The Internal Senses and Late Medieval Practices of Seeing." In *Visuality Before and Beyond the Renaissance*, ed. Robert S. Nelson, 197–223. Cambridge: Cambridge University Press, 2000.

Collareta, Marco. "Visibile parlare." *Prospettiva* 86 (1997): 102–4.
Curtius, Ernst Robert. "Argo." *Romanische Forschungen* 62 (1950): 1–27.
Denery, Dallas G. *Seeing and Being Seen in the Later Medieval World.* Cambridge: Cambridge University Press, 2005.
Frojmovic, Eva. "Giotto's Circumspection." *Art Bulletin* 89, no. 2 (2007): 195–210.
Giacomo da Lentini. *Poesie.* Edited by Roberto Antonelli. Rome: Bulzoni Editore, 1979.
Gilson, Simon A. "Dante and the Science of 'Perspective': A Reappraisal." *Dante Studies* 115 (1997): 185–219.
———. *Medieval Optics and Theories of Light in the Works of Dante.* Lewiston: Edwin Mellen Press, 2000.
Kablitz, Andreas. "Jenseitige Kunst oder Gott als Bildhauer: Die Reliefs in Dantes *Purgatorio (Purg. X–XII)*." In *Mimesis und Simulation,* ed. Andreas Kablitz and Gerhard Neumann, 309–56. Fribourg: Rombach, 1998.
Klemm, Tanja. *Bildphysiologie: Wahrnehmung und Körper in Mittelalter und Renaissance.* Berlin: Akademie Verlag, 2013.
Ledda, Giuseppe, ed. *La poesia della natura nella Divina Commedia.* Ravenna: Centro Dantesco Onlus, 2009.
Mocan, Mira. *La trasparenza e il riflesso: Sull'alta fantasia in Dante e nel pensiero medievale.* Milan: Mondadori, 2007.
Parronchi, Alessandro. "La perspettiva dantesca." *Studi Danteschi* 36 (1960): 5–103.
Pegoretti, Anna. *Dal "lito diserto" al giardino: La costruzione del paesaggio nel "Purgatorio" di Dante.* Bologna: Bononia University Press, 2007.
Pfisterer, Ulrich. *Donatello und die Entdeckung der Stile, 1430–1450.* Munich: Hirmer, 2002.
———. "Phidias und Polyklet von Dante bis Vasari: Zu Nachruhm und künstlerischer Rezeption antiker Bildhauer in der Renaissance." *Marburger Jahrbuch für Kunstwissenschaft* 26 (1999): 61–97.
Preimesberger, Rudolf. "Der Jahwist: Die erste Effigies (9. oder 8. Jhdt. V. Chr.)." In *Porträt,* ed. Rudolf Preimesberger, Hannah Baader, and Nicola Suthor, 65–69. Berlin: Reimer, 1999.
Stewart, Dana E. *The Arrow of Love: Optics, Gender, and Subjectivity in Medieval Love Poetry.* Lewisburg: Bucknell University Press, 2003.
Stierle, Karlheinz. *Das große Meer des Sinns.* Munich: Fink, 2007.
Tarud Bettini, Simone. *Luce, Amore, Visione: L'ottica nella lirica italiana del duecento.* Rome: Aracne Editrice, 2013.
Tollemache, Federigo. "Occhio." In *Enciclopedia Dantesca,* ed. Umberto Bosco, 4:117–21. Rome: Instituto della Enciclopedia Italiana, 1973.
Traska, Georg. *Die Gesellschaft der Räume: Laikale und bürgerliche Handlungsräume in der italienischen Malerei und Literatur um 1300.* Weimar: VDG, 2009.

7.

The Invisible Element in Art

Dürer, Shakespeare, Donne

CARLA MAZZIO

Hogs are so disturbed, affrighted, and strangely affected at the rising of Wind, that the Country People have a Notion, this is the only Creature that can see the Wind; and imagine it a frightful Sight.

—FRANCIS BACON

Is air thinkable?

—LUCE IRIGARAY

Toward a History of Air

In his "Catalogue of Particular Histories" required for the interpretation of nature, appended to the *Novum organum*, Francis Bacon suggested that someone write a book titled *The History of the Air.*[1] What might it have meant to constitute the air as a perceptual object—or simply to discern the very thing that one breathes, moves in, and depends upon for life but cannot see? This question, vexing as it

This essay has been adapted from an earlier article, "A History of Air: Hamlet and the Trouble with Instruments," *South Central Review* 26, nos. 1–2 (2009): 153–98. I thank the journal for permission to adapt the essay.

is, may help us expand our collective approach to the history of "vision and its instruments" in the early modern period. The sheer imperviousness of air to the faculty of sight posed problems for those who might otherwise aim to subject this element to epistemological or natural-historical consensus: "Air is strong, though not visible," thus only "known by its effects and our apprehensions," wrote the sixteenth-century physician Jean Feyens, who also observed that such a "substance, whether it be air, wind, or blast," is "to be heard or felt, though not to be seen."[2] Air could, of course, be approached by indirection or metonymy—gleaned through the movement of wind on leaves, water, and skin, through the liquid condensation of clouds, vapors, and fog, through the sounds of moving air, or through the smell of air, be it foul or fair.[3] But it was the resistance of air to *direct* visual perception as well as to visual instrumentation that led Bacon to assert, as late as 1622, that as far as natural philosophy was concerned, "the Nature and Power of the Air is *unknown*."[4]

Within and beyond the natural sciences, moreover, to contemplate the air could lead to sustained reflections on the status of affect at the limit point of mediated knowledge. "If I were asked," wrote John Donne in his *Devotions*, "what is a vapour, I could not tell, it is so insensible a thing; so near nothing, that it reduces us to nothing." The potential of air, as a natural "thing" that was also "nothing," to propel cognition into such an abyss could lead humans (and not simply hogs) toward states of being "disturbed, affrighted, and strangely affected." Indeed, the notion that such an unseen element, "not forced but breathed, should kill" led Donne to highlight at once the intractable coordinates of his own "melancholy" and the tyranny of nature in making the body an unwitting instrument of its own demise: "I do nothing upon myself, and yet am mine own executioner."[5] Whether thinking about "insensible" properties of the air evoked pleasure or pain, it could occasion highly charged expressions of that panoply of emotions ranging from petrification to giddiness (many of which we see in Robert Burton's own "Digression of the Air").[6]

By drawing on specific instances in the arts of engraving (Dürer's *Melencolia I*), verse drama (Shakespeare's *Hamlet*), and prose (from Donne's *Devotions* to treatises on meteorology and plague), I argue for a logic whereby an aesthetics of affect emerged from, and often managed to displace concerns about, the otherwise threatening power of an element that could not be directly seen, understood, controlled, or subjected to "capture." In the cases examined here, reckoning with the air—nature's own correlative for "that within that passes show"[7]—exposed forms of affect precisely at the limits of vision, visual instruments, and visual representation.

That the history of air involves the history of affect is no surprise. Air, the element long linked with blood in the humoral system, was part and parcel of a psychophysiological economy that was still operative in Shakespeare's England.[8] Moreover, that terms involving "air" included "ghost" (an English term for "breath") and "spirit" (from the Latin *spiritus*), and that various classical theories linked the air with psyche and soul, make the air's relationship to the question of inwardness perhaps all too explicit. To imply a separation between atmosphere and affect would thus be anachronistic.[9] What I want to stress is simply how some seminal innovations in the representation of affect, including Dürer's *Melencolia I* and Shakespeare's *Hamlet*, evince powerful traces of a "history of air" that could not be subjected to epistemology or mimetic representation, and that therefore led to extended reflections upon the "subjection" of both artist and instrument to the very air they might otherwise attempt to capture. For air could be reflected upon but not directly reflected, thus posing a challenge to those artists and scientists who might otherwise aim to mobilize various instruments, technologies, and forms of mediation in order to "hold," in Hamlet's words, "a mirror up to nature" (3.2.22).[10]

As an element that determined conditions of life and death without being seen, air could grip the imagination and dizzy the intellect. Writing about the sixteenth- and early seventeenth-century status of aerial investigation, Bacon himself foregrounded the cognitive vertigo that could be produced by attempts simply to subject the air to thought: "in the Inquisition of General winds, men have suffered and been in darkness, so they have been troubled with a Vertigo or giddiness concerning staid and certain Winds. Of the former they say nothing, of the latter, they talk up and down at random."[11] Further, he added the commonplace complaint that "men talk as if the wind were some body of it self and by its own force did drive and agitate the air. And when the wind changes its place, they talk as if it did transport it self into another place. This is the vulgar's opinion; yet the Philosophers themselves apply no remedy thereunto, but they likewise stammer at it."[12]

We might use these words to gloss Hamlet's "talk" in act 1, as he famously enters with "wild and whirling words" just after an encounter with an anthropomorphized entity, the ghost or "spirit" described, significantly, "as the air, invulnerable" (1.5.139, 1.1.150). This moment is followed by Hamlet's willingness to investigate the status of a "wondrous strange" phenomenon that, as he observes, exceeds the coordinates of natural "philosophy" (1.5.172–75). In Shakespeare's England, it would not necessarily take a ghost, or even a ghost story, to make otherwise reasoned speech seem to stammer, or to make a "philosopher" or "scholar" such as Horatio "tremble and look pale" in the face of inexplicable aerial phenomena (1.1.56). "The word *Ghost* in English," wrote Alexander Gill, "is as much as *athem*, or breath," and, just like "spirit," "*may* meane any being elementall, as we speak of the wind, or any subtle steam that may rise from a moist body."[13] This simple link between "ghost" and air (here, windy or vaporous) may prompt us to reconsider the relationship between the "ghost" of Hamlet's father and various encounters in the play between characters and the air at once around and in them, for air constituted one of the biggest problems for early modern as well as classical natural philosophy.[14]

But Bacon's observations and Hamlet's "wild" words also speak to a broader point about metaphysics: any attempt to integrate the air into a philosophy is likely to raise fundamental questions about philosophy itself. Addressing precisely this point in *The Forgetting of Air in Martin Heidegger,* Luce Irigaray posits the air as an often forgotten yet fundamental, if elusive, "ground" of being and intellection in postclassical philosophy. While focusing on the extent to which Heidegger "forgot" this detail and unwittingly privileged the element of *earth* as a ground of metaphysics, she argues that the air is all too easily forgotten in post-Enlightenment philosophy. For Irigaray, this is because to base metaphysics on the air would undermine some cherished ideals about the power of the visual and the autonomy of thought (99). It would also unsettle the privileged status of "things," "tools," and "technē" for the grounding of empirical science as well as metaphysics (86–88). Without air, she argues, philosophy is itself a kind of walking ghost: a form of deadened thinking grounded on the tangible dimensions of "earth"—as well as water, fire, and bounded space—rather than the "intangible, imperceptible, invisible, insensible, unintelligible" element of air out of which all thought and being proceed (4). She draws here from the classical philosophy of Anaximander, for whom air was not the "quintessence" (or "fifth essence," as the "aether" was for Aristotle) but the essence of man—the unknown "nothing" at the source of being. Like Donne before her, Irigaray asks what air might do to thought when it becomes the very basis of being, knowing, and navigating one's way in the world. Can air, which long existed at the limits of perception, be a "ground" or a "basis" of anything at all?

To ask the simple question "Is air thinkable?" is to propel philosophy toward the pre- and post-Cartesian as well as the pre- and post-Heideggerian—toward the idea that "je respire donc je suis." To think about air is to confront what it means to think *through* it, to acknowledge a condition of interdependence between mind and environment that thinking, in Irigaray's terms, likes to forget. In *Hamlet*, air is arguably *the* central element of the opening act. Yet the "air, invulnerable," becomes forgotten, displaced, philosophically exiled in an increasingly stifling atmosphere in which metaphysics finds a "ground" in the visible, tangible stuff of earth and, as we shall see, water. This leads, in the language of the play, to a final act rife with "people muddied, / Thick and unwholesome in their thoughts" (5.4.81–82). It also leads, conversely, to presumptions of the air as nothing. When Gertrude imagines Hamlet to "bend" his "eye on vacancy" and attempt to "hold discourse" with "th'incorporal air," we glimpse a moment in the play in which a substantial no-thing is diminished to "nothing," indeed "nothing at all" (3.4.117–18, 133). In the process, Shakespeare stages at once the limits of knowledge based on material, corporeal "things" and the limits of various technologies of observation through which "th'incorporal air" might be subject to epistemology and consensus.

"What if he who gives you air," asks Irigaray in a question central to *Hamlet*, "gives you air so rarefied, or compressed, or pure, or polluted . . . that he, in effect, gives you death?" (7). If such a question itself seems "rarefied," it would not have been so in Shakespeare's England. The daunting idea that humans do not simply live in but also *because* of the air was especially marked during periods of pestilence, in which many a mind, body, and soul were gripped by something frighteningly immune to detection.[15] Stressing the complete dependence of humans on that invisible element in a treatise printed in 1603, for example, Thomas Thayre invoked the classical commonplace that "the aire is one of the Elements wherof our bodies are composed; and without the inspiration, and respiration thereof we cannot liue." Yet because the "Plague or Pestilence, is no other thing than a corrupt and venemous aire, deadly enemie vnto the vital spirits," simply breathing during times of pestilence could be fatal.[16] In the plague-ridden atmosphere of Shakespeare's England, air thus carried a haunting metaphysics, with the plague itself based on an unknown and invisible "physics." If "some*thing*" is "rotten in the state of Denmark," that something could be nothing in natural-philosophical as well as moral terms, for the phrase "roten eyr" (rotten air) went back in England at least to 1330.[17] Further, however, to recall Irigaray's emphasis on the "deadening" of thought through fantasies of alternatively pure and rarefied or polluted and deadly air, it is notable that Hamlet is positioned between these two poles in a single speech: the rarefied "excellent canopy the air" in which man exists "like an angel" in action, "like a god" in apprehension, is also, in Hamlet's mind, as "but a foul and pestilent congregation of vapours" (2.2.299–303). Where does metaphysics stand between such poles? Moreover, where does science, in the general early sense of "knowledge," stand? In a word, nowhere: the ground of understanding is unearthed through the phenomena of the ghost, the impervious air, the "air, invulnerable."

Air could not be held or taken in the hand or "grasped" through the intellect any more than it could be caught by the eye. Nor could it be effectually subject—at least not in *Hamlet*—to material instruments and technologies that might aim to "mirror," "strike," or even "saw" it. As Marcellus puts it at the play's outset in regard to the attempt (sanctioned by Horatio) to "strike at" the ghost with instruments of war, "We do it wrong, being so majestical, / To offer it the show of violence. / For it is as the air, invulnerable, / And our vain blows malicious mockery" (1.1.148–51). Certainly, the process of "cutting the ayre with swords," according to Robert Burton's *Anatomy of Melancholy* (1621), was long considered a form of exorcism, one subject to much ridicule in Protestant polemic.[18] But the ghost raises questions well beyond those of Catholic rituals and superstitions; it raises questions about how one might understand the very air that we both inhabit and are inhabited by.

We might recall, at this point, all manner of technologies and material instruments available in the sixteenth and early seventeenth centuries through which the "air" might be manipulated. Bacon considered fans, bellows, windmills, guns, mines, sails, powder houses, and even dietary regimens for controlling the flow of air in and out of bodies in his quest to discern the properties of air.[19] Furthermore, as numerous critics have emphasized, Shakespeare's London was veritably exploding with new technologies and instruments for the manipulation of nature—including windmills, military and navigational instruments, and even heating devices.[20] Shakespeare's characters, as Adam Max Cohen cleverly puts it, often "turn tech," imagining themselves or others in terms of instruments, mechanical objects, or forms of "artificial life" (as Hamlet, of course, describes himself as a "machine").[21] For Jonathan Sawday, Hamlet's "machinations" align his ingenuity with the craft of the military engineer, and his language is "indebted to tactical uses in sieges of the period."[22]

When we attend closely to the matter of air, however, we find in *Hamlet* a world teeming with anxiety about the limits of "machines" and "instruments" in the face of the daunting presence of the air. The play alludes to numerous technologies through which air might be understood or manipulated, but air proves largely resistant to instrumentalization. Famously giving Rosencrantz and Guildenstern a recorder to play on, Hamlet says, "give it breath with your mouth, and it will discourse most excellent music" (3.2.349–50), a command that only leaves them to avow incompetence, staring blankly at the dumb or mute instrument. "Do you think I am easier to be played upon than a pipe?" asks Hamlet. "Call me what instrument you will, . . . you cannot play upon me" (360–63). Hamlet's body may be interpellated as an "instrument," but it is one that cannot, he stresses, simply be breathed into by another in order to produce "discourse," sweet airs, or, more to the point, recordable knowledge. Let us contextualize this curious dramatic moment by recalling, first, that Hamlet's much ado about "recorders"—inaugurated by the question "why do you go about to recover the wind of me?" (337)—occurs moments after he overestimates *his own* capacity to have effectually recorded, recovered, represented "the ghost's word" through the play within the play (280). Likewise, Hamlet's mocking request for his peers to but "govern these ventages" (350) is swiftly followed by his mocking of Polonius for seeing, along with him, various images of specific things in the air above (366–73)—a moment itself directly juxtaposed with Hamlet's solitary attention to the earth's ungovernable "ventages," through which "churchyards yawn" and "hell itself breathes out Contagion to this world" (380–81). These moments, situated in tight dramatic succession, foreground the limits of visual and acoustical technologies through which the unknown might be subject to epistemological capture and, no less important, the status of an element that proved difficult to "govern," never mind "mirror," "record," or decipher. Hamlet himself might be seen to "saw" at the air through his own playlet, since it is through theater that he attempts to verify a hypothesis about the meaning of the ghost, "th'air invulnerable," through collective observation, a topic to which we shall return.

The emphasis on air at the limits of governance and epistemology can be understood as a metaphoric subset of larger problems of illegitimate governance and vexed theology that dominate *Hamlet*. But such an emphasis also resonates with commonplace concerns within classical and early modern natural philosophy about the air's ungovernability: "moving air is an unconquerable thing," reads Seneca's *Naturales quaestiones*, and "the great force of air cannot be checked, nor does any compact structure hold this wind. For it loosens any bond and carries every weight away with it and makes a space for itself, pouring through the smallest fissures. By the indomitable force of its nature air frees itself, and especially when agitated it asserts its rights for itself."[23] Bacon, among others, repeated variants of this problem, emphasizing that "air" could not be properly governed until natural philosophy found a new experimental means through which to decipher it. Or, to borrow Hamlet's then already commonplace conceit, Bacon emphasized that as far as the air was concerned, nature exceeded anything yet "dreamt of in our philosophy" (1.5.174–75). Such a natural history of the air arguably animates, and is newly animated by, the metatheatrical dimensions of *Hamlet*, where the body of the actor comes into play as a possible but flawed instrument not only of controlled observation but also of controlled ventilation and experimentation.

The air in *Hamlet* is at once an eerie atmospheric haunt and an element integral to the technology of drama: "do not saw the air too much with your hand," Hamlet advises his players, "for in the very torrent, tempest, and, as I may say, the whirlwind of passion, you must acquire and beget a temperance that may give it smoothness" (3.2.4–8). This moment, "do not saw the air," evinces a direct transferal of that earlier image of incompetent atmospheric engagement—a sawing at the air with blunt force in the dark and chill of midnight—to the comparatively benign space and time of theater. Marcellus's "We do it wrong, being so majestical, / To offer it the show of violence. / For it is as the air, invulnerable, / And our vain blows malicious mockery" (1.1.148–51), transferred to the theater, brings air into play as a force that might be "mocked" through the visible as well as material dimensions of "show."

But the air in *Hamlet* is not easily "mocked," nor is it typically something to be effectually sawed at, penetrated, or cut into. Even in the military language of Claudius, the air is veritably "woundless": a "cannon" blast or "poison'd shot" of slander will only "miss our name / and hit the woundless air" (4.1.40–44)—itself a rhetorical revisitation of "th'invulnerable clouds" in *King John*, said to be immune to "cannons" blast.[24] If the air is largely "invulnerable" and "woundless" in *Hamlet*, however, it also produces spectacular and inexplicable effects that haunt, penetrate, and disrupt the lives of various characters in ways that scholars have not yet fully observed or accounted for. While the invulnerability of the air during Hamlet's encounter with the ghost is offset by the eerie materiality of the ghost's description of his own death, his crusted skin, and indeed his subjection to the fires of purgatory, this is part of a larger problem of understanding the air in and through recourse to other elements or material things that may do "violence" to both the physics and metaphysics of the air's sovereignty. That the word Marcellus uses to describe the air, "majestical," soon emerges in Hamlet's lexicon to describe nothing less than the "canopy the air" or "majestical roof" (2.2.299–303) subtly underscores the problem of a world without a legitimate "heir," or an "air" that might give health to the body politic.[25] But it likewise underscores the problem of air's sovereignty over the visible world, over that which can only "seem," for example, a kind of liquid constellation, or foul and pestilent congregation, of vapors.

For Bacon, the human eye, and even the most "exquisite instruments" available for extending its power, simply failed when it came to compassing the element of air—a concern that led to his privileging of "experiment" as an indirect means through which to approach the "thing itself."[26] We might consider *Hamlet* in similar terms: not as an "experiment" or even an "exquisite instrument" through which to test out various hypotheses of air circulating around 1600, but as an extended reflection on the limits of both. When Gertrude sees "nothing at all," while Hamlet sees a spirit, we have, to state the obvious, a failure of consensus. But this problem is integral to the status of theater. At this point, Hamlet has already experienced the ghost of "air." This has occurred not only through an intimate ear- and eyewitness encounter but also through a repeated structure of collective witnessing (including the sentinels and a fellow scholar at the outset)—and, indeed, by an artificial or dramatic experiment designed (through an orchestrated and ideally controlled structure of collective witnessing) to test a hypothesis about the significance of that ghost. "I'll take the ghost's word. Did's perceive?" says Hamlet in the wake of the play as a "thing" that might "trap" or "catch" what could not otherwise be directly seen. As scholars of the play well know, that experiment, the "Mousetrap," is itself compromised, designed to produce the effects it desired.

Thus what comes to the fore is the fallibility not simply of the visual but also of the experiment as a means by which to rationalize the significance of, or validate a hypothesis about, an aerial apparition. A reliable model of sensory verification is not, in *Hamlet*, enabled by the prosthetics of experiment or the production of "artificial life." Yet there remains a truth beyond experiment, and Shakespeare was confronting as well as dramatizing the limits of his own medium, or dramatic instrument, to subject that element of air to a form of "capture" capable of producing collective consensus.

Long before and long after *Hamlet*, to "hold discourse" with the air—or to attempt to "hold it" through discourse, or any other instrument—required a form of investigation that could smack of sheer madness. As late as 1670, Leibniz addressed the Royal Society, "Why does not the wind carry off our heads like balloons? Such questions, ridiculous as they are, are difficult to answer. People would think us mad if they should hear of such things being investigated, yet investigated they shall be."[27] Be that as it may, the air, well before Leibniz and Boyle, posed equally "difficult questions" and, in the sixteenth and early seventeenth centuries, made a series of particularly daunting demands. It made demands on artists who tried to represent this element through representations of wind and clouds.[28] It made demands on "school-men" who grappled with natural phenomena alongside revived classical theories, including the atomism of Democritus, Epicurus, and Lucretius, the miasmatic and climate theories of Hippocrates and Galen, and the natural histories of Aristotle, Seneca, and Pliny. It made demands on the "generality," as it was understood to produce devastating effects through weather, plague, disease, and environmental disaster. Perhaps—like the ghost of King Hamlet himself calling out from the dawn of the century, "Remember me"—the air made demands that could not, in most instances, be fully met.

As such, the air can be understood as a force of nature through which the limits of instrumentality—of the body and its tools—became a manifest problem with which to be reckoned. For the more the air "mattered"—and met with vigorous attempts in various artistic-artisanal and scientific domains to somehow approximate, capture, or reproduce it—the more of a challenge it posed to those instruments of body, eye, and mind. With this basic point that the "air invulnerable" made the body and its instruments vulnerable, let us now consider how we might put to use the trouble with instruments.

The Trouble with Instruments

In his influential analyses of the relationship between the body and its tools, the scientist and philosopher of science Michael Polanyi worked to break free of intellectual traditions in which bodies and psyches had been separated from the tools that they held and used. "When we use a language, or a probe, or a tool," he wrote, "we *interiorize* these things and *make ourselves dwell in them*."[29] Elsewhere, he wrote:

> We *use* instruments as an extension of our hands and they may serve also as an extension of our sense. We assimilate them to our body by pouring ourselves into them. And we must realize then also that our own body has a special place in the universe . . . we never attend to our body as an object in itself. Our body is always in *use* as the basic instrument of our intellectual and practical control over our surroundings. . . . Every time we assimilate a tool to our body our identity undergoes some change; our person expands into new modes of being.[30]

Polanyi's task of exposing the otherwise tacit or invisible life of technological and discursive "use" is generative and laudable. Yet to begin to expand on his already dynamic model of instrumentality, let us simply note that the bias in the passage above is on *effectual* use: on the body and its instruments effecting "intellectual and practical control over . . . surroundings." Indeed, let us ask whether the opposite may also be the case, whether every time we *fail* to assimilate a tool to our body, our identity undergoes some change, our person expanding into new modes of being.

Instruments that don't work—or that cannot be made to work—can produce a whole panoply of thoughts and emotions that effectual forms of use can otherwise manage or render less visible: frustration, anger, irritability, disappointment, abjection, self-loathing, melancholy, or a sense of absurdity, to mention just a few. Since material "instruments" themselves, be they printed books, writing implements, or trade or scientific tools, have accrued a great deal of capital within various disciplines in the present-day humanities, a focus on something that we might call the *productiveness of failed instrumentality* may be in order.[31]

A book that approaches, in historical terms, what we might call the poetics of failed instrumentality is James Elkins's *The Poetics of Perspective.* Of the many instruments of vision long held to be pivotal for the Renaissance of visual arts, perspective long reigned as a central tool or instrument, leading, for many, to the "rationalization" of space and sight.[32] Elkins, however, exposed the conceptual blinders put on historians of art by Enlightenment and post-Enlightenment ideals, blinders that led to anachronistic treatments of Renaissance perspective as rational, unified, and indeed proto-Cartesian. More centrally, he mined the many examples in the history of perspective when geometers and visual artists confronted the manifest limits of that tool, perspective, at work in their own hands, eyes, and minds. The limits of perspective as a readily usable tool, he argued, generated increasingly visible forms of subjectivity—of psyches and of artistic representations of those psyches—experienced as fragmented, partial, disoriented, disunified, or, in a word, melancholic. As Elkins historicized the art of perspective anew as an "experiment in the ruins of mathematics," he also excavated how such ruins were internalized, dwelled in, felt, and expressed: "The association between perspective and melancholy strengthened in the sixteenth century as treatises increasingly turned to instruments and constructions detached from practical use."[33]

To bring the issue of failed instrumentality to bear on the topic at hand, the elusive history of the air, let us now consider Dürer's *Melencolia I* (fig. 7.1), an engraving that Victor Hugo in fact once linked to the melancholic dimensions of Shakespeare's character Hamlet. As Hugo wrote of the alignment of the verbal with the plastic arts: "Shakespeare, wonderful plastic poet, makes the grandiose pallor of [Hamlet's] soul almost visible. Like the great spectre of Albrecht Durer [*sic*], Hamlet might be named 'Melancholia.' Above his head, too, there flits the disemboweled bat; at his feet are science, the sphere, the compass, the hourglass, love; and behind him, at the horizon, a great and terrible sun which seems to make the sky but darker."[34] From my perspective, both *Melencolia I* and *Hamlet* converge through a common concern, not simply with melancholy or science in general but with the air itself existing at the limit point of mediation—for Dürer's famous engraving is arguably about its own inability to compass the air, about its own inability to breathe.

Klibansky, Panofsky, and Saxl long ago argued for this engraving's explicit engagement with iconographic traditions linking measurement with "Saturn," or melancholy, and with philosophical traditions through which melancholy was understood to be part and parcel of genius.[35] What I want to emphasize is simply the all too ubiquitous element of air in the engraving as something more than a central element in the humoral tradition of melancholia (melancholia produced through bilious

7.1 Albrecht Dürer, *Melencolia I*, 1514. Engraving, 31 x 26 cm. Kupferstichkabinett, Staatliche Museen, Berlin.

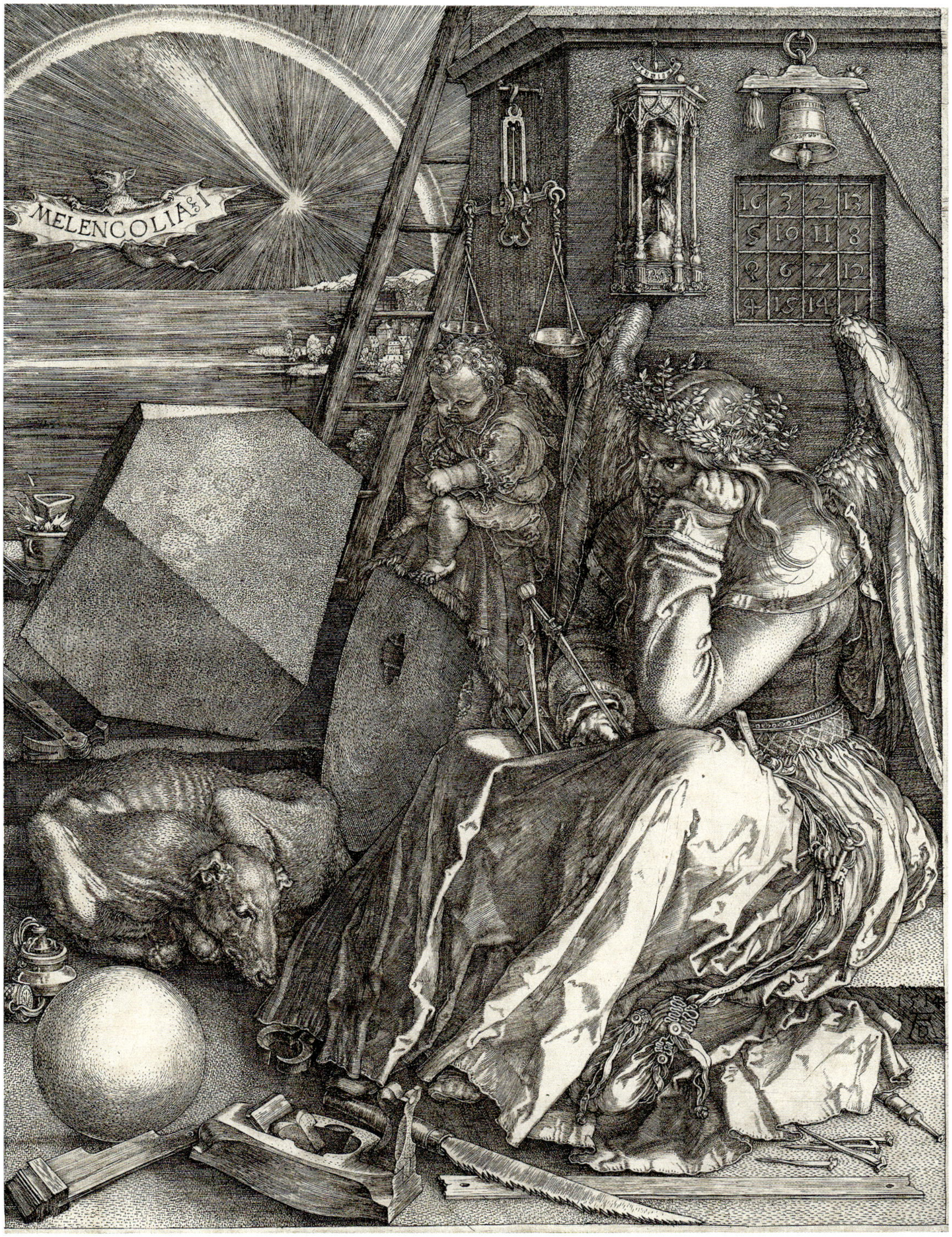
MELENCOLIA I
16 3 2 13
5 10 11 8
9 6 7 12
4 15 14 1

"vapors," climatological conditions, elemental imbalance, and so on); it is also an element that, precisely by proving resistant to instruments of apprehension and representation, inspired—or, rather, breathed its own form of life into—aesthetic innovation.

The specter of geometric disarray and instrumental uselessness in *Melencolia I* is positioned, in large part, as a product of aerial phenomena beyond the grasp of both engraver and engraved. As Susan Stewart has written in another context altogether, "Wind and water press against and move us, but they cannot be grasped. By flying and floating we can have some sense of mastery over our contact with these elements, but we cannot contain them in our hands for more than a moment."[36] This insight provides a nice gloss on a particular aspect of Dürer's engraving. Beneath the images of aerial motion and flight (not only the comet but also the violently propulsive lines of sky, doubled with those of sound coming from the conspicuously open mouth of the airborne creature), the motionless bell and the nozzle of the abandoned bellows call attention to the silence that would be sound, to the air that would be wind. The air on earth is relatively still, yet through images of motion stopped dead in its tracks—including the otherwise winged and airborne but now stationary, downcast *putto* and Melancholy herself; the motionless bell; the idle bellows; and that Platonic prototype for unimpeded motion, the sphere, all brought to a manifest halt—the limits of the engraver's medium to depict sound and motion in pictures, as nature could in reality, is on full display here. Additionally, the spaciotemporal flux implied by the hourglass and the setting sun is counterbalanced by the "1514," which, like the title *Melencolia I*, gestures toward the limits of "still life" as a medium stuck in time.

The truncated geometrical solid, existing as if whacked by the hammer or cut off by the saw below, reflects on the injury that a two-dimensional space might do to a three-dimensional one. This "octahedron," with an illuminated triangle marking an empty space, arguably alludes to the original construction of air, like all of the elements in Plato's *Timaeus*, through a combination of triangular forms, but also of Plato's specific alignment of air, alone among the elements, with the octahedron.[37] (Here we find a visual correlative to what it might have meant, in *Hamlet*, to attempt to "saw the air" through manual or instrumental technology.) Finally, the overarching rainbow indexes a natural occurrence produced in the air through a mixture of liquid drops and light that in some respects exceeds the instruments of the engraver. Even as shades of black and white were used in this period to signify or suggest color, and even as rainbows were made visible against the darkened sky, Dürer's rainbow is placed less as the obverse of than on a continuum with the upward curving banner.[38]

What is suggestive for my purposes is the extent to which the air forms the elemental nexus of such limit points of mediation. The air, even as invisibly and inexplicably present through the eerily empty but still unbalanced scale, proves resistant to instruments in ways that bring "melancholy" into being. Interpretations of the engraving as an illustration of failed "inspiration," humoral imbalance, and, more generally, the elusiveness of nature to the instruments of scientific knowledge are by now familiar. But the specific limits of the medium to compass the complexity and substance of the air seems to me to inform a "dispirited" engraving that follows in the wake, as it were, of the engraver's own art, hand, and eye.

The limits of body, eye, and implements for self-extension in the particular matter of aerial or meteorological phenomena were not, of course, new. But reflecting further on this topic can help to counterbalance the accrued capital attributed to concepts of "instrumentality" in its various academic fields and discursive registers—as well as the power of scientific instruments and scientific objects as vehicles of making meaning. To simply highlight the impotence of instruments and instrumentality as it occasioned or "made visible," in always historically specific ways, arenas of subjective experience can help us to reassess our own investments in sometimes forgetting—at times strategically and at times unwittingly—the affective as well as the material dimensions of the "history of air."

The body's apparent uselessness as an instrument of agency or rational apprehension, as well as its status as something at once used and abused by the very air through which it lived, was never more apparent, in early modern Europe, than during periods of plague or serious illness. Yet it is precisely for this reason that the task of the artist, engraver, poet, dramatist, and essayist was sometimes less to "remember" than to somehow "forget" the air as a necessary and terrifying space of internal and external dwelling—and to dwell rather on the domain of feelings and imaginative resources *produced* at the limit of the instrumentality.

The configuration of the artist-philosopher as his "own executioner"—as powerfully represented in Dürer's *Melencolia I* as in Donne's *Devotions*—is one way to take air's power over life and death into one's own hands. For Donne, the combined power of technologies, such as the "bullet," the "cannon," and other explosives or "scornful instruments" used to injure or kill, simply paled in comparison with that stealth and "insensible" element, the air, which "disables us from seeing our danger."[39] That the "air that nourishes us should destroy us" led Donne to ask, in several essays in his *Devotions*, composed during a period of illness, "What will not kill a man if a vapour will?" (77). At once to express and to compensate for the disconcerting power of "so insensible a thing" to reduce him to "nothing," Donne turns to a kind of ferocity of *ekphrasis*: "who would not think himself miserable to be put into the hands of Nature, who does not only set him up as a mark for others to shoot at, but delights herself to blow him up like a glass, till she see him break, even with her own breath?" Here, with Nature's own "breath" and "hands" making the body feel as fragile as glass, the avowal of air's power could altogether undermine fantasies of human autonomy, agency, and instrumentality.

Yet because imagining the power of Nature's breath to inflate, suffocate, and destroy—and the comparative fragility of man as blown and shattered glass—could, as Donne points out, smack of heresy (77), he quickly internalizes the logic of vaporous production, or tests out the possibility that he has somehow infected the air within him. On how we kill ourselves with our own vapors, he writes, "But what have I done, either to breed or to breathe these vapours? They tell me it is my melancholy; did I infuse, did I drink in melancholy into myself? Is it my thoughtfulness; was I not made to think? . . . I have done nothing willfully, perversely toward it, yet must suffer in it, die by it" (78). The avowedly blurred lines between inside and outside, body and world, victim and executioner, air and liquid, and intellection and melancholy—so central to Donne's philosophical vexation at the air's power—will resurface when we return to our discussion of *Hamlet*. So too will the logic of analogy, for it is to analogy that Donne turns in order to make *something* of that insensible substance through which he might otherwise be reduced to "nothing."

As Donne works to "extend this vapour, rarefy it; from so narrow a room as our natural bodies, to any politic body, to a state," it becomes evident that the substance of vapors in and around the body so resists sense and material localization that Donne can only gesture toward it through analogy, where it emerges as a form of discourse (79). Thinking of the toxic and infected world of *Hamlet*, it is suggestive that he writes, "That which is fume in us is, in a state, rumour; and these vapours in us, which we consider here pestilent and infectious fumes, are, in a state, infectious rumours . . . and therefore, when these vapours, these venomous rumours, are directed against these noble parts, the whole body suffers" (79). Directing hyperbole against social agents, Donne diminishes the material force of environmental toxins: "What ill air that I could have met in the street, what channel, what shambles, what dunghill, what vault, could have hurt me so much as these homebred vapours?" (79). Significantly, Donne lists here a series of places where the plague was commonly thought to be transmitted, but he renders such materialism secondary to the more toxic vapors produced by human agents through sin, as he moves swiftly from environmental analogies of "homebred vapours" to the theological origin of smoke as that which brings disease and death (81). From Genesis—where life itself is but "a vapour, that appeareth for a little time, and then vanisheth away" (80)—to the smoke, clouds, incense, and fumes of the Psalms, James, Leviticus, Isaiah, Ezekiel, and Revelation, where "out of the bottomless pit arose a smoke that darkened the sun" (81), Donne tries to reckon with the agency of air, ultimately turning to theology, pleading to God to "afford us [a] . . . remedy," to expel the toxic air from body, soul, and world. "For what is sin but a vapour," he asks, "but a smoke, though such a smoke as takes away our sight, and disables us from seeing our danger, and it is just that thou punish us with vapours too" (81).

Through scripture, then, with vision disabled and insight precluded by sin, a logic of occluded visuality is restored where none may otherwise be found. By turning away from the mire of invisible materialism to that other invisible agent, God, and to the power of language itself, Donne counters the palpable impotence of his instruments of body and mind ("If I were asked again what is a vapour, I could not tell, it is so insensible a thing; so near nothing, that it reduces us to nothing"). Through language, he turns the air into something other than itself, finding a way, in Irigaray's terms, to forget the air. Yet before the leap to, or between the leaps of, affirmation, there are spaces in which Donne mourns the limits of his intellect, the loss of control over his bodily and intellectual surround, and the limits of his own powers of insight. If, as Irigaray put it, "not being perceived, air can serve as the base for mourning" (43), we find traces of such mourning in the vividly memorable power of Donne's prose.

Let us turn again to what has become one of the most memorable dramas of the English Renaissance, *Hamlet*. To develop an issue already hinted at, we might simply ask, what makes this play so memorable? Other than that the accumulated capital through which *Hamlet* has become something of a touchstone of cultural memory, all manner of possibilities may swiftly come to mind: the combined dramas of ghost, incest, regicide, suicide, spying, melancholy, madness, or what many have seen, however hyperbolically, as a turning point in the representation of modern subjectivity, to mention a few. Among these and many other keywords through which *Hamlet* has come to mean so much, we would most definitely *not* find much memorable about the "air," the "plague," or the "climate," not to mention the "weather."[40] That *Hamlet* is thick with allusions to each of these things is, in and of itself, worth dwelling upon. But that *Hamlet* also manages *not* to be about an element that informs so much of its language, imagery, and action is the more interesting issue at hand, particularly given its initial staging at a time when the air was a powerful object of fascination and terror for the diverse inhabitants of England. Let us now consider *Hamlet* in terms of surrounding discourses dealing with the air, but also pay close attention to its own language, which contains traces of a now forgotten "history of air."

As in Dürer's *Melencolia I*, but now in the form of a sound and motion picture, the "matter" of air in *Hamlet* is so central as to be ubiquitous, yet it is also strategically subordinated to a series of reflections on the limits of the medium as an instrument through which the air might be subject to capture.

Holding the Mirror: Hamlet and the Air Apparent

> Surely nothing is more reproachful to a being
> endowed with reason, than to resign its powers
> to the influence of the air, and live in dependence
> on the weather and the wind.
> —SAMUEL JOHNSON

Hamlet has been mined, indeed gutted, in recent years for its preoccupation with the essence, or quintessence, of dust: dirt, land, and soil; flesh, bone, and brain; worms, viscera, and human remains—in short, matter.[41] But matter was, in both play and period, in the air as well as on the ground: "there's matter in these sighs . . . you must translate" (4.1.1–2); "I eat the air, promise crammed" (3.2.93–94). Whether, as Hippocrates suggested, air was itself a kind of food, something that might actually be eaten, air "mattered" a great deal in this period.[42] Air mattered as an element packed with atoms, "motes," invisible particles knocking about and unsettling those who couldn't see, or rather observe, such matter. It mattered as a medium through which other elements—earth, water, and fire—could move or be moved, and could be transformed, becoming something other than themselves. It mattered because it was a central medium of intellection and communication. And it mattered all the more during periods of disease and environmental disaster, and also periods in which the air became, *through indirect or secondary effects*, visible in ways that were, at times, astonishing.

A detail from Giulio Romano's enormous fresco in Mantua (fig. 7.2) stages the subjection of giants to the power of the elements and captures a form of collective astonishment and fear at the power of the air. As in much early modern visual art, air enters into the field of perception or visible "matter" through secondary, anthropomorphic, or indirect means: the personified winds and the specter of Aeolius himself merge with the otherwise naturalistic dimensions of the fresco's ominous cloud formations. The giants stare fearfully into a space of meteorological chaos that is also a space of personified forces, indexing a common form of double vision informed by myth. Beyond the visual arts, however, we can find in early modern meteorological discourse evidence of the often spectacular ways in which the air could seem to "matter," again through secondary

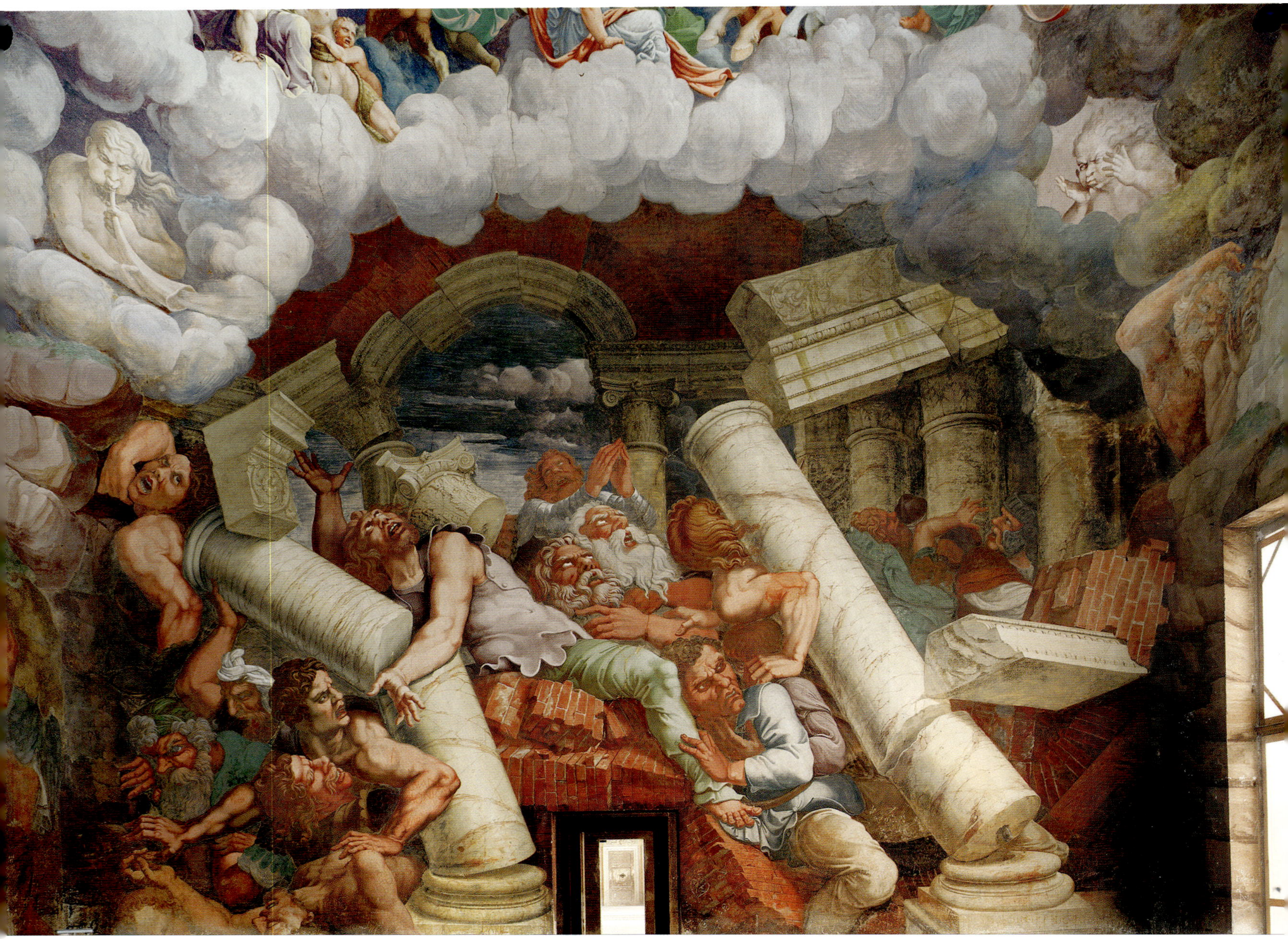

7.2 Giulio Romano, *Fall of the Giants*, detail, 1530–32. Fresco. Palazzo del Te, Mantua, Italy.

effects, in ways that can help us further grasp the significance of the element at work in *Hamlet*.

A 1563 meteorological text by William Fulke, for example, recounts various first-, second-, and third-hand accounts of historical persons who witnessed something like the air (a combination of liquid, air, and light that produced fog, clouds, steam, or vapors) moving as if of its own accord and in ways that could be "very fearefull to loke vpon."[43] The weather itself, that is, could generate visual spectacles in which signs were taken for wonders, ghosts, moving pictures, or spectacular forces capable of acting on their own. With *Hamlet* in mind, it is significant that in early meteorology the air was understood to be particularly "thick" around the bodies of the recently dead, which emanated vapors or, as Fulke puts it, oily or "unctuous and fat matter" (12). Thus to be in proximity to the dead was to breathe them in ("Pah!" says Hamlet of even the long-buried skull of Yorick that he holds close enough to smell).[44] It was also, for Fulke, to see the dead—or rather to imagine seeing them, especially in the dark of night, when light flickered on airborne "unctuous and fat matter" so as to produce illusions of movement, leading "ignorant & superstitious fooles" to think they see "soules tormented in the fyre of purgatorie" (12). This apparent movement of the air could "leade men out of their waye," appear to "follow men and go before them," and lead such eyewitnesses to "tell a greate tale, how they haue béen lead about by a spirite in the lykenes of fyre" (12).

Such an account in this meteorological text is no more surprising than the vulgar errors that pepper Bacon's prose, for the use of folklore and popular knowledge in such scientific texts could both entertain and elaborate on supernatural "fictions," against which developing forms of natural history could be pitched and authorized. Accordingly, such accounts could also, for avowed Protestants like Fulke and Bacon, occasion pointed stabs at the iconophilic traditions, beliefs, and practices of Catholicism—against which a more sensible vision of religion and nature could emerge. Yet as conventional as Fulke's text was, it is interesting as a text that may inform, as well as encode questions central to, *Hamlet*. As Fulke writes toward the close of his section on "apparitions" in the air: "thoughe I haue enterprysed to declare these by naturall reason, yet beleuing y^t not so much as on sparrow falleth to ye grounde, without Gods prouidence, I doe also acknowledge Gods prouidence bryngeth these to passe . . . vsing these causes, as meanes and instrumentes to doe them" (46). This trajectory in Fulke—from recounting and debunking phenomena such as purgatorial "spirits," to attempting to establish proper "causality," and, finally, to restoring divine agency to the matter of nature—mirrors the trajectory of *Hamlet*. We might recall Hamlet's famous "Not a wit, we defy augury: there's a special providence in the fall of a sparrow," a line delivered, interestingly enough, toward the end of the play, precisely as he gives up his own instrumentality in effecting or carefully observing the world around him (5.2.215–16). Fulke's text includes an extended treatment of a meteorological formation that was seen to fly over the Thames and hit on Stratford—the home, of course, of Shakespeare's birth and youth—which, Fulke demonstrates, became a focus of cultural speculation, fascination, and reporting for more than a decade (10–11).

I am not interested in arguing for Fulke's text as a "source text" for *Hamlet*, though it may well have been (the text had been printed three times by 1602). Instead, I am interested in how such *meteorological* debates of the period can help us to understand the ghost in *Hamlet* as something other than a haunt of Catholicism or a phantasmatic product of a son unable to mourn his father. Neglecting to examine this dimension of the play would underestimate the haunt and power of the air itself as it was understood to affect lives and souls, as it was during this time made "visible" through a range of medical treatises, natural histories, and cultural fictions, and as it was likewise subjected to new modes of experimentation and observation in the years leading up to the play.

When Fulke's treatise on the weather observes, as many meteorological texts did, that "within the globe of the earth, be wonderful great holes, caues, or dongeons, in which the ayer" is bound yet "cannot abide to be pined in," and escapes with "force & vehemens" (18)—or when Bacon, noting the same phenomenon of the "winds prison," suggests ways to "make" the wind by "by letting forth of winds out of subterraneal places," or simulating the release of air from "imprisoned places" to examine the sound as well as force of moving air[45]—we come intriguingly close to an Ariel released from years of imprisonment in a cloven pine in *The Tempest*, or to a ghost of King Hamlet emerging from a subterranean space with frightening force. When Prospero not only enacts mastery over "Ariel" as a means of extending his powers, but also ultimately says of the theatrical medium, "These our actors, / As I foretold you, were all spirits, and / Are melted into air, into thin air," we find the limits of theatrical and scientific experimentation on an explicit continuum.[46]

The science of air—even or especially when exposed as a kind of fiction or product of unreliable sensory information, "such stuff as dreams are made on"—was integral to the medium of drama as an art in more ways than one. Certainly, in many of his plays Shakespeare made "visible," through aerie spirits, apparitions, *ekphrastic* scene painting, and meteorologically imagined atmospheres, the element of air. Such imaginative fictions, theatrically produced, at once staged and enabled forms of collective observation that concerned the nature and power of the air. The historical context in which some of these fictional experiments took place is important, particularly if we consider the place of collective eyewitnessing in sixteenth-century meteorological matters. This is, in part, because, on particular days when the weather was just right, nature seemed capable of producing vividly and carefully detailed motion pictures that were verified by collective consensus in ways that might strike us, in the twenty-first century, as startling.

Fulke reports, for example, that one day in London, vivid images of persons, places, and things, all in action, could be seen in the upper region of the air. Specific and vivid representations of life on earth were witnessed not by one person, nor by what Bacon would call the "Country people," but by a socioeconomically diverse collective who all saw the same thing at once. "The very similitude of persones knowen to the beholders," he writes, "as of late, was séen the very image of the Emperour Charles, insomuche that they whiche behelde it, put of[f] their cappes, thinking verely it had béen he." Likewise with "Frederick prince Elector of Saxon" (45). Such communal deference to airborne rulers might well index an overactive political imagination, but the air was a locus of representations that expanded beyond the political to the everyday, from the grand to the minute. Other accounts collectively witnessed and recorded included.

> sites of armies fighting, in the ayre, of Castels, Cities, and Townes, with whole countries, hauing in them hills, valies, ryuers, woodes, also beastes, men, and foules, monsters, of whiche ther are no suche kyndes

> on the earth, and fynally all maner of things and actions, that are on the earthe, as burialles, processions, iudgementes, combates, men, women, children, horses, crownes, [an] arme of certayne noble men, . . . many thynges and gesturs done by men or beastes. (45)

As a representational medium, the sky becomes a vivid motion picture here, structurally akin to drama in representing both process and persons. The sightings on this particular day in London, as the historian Alexandra Walsham has shown, were by no means an isolated event. Different but similarly spectacular aerial phenomena were witnessed, again collectively, in Nuremberg (1554), Montpellier (1573), Ghent (1579), Poland (1581), Normandy (1590), Frankfurt (1600), and Croatia (1605).[47] So too, as Fulke points out in 1563, "In Germany . . . such many things, as it were long stories, [were] séen in the ayre" (45). This brings us into a historically specific cultural climate in which "cloud gazing" could form the basis of cultural consensus.

Juxtaposed with this, we might recall the moment in *Hamlet* when Polonius, looking skyward, agrees that various cloud formations are shaped exactly as Hamlet says they are:

HAMLET: Do you see yonder cloud that's almost in shape of a camel?
POLONIUS: By th'mass and 'tis—like a camel indeed.
HAMLET: Methinks it is like a weasel.
POLONIUS: It is backed like a weasel.
HAMLET: Or like a whale.
POLONIUS: Very like a whale. (3.2.367–73)[48]

One might be counted a fool, of course, to agree with Hamlet at this moment, but what if what one saw was an exact "similitude," as Fulke put it, "like the King that's dead" (*Hamlet*, 1.1.44)? And what if that similitude was witnessed, as in the opening act of *Hamlet*, collectively? Indeed, before jumping to either psychoanalyze Hamlet for seeing an apparition of his father in—and as—the air, or placing the ghost exclusively in the domain of Catholic superstition,[49] we might consider just how seriously "marvelous apparitions" were taken, and taken as historical phenomena, in the period leading up to and surrounding the production of *Hamlet*.

Like many others who aimed to naturalize celestial apparitions, Fulke drew heavily on Aristotle and Pliny to produce a range of explanations for the remarkable artistry of nature. Those collectively witnessed "apparitions," for example, were the product of times "when the disposition of the ayre, hath béen suche, that it hath receiued the image of manye thinges placed and done on the earth" (46). This theory of celestial impressionability positions the air as a mimetic medium or mirror, reflecting things exactly as they are, or once were, on earth. What Fulke does *not* do in this instance is discredit the idea that what was seen was in fact what was there. Instead, to account for eyewitness accounts of images of monsters that don't exist on earth, he describes the natural cause of such phenomena—and invokes Horace's famous dictum in the process: "because [air] is apte to receyue dyuerse images, as well in one place as in another, these monstruous formes and straunge actions, or stories proceade of the ioyninge of dyuerse formes and actions," mix and comingle in unpredictable ways, "as if twoo histories, were confusely paynted in one, the whole picture would be straunge or (as the Poet sayeth) if a paynter to a mans head, should set a horses neck, & after dyuerse fethers" (46). At the same time, Fulke observes that through the use of catoptrics, "artificially by certein glasses, and instruments," artisans could produce the very same effects of nature (45–46).

In turning to the "bad air" in *Hamlet*, a text marked by the very constellation of vapors we encounter in Donne, I want to emphasize that one of the central symptoms of both melancholy and the plague was breathing problems caused by the quality of the air. The "worst" kind of air, wrote Robert Burton, "is a thicke, cloudy, misty, foggy Ayre, or such as come from fennes, Moorish grounds, lakes, muckhils, draughts, sinkes, where any filthy carcasses or carrion lies, or from whence any stinking fulsome smell comes . . . much Air is unwholesome, and *ingenders melancholy, plagues, and what not*."[50] Thomas Phayre's *Regiment of Life*, printed eight times between 1544 and 1600, blames "the stynch and filthy sauours that corrupte that ayre whiche we lyue in: for we can not liue wythout drawynge of the breath, and we haue none other breth but of the ayre rounde aboute vs." Of the plague, he writes, "many one is so strayte wynded, that he can not speake, and when he breatheth it is with great laboure and difficultie."[51]

When, in the final act of the play, Hamlet speaks of the "fatness of these pursy times"—pursive meaning "short-winded"—he not only anticipates his own condition of being "fat and scant of breath" (5.3.290); he also articulates the breathing problems that mark the "rotten . . . state of Denmark" (1.5.90). From the performance of woe through "windy suspiration of forc'd breath" (1.2.79) to Ophelia's death by drowning, this is a play marked by characters who not only have trouble breathing but try to take the air into their own hands. Although scholars have long diagnosed Hamlet's symptoms as signs of melancholy, the play returns again and again to signs of the plague and, specifically, the breathing of toxic air. "A man may know the infection of the aire which threatneth vs with Pestilent sicknesses," wrote Thomas Lodge in 1603, not only by "thicke, cloudy, moyst, and ill smelling vapours, the Skie vnaccustomed to Northren windes, but sollicited with Southerly blastes," but also by a proliferation of vermin, dead bodies, and conspicuous shifts in temperature and air quality.[52]

Throughout *Hamlet*, the fumes and vapors of a decayed world align melancholy with both religious and climatological forms of corruption. If, as Hamlet says, directly after the scene of cloud gazing with Polonius, "tis now the very witching time of night, / When churchyards yawn and hell itself breathes out / Contagion to this world" (3.2.379–81), then this toxic air is embodied in the ghost, imagined as a product of the earth's breath after consuming the dead. This issue is further emphasized in the 1603 Quarto of *Hamlet*, where Hamlet is allegedly sent away to get him out of the bad air in Denmark: "the aire and climate of the Country / May please him better than his natiue home."[53] "This most excellent canopy the air," observes Hamlet, "why it appeareth nothing to me but a foul and pestilent congregation of vapours" (2.2.299–303). However poignantly such words configure melancholic perception, the very same conceit emerges in Dekker's *The dead tearme*, where the breath of a ghost, an embodiment of plague, hangs in the London air: "Dead vnto vs," he writes, "are our liueliest dayes, whilest this pestilent vapour hanges ouer our heades."[54] The ghost of King Hamlet, so familiar to us now as a haunt of purgatory, a sign of melancholy, or an illusion of the devil himself, seems a striking embodiment of environmental blastments and the coming of plague. For Hamlet is not only blasted with the air of that questionable shape, but also ultimately lurks in places of the dead where bad air, melancholy, and the plague were thought to be "engendered": he lugs the guts of Polonius, holds human remains in the graveyard close enough to smell ("Pah!"), and wishes to be buried alive alongside the corpse of Ophelia. As such, we might think more literally about Walter Benjamin's sense that "Hamlet wants to breathe in the suffocating air of his fate in one deep breath."[55]

But this sigh or deprivation of air that might "end ones being" is fatally imagined, not only in Hamlet—in some sense literally walking "out of the air" and into his "grave"—but also in Ophelia's drowning. The fleeting fantasy that one might live without air and, thus, in the water—as Ophelia is, for a moment, "like a creature native and indued / Unto that element" (4.8.178–79)—is counterpoised with the fear that to walk in the air, to "eat the air," to breathe in this play, where "rank corruption, mining all within / Infects unseen" (3.4.150–51), is to be vulnerable to what Laertes calls "contagious blastments . . . most imminent" (1.3.42). And this is a language of the plague that is further echoed as Hamlet gives Ophelia "this plague for thy dowry" (3.1.136–37). Air, then, as the mythic and Galenic medium of thought, intellect, communication, is, like the vapors of Donne's world, "infected" in *Hamlet*. It is also, we might note, rather wet. We might transfer Donne's words to Hamlet: "But what have I done, either to breed or to breathe these vapours? They tell me it is my melancholy; did I infuse, did I drink in melancholy into myself? Is it my thoughtfulness; was I not made to think?" The uneasy mingling of air and water (the conflation here of breathing and drinking) that constitutes being and thinking is in fact central to *Hamlet*. In perhaps all too literal terms, given the dominant trope of alcoholic "spirits" in the play, it is at least minimally suggestive that Lodge stresses that drinking "wine" in times of pestilence would only hasten fatality, making the body "more capable to receiue the euill influence of the ayre."[56] Further, the relation of liquids to air is writ both large and small in *Hamlet*. In small, indeed very small, recall the toxic wine—designed to "drink to Hamlet's better breath" (5.2.268)—that Gertrude imbibes, only to swiftly expire; the liquid "unction" or "contagion" on the tip of a sword that "quite o'er crows my spirit," says Hamlet, as if he has become his own father's ghost (4.7.140, 156, 5.2.358); and the famous staccato of "O, o, o, o" in the 1603 Quarto, which stresses the fatal effect of even a small drop of liquid.

In more general terms, the plague itself was classically understood as originating in an "earth ouerflowed with *too much moysture and filled with grosse and euill vapours*, which by vertue of the Sunne béeing lifted vppe into the ayre, and mixed with the same, corrupteth the nature and complexion thereof," so that they who drink or "sucke this infected aire are in daunger to be attainted with this contagion and sicknesse of the Pestilence."[57] Since "*Water*," as Bacon put it, is "contiguous with *Aire*," and since "vapours" moistened the atmosphere, to breathe infected air during times of plague constituted a form of involuntary and fatal drinking.[58] This simple point may gloss Donne's conflation of breathing and drinking, but it may also compel us to take what has long been understood as the gravedigger's burlesque of legal argumentation in *Hamlet* at least a bit more seriously: "If the man go to this water and drown himself, it is, will he nill he, he goes, mark you that. If the water come to him and drown him, he drowns not himself" (5.1.18–19). While "water is a sore decayer of your whoreson dead body," entering through the "hide" or skin (5.1.163–65), so too it seems to pose a threat to those living in the air, as "churchyards yawn and hell itself breathes out / Contagion to this world," constituting a wet and eerily mobile atmosphere of airborne vapors.

The problem of liquid was of course integral to purgatory—itself a place where fire purged impure spirits of sin (hence the unpurged form of the ghost, subject to purgatorial fires).[59] But it was equally integral to climate theory and the frustrations of inhabiting an unhealthy atmosphere that seemed, in London, constitutionally resistant to the "melting," "distilling," or "resolution" of vapors. As Jean Feyens put it in the 1580s, following Hippocrates, "A cold and moist air, as it is in the North, is like the Winter, stupifies the spirits, stops the pores, and burdens the body with many superfluous humours."[60] The North here, a perpetual winter of discontent, stops up pores, making for "too too solid flesh" unable to evaporate, "melt, thaw, or resolve itself."

In addition to engendering, as Mary Floyd-Wilson has recently observed, a famously melancholy region,[61] such a climate was also understood to breed far more fatal vapors, which entered in at the mouth but could not exit through the pores of the skin. During plague time especially, closed pores could inflame the interior of the body and hasten the progress of infection. At the same time, many, such as Thomas Moulton, warned that during periods of heightened atmospheric corruption, one should not exercise or sweat "to much, for all openeth the pores of a mans bodye, and maketh the venomous ayre to enter, and for to infect the blood, & the spiritual members of man, & for to destroy al lyuely spirits in man."[62]

Hamlet's dying wish returns to the domain of transmutation, not only in turning to God, where analysis, hypothesis, and human agency reach their limit, but also in imagining his own afterlife as a form of breath issuing from Horatio's mouth. The afterlife of *Hamlet*, as play, depends of course on the medium of air as the medium of speech and theatrical performance—a medium that will, at least ideally, move, liquefy, melt, purge, and resolve the attentive auditor. Performance itself, in Hamlet's own advice to the players, is a delicate technology of the air, one that should not blast the ears but rather bring one to tears. Within Shakespeare's plays, in more general terms, a transmutation is always effected that transforms potential atmospheric casualties into dramatically situated social agents. To state the obvious, his characters don't die from the plague or venomous air but from error, mischance, or love. Indeed, what is interesting is that in Shakespeare potential victims of the plague often become lovers or melancholics who might die for love. Given the imagery of disease and plague that riddles *Romeo and Juliet*, for example, it is suggestive that in an early version of the story Juliet suffocates herself to death, and in Shakespeare's play Lady Montague simply stops breathing: her grief "hath stopped her breath."

Remembering Irigaray's statement that "not being perceived, air can serve as the base for mourning," for Shakespeare, the recoding of "air" over the course of his plays may facilitate a move from melancholy to mourning, opening up representational processes that encode affective and paradoxically living forms of death. Shakespeare transmuted the infected air into something altogether different, turning fears of breathing into inspired dramas, where the "fog and filth" produced, in *Macbeth*, "sighs, and groans and shrieks that rent the air"; in *Twelfth Night* and *Romeo and Juliet*, love that might "purge the air of pestilence"; and in *Antony and Cleopatra*, poison itself that might be "as soft as air, as gentle."[63]

In *Hamlet*, all manner of instruments through which the air might be manipulated tend to falter. We see this in the language of navigation, as Claudius puts Hamlet on a ship where the "bark is ready, and the wind at help," yet the ship proves "too slow of sail," easily overtaken by pirates (4.3.44, 4.7.16). We see this in the language of music, with Hamlet as a self-imagined "instrument," "pipe," or "recorder" through which his fellow students might "go about to recover the wind of" him but promptly fail (3.2.337–64). We see it in the language of meteorology, from weather "talk," to ghost sightings, to cloud gazing, where the instrument of the eye proves all too fallible in discerning the air. And we see it in the language of mechanical artisanry, from Polonius's metaphorical and ineffectual "windlasses"—a mechanical device used in mining as well as in navigation to lure objects out of the depths—to the less than effectual mechanics of the players. Indeed, the dramatized inability of artists to manipulate the air with skill, as in the case of players who might "saw the air" or characters who might stare blankly at a simple "pipe" without knowing how to use it, is not unrelated to the antiquated dramatic technology of staging a ghost emerging from the mouth of hell, calling for revenge—itself a form of theatrical transparency explicitly mocked as mere buffoonery or child's play in the induction to *A Warning for Fair Women* (1599), a play performed by Shakespeare's own company. Finally, we see the limits of instrumentalizing the air in the language of military engineering, where the very idea of an "enginer hoist with his own petard" returns in Horatio's recasting of the entire play of *Hamlet* as one "of deaths put on by cunning and forc'd cause, / And, in this upshot, purposes mistook / Fall'n on the inventors' heads" (5.2.388–90).

These multiple dramas of failed instrumentality, with various engineers hoisted with their own petards, transposes a larger concern about the body itself as a vulnerable "instrument" through which the air might breathe or play upon at will, particularly as a nonhuman entity or haunt that "breathes out contagion to this world." It is in this transposition, however, that Shakespeare creates an instrument of tragic dramatic—as opposed to common atmospheric—consequence. As we have seen, Dürer, Shakespeare, and Donne deploy their respective artisanal technologies to encode the problem of subjecting air to capture, and yet reimagine technological limit as artistic and affective potential. The resistance of this invisible or quasi-visible element to intellection and instrumentalization stimulated creative alternatives in both the "arts" and "sciences" through which the air could be taken in anew. For if it is true that, to recall the quotation from Bacon, "the Nature and Power of the Air is unknown," then Shakespeare, along with Dürer and Donne, managed to transform that terrifying unknown into something that might—even or especially in its elusiveness to capture—be absorbed, experienced, and willfully taken in.

Notes

1. Bacon, *Catalogvs Historiarvm Particvlarivm*, D1v. The epigraphs to this chapter are taken from Bacon, *History of Winds*, 112; and Irigaray, *Forgetting of Air*, 12 (hereafter cited parenthetically in the text).

2. Feyens, *New and Needful Treatise of Wind*, 3, 1.

3. On the smell of air in Shakespeare, see Nagler, "Toward the Smell of Mortality." On the representation of wind and air from ancient to modern times, see Alesandro Nova's invaluable *Book of the Wind*.

4. Bacon, preface, *Natural and Experimental History of Winds*, *b1v (emphasis added). Unless otherwise noted, all references to Bacon's *Historia Ventorum* are to this edition.

5. Donne, *Devotions upon Emergent Occasions*, 78–79.

6. Burton, *Anatomy of melancholy*, 318–37.

7. *Hamlet*, 1.2.85. Unless otherwise noted, all references to *Hamlet* are to Jenkins's Arden Shakespeare edition, hereafter cited parenthetically in the text.

8. See Gail Kern Paster, *Humoring the Body*, 25–76.

9. On environment as a crucial interface between early modern body and world, see especially ibid.; Floyd-Wilson and Sullivan, *Environment and Embodiment in Early Modern England;* and Kenda, *Aeolian Winds and the Spirit*.

10. The idea that elemental mixtures productive of "heavy air" could function as "mirrors" goes back at least to Aristotle's *Meteorologica*. See Clark, *Vanities of the Eye*, 213. On both Shakespeare and Hamlet as finding a suitable Protestant aesthetic through which to hold "the mirror up to nature," indeed, to "develop a drama that mirrors the natural world," see Diehl, *Staging Reform, Reforming the Stage*, 67–93, 92.

11. Bacon, *Natural and Experimental History of Winds*, 27.

12. Ibid., 9. For a similar complaint, see also Fracastoro, *De Contagione*: the presumed power of air to injure or kill during times of plague and disease often met with anthropomorphic fantasies in which "particles imperceptible to our senses" were treated like "enemies," "as though they possessed cognition and will" in trying "above all to make for and attack the heart" (43, 25).

13. Gill, *Sacred Philosophie of the Holy Scripture*, chap. 33, p. 113 (emphasis added; Gill was less concerned with the elemental materiality of "ghosts" than, as he stressed, the spirituality of the Holy Ghost).

14. As Seneca put what became a long-standing commonplace, "the great force of air cannot be checked." *Naturales quaestiones*, 2:178–81.

15. As Gabrielle Walker writes in *Ocean of Air*, "We don't just live *in* the air. We live *because* of it" (xv). On the disruption of drama by plague, see Barroll, *Politics, Plague, and Shakespeare's Theater*. On miasma and contagion through touch, see Healy, "Anxious and Fatal Contacts." On the Second Quarto of *Hamlet* in relation to the historiography of plague and Jacobean politics, to which I am indebted, see Mallin, *Inscribing the Time*, esp. 62–105.

16. Thayre, *Treatise of the pestilence*, 14, 1.

17. *OED*, s.v. "rotten," 1a and 1b.

18. Burton, *Anatomy of melancholy*, 290–91.

19. See especially Bacon's *Natural and Experimental History of Winds* (*Historia Ventorum*) and *Sylua syluarum*. For the 1660s, a period later than the one I am addressing, see Shapin and Schaffer, *Leviathan and the Air Pump*.

20. Among recent studies of Shakespeare focused on scientific instruments and technologies, see Cohen, *Shakespeare and Technology*, which explores Shakespearean drama in terms of an array of instruments, including mirrors (convex, concave, cylindrical, cheap), maps, clocks, and books. See also Arthur F. Kinney's approach, in *Shakespeare's Webs*, to these same instruments from the perspective of cognitive science; and Jonathan Sawday's approach to concerns as well as ideals with regard to machine technologies in the early modern period, in *Engines of the Imagination*. On windmills, heat production, and other technologies and instruments available in Elizabethan London, see especially Harkness, "'Strange' Ideas and 'English' Knowledge."

21. As Cohen suggests of instruments and technologies, "authors like Shakespeare often depicted them in human terms, and they also described human beings themselves as technologies." *Shakespeare and Technology*, 177. On Shakespeare's art being informed by "modes of artificial life," see Johnson's "Preface to Shakespeare," 11.

22. Sawday, *Engines of the Imagination*, 99–100. In more general terms, for Pauline Kiernan, Shakespeare's central "instrument" was the body of the actor itself, upon which he made his art. See her *Shakespeare's Theory of Drama*, esp. 11, 15, 193. For Bruce Smith, early modern theaters were themselves architecturally engineered "sound-devices": "the South Bank amphitheaters were, in fact, instruments for producing, shaping, and propagating sound." *Acoustic World of Early Modern England*, 206.

23. Seneca, *Naturales quaestiones*, 2:178–81.

24. *The Life and Death of King John*, 2.1.251–53, in Greenblatt's *Norton Shakespeare*.

25. On the "majestical roof" as a specific allusion to air, see Horace Howard Furness's note in the New Variorum Shakespeare edition of *Hamlet*, 3:158–59n293. On the constellation of homonyms, including "heir" and "air," in *Hamlet*, see Kokeritz, *Shakespeare's Pronunciation*, esp. 90–91, 111.

26. Bacon, *Novum organum*, 58.

27. Quoted in *Correspondence of Henry Oldenburg*, 7:167.

28. See especially Cole, "Demonic Arts." Cole examines two Renaissance "models on which both demonologists and artists understood the medium: as the usually invisible yet always perceptible 'air' that joins all things, and as the 'binds' that could ensnare and move the targets of one's spells" (621).

29. Polanyi, *Knowing and Being*, 148.

30. Polanyi, *Study of Man*, 31.

31. On failure and subjectivity, see especially Yates, *Error, Misuse, Failure.*

32. For various theories on the "rationalization" of space and sight, see my "Three-Dimensional Self," esp. 47–48.

33. Elkins, *Poetics of Perspective*, 116, 120.

34. Hugo, *William Shakespeare*, 236–37.

35. See Klibansky, Panofsky, and Saxl, *Saturn and Melancholy.* On the anachronism of this long-standing postulate of "genius," see Doorly, "Dürer's *Melencolia I.*"

36. Stewart, *Poetry and the Fate of the Senses*, 165.

37. As Plato's *Timaeus* reads, "To the earth . . . may be assigned the form of a cube; to fire the form of a pyramid, and the remaining forms to the other element: to air the octahedron, and to water the icosahedron." *Dialogues of Plato*, 2:472. On the geometry of the elements in Plato, see Harte, *Plato on Parts and Wholes;* on air and the octahedron, see esp. 238. Finally, although Patrick Doorly's "Dürer's *Melencolia I*" does not discuss "air" in particular, I am indebted to it for leading me to Plato's *Timaeus.*

38. The rainbow and comet, of course, remain unobserved by the downward-looking figures, as if these figures have not managed to see or perceive properly—neither in and through their eyes nor with the array of instruments strewn about. Christopher Wood has argued that *Melencolia I* evinces both the "blindness of drawing" and the "blindness of theory," a problem of composition between spaces of "outdoor" and "indoor" and between nature and the compositional spaces, such as the studio integral to contemplation. See his "Indoor-Outdoor." For a pointed challenge to Wood's argument, see Beth Holman's review of the volume in which Wood's essay appears, Cole and Pardo's *Inventions of the Studio*, in *Art Bulletin* 88, no. 3 (2002): 601–5.

39. Donne, *Devotions upon Emergent Occasions*, 77, 79, 81 (hereafter cited parenthetically in the text).

40. By contrast, Tom Stoppard often highlights the weather, the wind, and the elements in his *Rosencrantz and Guildenstern Are Dead.*

41. The epigraph from Johnson is from the *Idler* 11 (24 June 1758), *Works of Samuel Johnson*, 367. The quotation continues: "To look up to the sky for the nutriment of our bodies, is the condition of nature; to call upon the sun for peace and gaiety, or deprecate the clouds lest sorrow should overwhelm us, is the cowardice of idleness, and the idolatry of folly." On *Hamlet* and matter, see especially de Grazia, "Weeping for Hecuba"; Guillory, "'To Please the Wiser Sort'"; Hillman, "Visceral Knowledge"; and Garber, "Out of Joint." See also Hillman, *Shakespeare's Entrails.*

42. On the chameleon eating air, see Bacon, *Sylua syluarum*, 96–97. On the air in Shakespeare, see especially Jonathan Gil Harris's suggestive essay "Atomic Shakespeare."

43. Fulke, *Goodly gallerye*, 10 (hereafter cited parenthetically in the text). While this text went through seven printings between 1563 and 1671, all references are to the original 1563 edition.

44. The "absolute equivalent between stench and death," writes Annick Le Guerer in *Scent*, was especially marked during periods of plague (38, quoted in Nagler, "Toward the Smell of Mortality," 44). For multiple views on this matter, see Bacon, *Sylua syluarum*, esp. 244, 246.

45. See Bacon, *Natural and Experimental History of Winds*, 17, 46. While "there is great quantity of subterraneal spirit required to shake or cleave the earth," wrote Bacon, the matter of finding "subterraneal places" or "prisons" of locked air was no easy matter: "to know such places of prisons, is very hard and difficult."

46. Shakespeare, *Tempest*, 4.1.148–50. For a more extended discussion of Prospero's "art" as a model of "science," see Spiller, "Situating Prospero's Art."

47. See Walsham, "Sermons in the Sky." For classical accounts of similar phenomena, see especially Agrippa, *De occulta philosophia.*

48. What one sees here is, to an absurd degree, conditioned by what one hears. The same situation applies, of course, in the "apparition" of the ghost, who famously appears after numerous *verbal* descriptions early on.

49. See especially Greenblatt, *Hamlet in Purgatory.*

50. Burton, *Anatomy of melancholy*, 236 (emphasis added). On the status of plague in the Second Quarto of *Hamlet* specifically, see Mallin, *Inscribing the Time*, 62–105.

51. Phayre, *Regiment of life*, N1r, P5v.

52. "If . . . there appeare any increase of such creatures as are engendred of putrifaction, as wormes of the earth, flies, gnattes, eales, serpents, toades, frogs, and such like foretokening corruption and putrifaction in the earth and waters, and *when the aire the same day chaungeth from faire to foule*, and from cleare to cloudy, when the Sunne shineth and afterwards hideth his head in cloudes, in one and the same day, it is a signe that the temperature of the aire is altered. And when as Rats, *Moules, and other creatures, (accustomed to liue vnder ground) forsake their holes and habitations*, it is a token of corruption in the same, by reason that such sorts of creatures forsake their wonted places of aboade." Lodge, *Treatise of the plague*, C1v (emphasis added).

53. Shakespeare, *Tragicall historie of Hamlet*, G3v.

54. Dekker, *Dead tearme*, G1r–v. "Fiue yéeres hath shee beene a troublesome *Guest* vnto me," Dekker continues, "she doth now walks still in a ghostly and formidable shape vppe and downe my stréets. Shee (with her *Ill Company*) infecteth my Sonnes and Daughters, and leades them dayly into such daungers, that (in hundreds at a time) doe they loose theyr liues."

55. Benjamin, *Origin of German Tragic Drama*, 137.

56. Lodge, *Treatise of the plague*, E3r.

57. Ibid., B4r–v (emphasis added).

58. See Bacon, *Sylua syluarum*, where air and water intermingle: "*Water*, being contiguous with *Aire*, Cooleth it, but Moisteneth it not, except it *Vapour*" (231).

59. As Arden editor Harold Jenkins observes, "Ficino explained that at the death of the body, while pure souls ascended immediately to the eternal regions, impure souls, until they had undergone purgation, stayed close to the earth and, assuming forms of air, were visible to men" (*Hamlet*, 454n).

60. Feyens, *De flatibus humanum corpus molestantibus*, 5–6.

61. See Floyd-Wilson, *English Ethnicity and Race.*

62. Moulton, *Mirrour or glasse of health*, 14.

63. Shakespeare, *Macbeth* (4.3.169); *Twelfth Night* (1.1.22); *Antony and Cleopatra* (5.2.303), in *Norton Shakespeare.*

Bibliography

Agrippa, Henry Cornelius. *De occulta philosophia, or Three Books of Occult Philosophy: A Complete Edition.* Hastings, UK: Chthonios Books, 1986.

Bacon, Francis. *Catalogvs Historiarvm Particvlarivm.* Appended to *Francisci de Verulamio, Summi Angliae Cancellarii, Instauratio magna.* London, 1620.

———. *The History of Winds.* In *The Works of Francis Bacon, Baron Verulam, Viscount St. Alban, and Lord High Chancellor of England*, 12:1–124. London: M. Jones, 1815.

———. *The Natural and Experimental History of Winds.* Translated by R.—G. Gent. London, 1671. (Originally published in 1622 as *Historia Ventorum* and translated by Gent in 1653.)

———. *Novum organum.* 1620. In *The Works of Francis Bacon*, ed. James Spedding, Robert Leslie Ellis, Douglas Denon Heath, and William Rawley, 4:39–248. London: Longman, 1858.

———. *Sylua syluarum: or A naturall historie In ten centuries.* London: 1627.

Barroll, Leeds. *Politics, Plague, and Shakespeare's Theater.* Ithaca: Cornell University Press, 1991.

Benjamin, Walter. *The Origin of German Tragic Drama.* London: Verso, 1977.

Burton, Robert. *The anatomy of melancholy.* Oxford, 1621.

Clark, Stuart. *Vanities of the Eye: Vision in Early Modern European Culture.* Oxford: Oxford University Press, 2007.

Cohen, Adam Max. *Shakespeare and Technology: Dramatizing Early Modern Technological Revolutions.* New York: Palgrave, 2006.

Cole, Michael. "The Demonic Arts and the Origin of the Medium." *Art Bulletin* 84, no. 4 (2002): 621–40.

de Grazia, Margareta. "Weeping for Hecuba." In *Historicism, Psychoanalysis, and Early Modern Culture*, ed. Carla Mazzio and Douglas Trevor, 350–75. New York: Routledge, 2000.

Dekker, Thomas. *The dead tearme.* London, 1608.

Diehl, Huston. *Staging Reform, Reforming the Stage: Protestantism and Popular Theater in Early Modern England.* Ithaca: Cornell University Press, 1997.

Donne, John. *Devotions upon Emergent Occasions.* New York: Vintage, 1999.

Doorly, Patrick."Dürer's Melencolia I: Plato's Abandoned Search for the Beautiful." *Art Bulletin* 86, no. 2 (2004): 255–76.

Elkins, James. *The Poetics of Perspective.* New ed. Ithaca: Cornell University Press, 1996.

Feyens [Fienus], Jean. *De flatibus humanum corpus molestantibus.* Antwerp, 1582. Translated into English by William Rowland as *The New and Needful Treatise of Wind Offending Mans Body.* London, 1676.

Floyd-Wilson, Mary. *English Ethnicity and Race in Early Modern Drama.* Cambridge: Cambridge University Press, 2006.

Floyd-Wilson, Mary, and Garrett Sullivan, eds. *Environment and Embodiment in Early Modern England.* New York: Palgrave Macmillan, 2007.

Fracastoro, Girolamo. *De Contagione.* 1546. Edited and translated by Wilmer Care Wright. New York: G. P. Putnam's Sons, 1930.

Fulke, William. *A goodly gallerye. . . .* London, 1563.

Garber, Marjorie. "Out of Joint." In *The Body in Parts: Fantasies of Corporeality in Early Modern Europe*, ed. David Hillman and Carla Mazzio, 22–51. New York: Routledge, 1997.

Gil Harris, Jonathan. "Atomic Shakespeare." *Shakespeare Studies* 30 (2002): 47–51.

Gill, Alexander. *The Sacred Philosophie of the Holy Scripture.* London, 1635.

Greenblatt, Stephen. *Hamlet in Purgatory.* Princeton: Princeton University Press, 2002.

Guillory, John. "'To Please the Wiser Sort': Violence and Philosophy in *Hamlet*." In *Historicism, Psychoanalysis, and Early Modern Culture*, ed. Carla Mazzio and Douglas Trevor, 82–109. New York: Routledge, 2000.

Harkness, Deborah. "'Strange' Ideas and 'English' Knowledge: Natural Science Exchange in Elizabethan London." In *Merchants and Marvels: Commerce, Science, and Art in Early Modern Europe*, ed. Pamela Smith and Paula Findlen, 137–60. New York: Routledge, 2002.

Harte, Verity. *Plato on Parts and Wholes: The Metaphysics of Structure.* Oxford: Oxford University Press, 2005.

Healy, Margaret. "Anxious and Fatal Contacts: Taming the Contagious Touch." In *Sensible Flesh: On Touch in Early Modern Culture*, ed. Elizabeth Harvey, 22–38. Philadelphia: University of Pennsylvania Press, 2003.

Hillman, David. *Shakespeare's Entrails: Belief, Scepticism, and the Interior of the Body.* New York: Palgrave Macmillan, 2007.

———. "Visceral Knowledge: Shakespeare, Skepticism, and the Interior of the Human Body." In *The Body in Parts: Fantasies of Corporeality in Early Modern Europe*, ed. David Hillman and Carla Mazzio, 80–105. New York: Routledge, 1997.

Hugo, Victor. *William Shakespeare.* Translated by Melville B. Anderson. Chicago: A. C. McClurg, 1886.

Irigaray, Luce. *The Forgetting of Air in Martin Heidegger.* Translated by Mary Beth Mader. Austin: University of Texas Press, 1999.

Johnson, Samuel. "Idler 11." In *The Works of Samuel Johnson*, ed. Arthur Murphey, 41–44. New York: Alexander V. Blake, 1842.

———. "Preface to Shakespeare." 1765. In *Johnson on Shakespeare*, ed. Walter Raleigh, 8–63. London: Henry Frowde, 1908.

Kenda, Barbara, ed. *Aeolian Winds and the Spirit in Renaissance Architecture.* New York: Routledge, 2006.

Kiernan, Pauline. *Shakespeare's Theory of Drama.* Cambridge: Cambridge University Press, 1998.

Kinney, Arthur F. *Shakespeare's Webs: Networks of Meaning in Renaissance Drama.* New York: Routledge, 2004.

Klibansky, Raymond, Erwin Panofsky, and Fritz Saxl. *Saturn and Melancholy: Studies in the History of Natural Philosophy, Religion, and Art.* New York: Basic Books, 1964.

Kokeritz, Helge. *Shakespeare's Pronunciation.* New Haven: Yale University Press, 1953.

Le Guerer, Annick. *Scent: The Mysterious and Essential Powers of Smell.* London: Trafalgar Square, 1993.

Lodge, Thomas. *A treatise of the plague.* London, 1603.

Mallin, Eric. *Inscribing the Time: Shakespeare and the End of Elizabethan England.* Berkeley: University of California Press, 1995.

Mazzio, Carla. "A History of Air: Hamlet and the Trouble with Instruments." *South Central Review* 26, nos. 1–2 (2009): 153–98.

———. "Three-Dimensional Self: Geometry, Melancholy, Drama." In *Arts of Calculation*, ed. David Glimp and Michelle R. Warren, 39–65. New York: Palgrave, 2004.

Moulton, Thomas. *The mirrour or glasse of health.* London, 1580.

Nagler, Danielle. "Toward the Smell of Mortality: Shakespeare and Ideas of Smell, 1588–1625." *Cambridge Quarterly* 26, no. 1 (1997): 42–58.

Nova, Alesandro. *The Book of the Wind: The Representation of the Invisible.* Montreal: McGill-Queen's University Press, 2011.

Oldenburg, Henry. *The Correspondence of Henry Oldenburg.* Edited by Rupert A. Hall and Marie Boas Hall. Vol. 7, *1670–1671.* Madison: University of Wisconsin Press, 1970.

Paster, Gail Kern. *Humoring the Body: Emotions and the Shakespearean Stage.* Chicago: University of Chicago Press, 2004.

Phayre, Thomas. *The regiment of life, whereunto is added a treatise of the pestilence, with the boke of children, newly corrected and enlarged by T[homas] Phayre.* London, 1546.

Plato. *The Dialogues of Plato.* Edited and translated by Benjamin Jowett. 5 vols. New York: Charles Scribner and Sons, 1907.

Polanyi, Michael. *Knowing and Being.* Chicago: University of Chicago Press, 1969.

———. *The Study of Man: The Lindsay Memorial Lectures.* Chicago: University of Chicago Press, 1959.

Rowland, William. *The New and Needful Treatise of Wind Offending Mans Body.* London, 1676.

Sawday, Jonathan. *Engines of the Imagination: Renaissance Culture and the Rise of the Machine.* London: Routledge, 2008.

Seneca. *Moral Letters to Lucilius.* Translated by Richard Mott Gummere. 3 vols. London: William Heinemann, 1919–25.

———. *Naturales quaestiones.* Translated by Thomas H. Corcoran. 2 vols. Cambridge: Cambridge University Press, 1972.

Shakespeare, William. *Hamlet.* The Arden Shakespeare. Edited by Harold Jenkins. London: Methuen, 1966.

———. *Hamlet.* The New Variorum Shakespeare. Edited by Horace Howard Furness. Philadelphia: J. B. Lippincott, 1877.

———. *The Norton Shakespeare.* 2nd ed. Edited by Stephen Greenblatt. New York: W. W. Norton, 2008.

———. *The Tempest.* Edited by Gerald Graff and James Phelan. Boston: Bedford / St. Martin's Press, 2000.

———. *The tragicall historie of Hamlet.* London, 1603.

Shapin, Stephen, and Simon Schaffer. *Leviathan and the Air Pump: Hobbes, Boyle, and the Experimental Life.* Princeton: Princeton University Press, 1986.

Smith, Bruce R. *The Acoustic World of Early Modern England: Attending to the O-Factor.* Chicago: University of Chicago Press, 1999.

Spiller, Elisabeth. "Situating Prospero's Art: Shakespeare and the Making of Early Modern Knowledge." *South Central Review* 26, no. 1 (2009): 24–41.

Stewart, Susan. *Poetry and the Fate of the Senses.* Chicago: University of Chicago Press, 2002.

Stoppard, Tom. *Rosencrantz and Guildenstern Are Dead.* New York: Grove Press, 1967.

Thayre, Thomas. *A treatise of the pestilence.* London, 1603.

Walker, Gabrielle. *An Ocean of Air: Why the Wind Blows and Other Mysteries of the Atmosphere.* New York: Houghton Mifflin Harcourt, 2007.

Walsham, Alexandra. "Sermons in the Sky: Apparitions in Early Modern Europe." *History Today* 51, no. 4 (2001): 56–63.

Wood, Christopher S. "Indoor-Outdoor: The Studio Around 1500." In *Inventions of the Studio, Renaissance to Romanticism*, ed. Michael Cole and Mary Pardo, 36–72. Chapel Hill: University of North Carolina Press, 2005.

Yates, Julian. *Error, Misuse, Failure: Object Lessons from the English Renaissance.* Minneapolis: University of Minnesota Press, 2002.

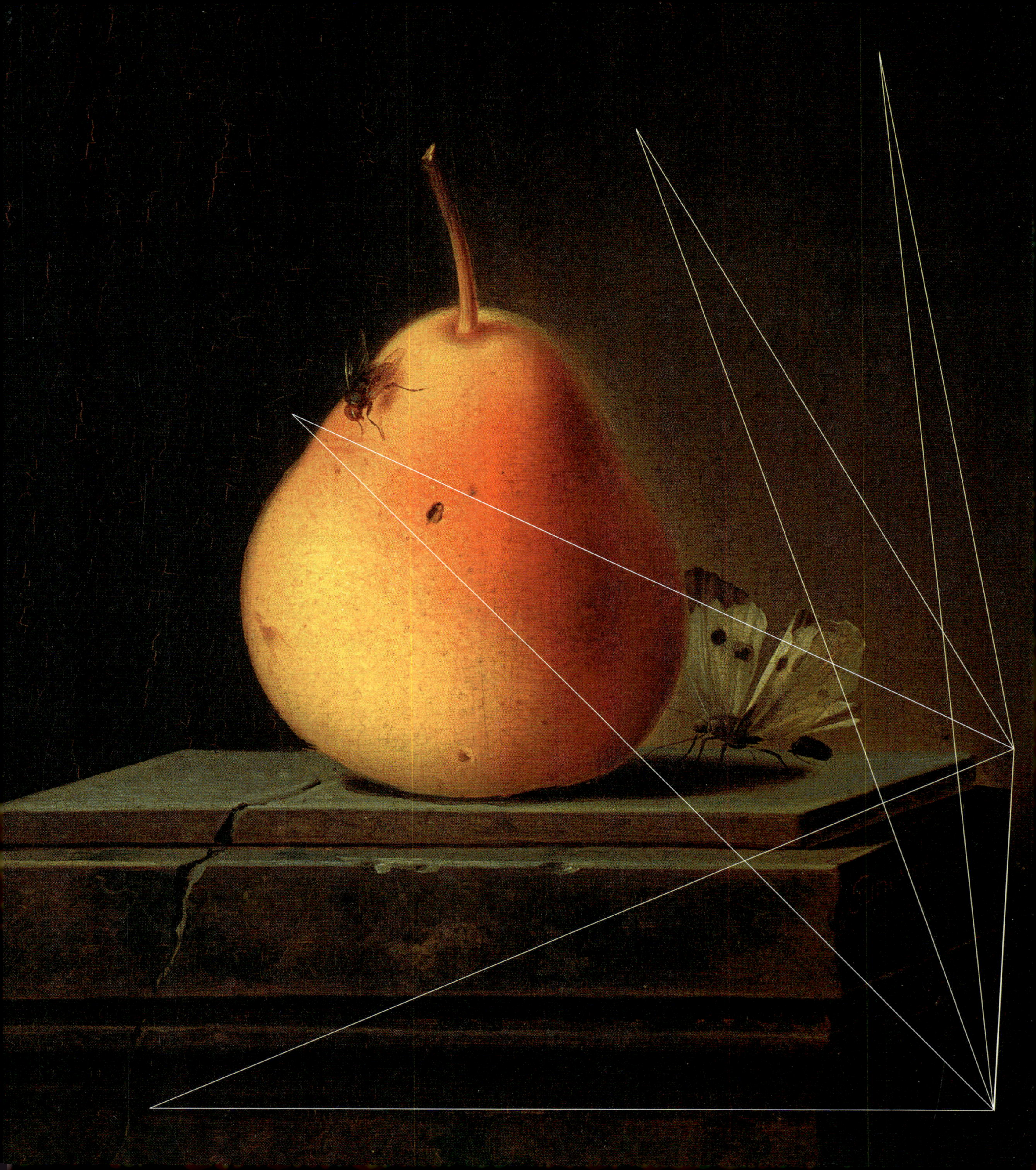

III.

THE PAINTER'S BRUSH AND THE MIND'S EYE

8.

"Art on the Tip of the Brush": A Blind Manœuvre?

Reflections on Correggio's Brush, Arent de Gelder's Spatula, and Pietro Testa's Figure of Practice

NICOLA SUTHOR

In his essay "Recueil des essaies des merveilles de la peinture," written in 1635, Pierre Lebrun argues that "poets have their inspiration in their heads, which is where their poetic nerves are located, and painters at the end of their fingers and on the tips of their paintbrushes."[1] The basic difference between the creative processes at work in poetry and painting, then, is that in the former the imaginative faculty is the seat of invention, while the *poesis* of painting is released in the act itself. While the writer's hand is not granted a role in the creative process, for the painter, fingers—and their extension, the paintbrush—are understood to stimulate the development of the pictorial ideas.

Lebrun's remark stands opposed to the predominant view articulated in early modern art theory according to which the creation of art is only a product of contemplation and thus purely spiritual. This attitude is reflected programmatically in Marcantonio Raimondi's engraving *Raphael in His Studio* (ca. 1520), showing the hero of academic art theory absorbed in the thought process required for creation (fig. 8.1): the painter sits in a contorted and crouched position on a step. His knees are turned to the right, whereas his slightly shadowed gaze faces the other direction. A second step serves as a shelf. On Raphael's left side, behind his back, stands a blank panel.

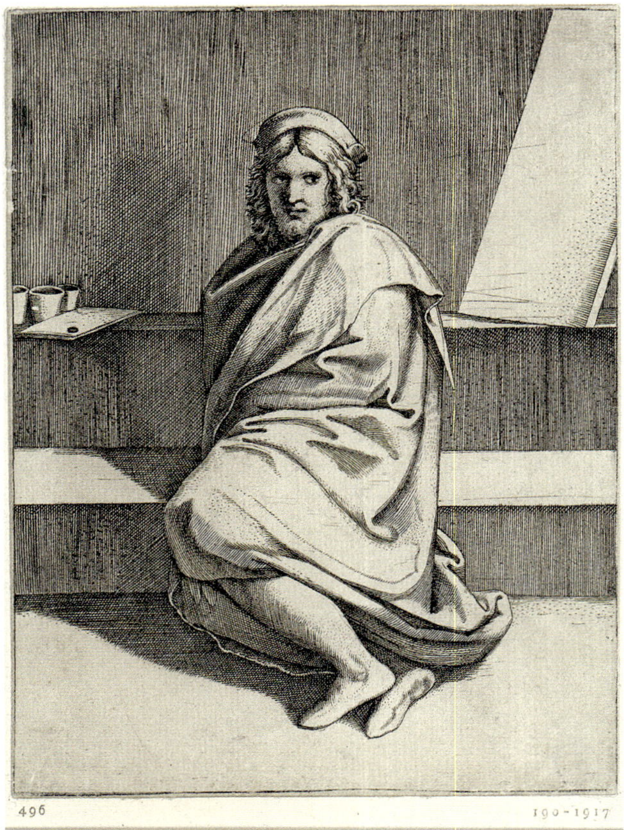

8.1 Marcantonio Raimondi, *Raphael in His Studio,* ca. 1520. Engraving, 13.9 x 10.7 cm. bpk, Berlin / Kupferstichkabinett, SMB / Volker-H. Schneider.

The painter's tools, palette and color cups, are placed on his right. The most stunning element of the engraving is the wide cape that completely covers Raphael's hands and arms and thus isolates him from his external surroundings.[2] The cape stresses the concentration of his gaze on, as he put it in a letter to Baldassare Castiglione, "a certain idea that comes to the mind,"[3] and gives at the same time the strange impression that he is freezing. The fire of enthusiasm with which the artist will take up his work seems not yet to burn.

I argue that in the course of the seventeenth century, an understanding of artistic creation as a purely inner process became increasingly contested. The opposing idea that the desired image emerges palpably during the process of drawing or painting—and not beforehand in the imagination—is tied to an acknowledgment of joy and curiosity as feelings experienced during artistic practice. Eighteenth-century academic art theory readopted the sixteenth century's critical perspective and condemned an art concept based on practice; but, while reactionary, that theory's stance is also insightful, furnishing us with an alternative viewpoint, thus giving the concept of art discussed here more plasticity.

Correggio, who in this respect served as a counterpart to Raphael in academic discourse, is the joyful artist par excellence, and for that reason was considered problematic. A first point we need to consider is thus Correggio's ambivalent reputation in art theory, especially in the eighteenth century. The critical reactions to his unique delight in rendering textures like skin and fur palpable are telling. They reveal a rather direct repression of sensual excitement, considered dangerous in many respects.

The academic reproach that this practice was superfluous because it involved a preoccupation with mimesis was neither new nor specific to Correggio, and this leads us to the next point we must consider. Commonly aimed at Dutch art, the reproach juxtaposes skillful mimesis with genuinely artistic invention. But the highly original application of paint to create stunning mimetic effects in the golden age of Dutch art must be understood in the broader context of seventeenth-century Dutch experimental culture. In this respect, it is no coincidence that the philosopher and central figure in the rise of rationalism and the Enlightenment, Descartes, although he emerged as a key reference point in French academic discourse, developed his revolutionary ideas in exile in Amsterdam. Against this backdrop, I conclude by considering the problematic role of the personification of practice in Pietro Testa's engraving *The School of Painting*, published the same year as Descartes's *Discourse on Method*. Here, the blind man's stick will bring us back to Correggio's idea of "having one's drawing on the tip of the paint brush."

Correggio's Self-Love

The idea not only that painting is executed corporally but that creativity is generated in a quasi-physical fashion reappears in the *Microcosmo della pittura* of 1657, written by Francesco Scannelli, artistic advisor to the Duke of Modena. For Scannelli, the human physical organism serves as a model for explicating the stylistic complexity of painting. In Scannelli's microcosm, Raphael forms the liver, Titian the heart, Veronese the genitals, Correggio the brain, and Michelangelo the spine, which supports the inner organs and guarantees the vitality of the whole. As the liver, Raphael draws the substance of knowledge from "Mother Antiquity" and extracts what is fine and tender in painting from the hard materials of antiquity's legacy in stone and bronze. Scannelli describes the heart, personified by Titian, as the origin of life, invigorating the spirits through the circulation of blood. He ties this physiological observation to a stylistic one in Titian's "bold and true style" (gagliarda, e vera maniera). Correggio's ability "to harmonize exaggeration in every area thanks to an unusual and subtle procedure" (con modo d'insolita, è più fina operatione ha saputo temperare per ogni parte i vitiosi eccessi) is equated with the brain.[4] Scannelli's analogy of the well-functioning human organism was thus offered in view of the productivity of the process of incorporation.

Interestingly, Scannelli was expressing a view of painting essentially the same as Pierre Lebrun's twenty years earlier. Scannelli informs us that Correggio describes himself as "having his drawing at the tip of his brush." This self-praise is preceded by the remark that Correggio—to whom Scannelli grants "the most exquisite excellence in painting"—detested the practice of drawing.[5] This cannot be intended to indicate that Correggio would have preferred not to practice drawing—after all, more than a hundred of his drawings are extant. Yet Scannelli's statement, written more than a century after Correggio's death, should not be understood simply as a rhetorical trope but as reflecting on a particular drafting technique that, as David Ekserdjian has observed, Correggio developed largely between 1522 and 1525, that is, on the eve of his most productive years.

Let us now turn from the rhetoric of art theory to a concrete object. In a preparatory study of the Annunciation that not only comes very close to the executed fresco (demounted and exhibited in the Galleria nazionale di Parma) but also already exhibits the *quadrettatura* for transferring the study to the larger format, we can see how Correggio used the brush tip (figs. 8.2 and 8.3).[6] The broad use of pen on pink-dyed paper appears to dissolve in the exorbitant light spilling over the archangel and Mary. Lead white highlights combine with the grayish-brown watercolor in order to create the shadow that spreads out over the drawing. Mary's cloak, draped over the right side of her upper body, is laid out over her lectern like a tablecloth. Pulled downward, it voluminously distends her kneeling leg so that it can connect with the cloud on which Gabriel appears. The thick application of lead white emphasizes the materiality of the depicted light stretched out behind her back. The figures, with the large heads so specific to Correggio, are subjected to the rule of light and seem to become its plaything. The dominant role of chiaroscuro in clarifying the form conforms to the image of Correggio's use of the tip of the brush as both brush and pencil. After all, chiaroscuro is a contested area, claimed sometimes by drawing, sometimes by color. In addition, the shadowy zone, which is also incorporated into the fresco to complete the contour of the archangel's profile, appears to result from a rethinking of the composition—it seems a late decision if we take account of the *quadrettatura*. An earlier study is hidden in this dark cloud: a roughly sketched face seen in approximately three-quarter profile, covered over with white lead and vanishing into the mist of light. In the corresponding entry in the catalogue for the Caravaggio exhibit of 2008, Mary Vaccaro was the first to point to the traces of rethinking in this drawing.[7] As she indicates, we also find an alternative positioning—again covered over with white lead—beneath Gabriel's left arm. The angel's profile is established solely through the thick application of white lead, with few markings (eyes, brows, nose, mouth) (fig. 8.4). A *putto* hides in the shadow of the billowing cloak; his frontal facial features are drawn on the pink base with a few brush strokes, and his arm is hardly materially distinguishable from the cloak's raised crease edge. A second *putto* joins him in the fresco, replacing a still nonconcrete white patch in the drawing. Likewise with the *putto* that braces the cloud from below: is his trace already present in a brush mark? A *putto* with lily appears at Gabriel's side; the Holy Spirit's dove, pouring warm light over Mary, will appear only when the picture takes on color in the fresco.

Scannelli distinguishes Correggio's manner of drawing from that of Raphael and Parmigianino, from whom we now have ten times more drawings. Raphael and Parmigianino, he indicates, developed their finished pictures with great care, "con longhe fatiche di studiosi esercitij," which is to say, on the basis of detailed preliminary studies executed "with extreme perfection."[8] For his part, Correggio simply captured "his inspired ideas" (pensieri accenati) in sketches executed very quickly, as Scannelli puts it, as if they were drawn out of himself.[9] In this respect, Correggio's metaphor of drawing "at the tip of the brush" is striking, aimed as it clearly was at the general expectation of painstaking preliminary drawing. The tip of the tool is the contact point where the pictorial idea takes material form. The traditional hierarchy—drawing as the paramount activity, which coloring is meant to complete in a second, subordinate step—is reversed here.

However, the experimental field opened up here to praxis was highly controversial—a critique of it already being inscribed, in fact, in Correggio's remark about the tip of the brush. Indeed, the remark, as recorded by Scannelli, is actually the direct quotation of another artist, who in Vasari's *vita* of Perino del Vaga reveals his self-satisfaction in precisely this manner. "What cartoons and yet not cartoons!" Girolamo da Treviso exclaims. "I, I have art at the tip of my brush." (Che cartoni e non cartoni! Io, io ho l'arte su la punta del pennello.) The artist was using the metaphor here to elevate himself over the intensive preliminary study pursued by his competitors, who were working with him in 1527–28 on the fresco of Casa Doria in Genoa. As Vasari records, having received word of this boast, Perino del Vaga responded by bringing his cartoons to an agreed-upon place. Girolamo now

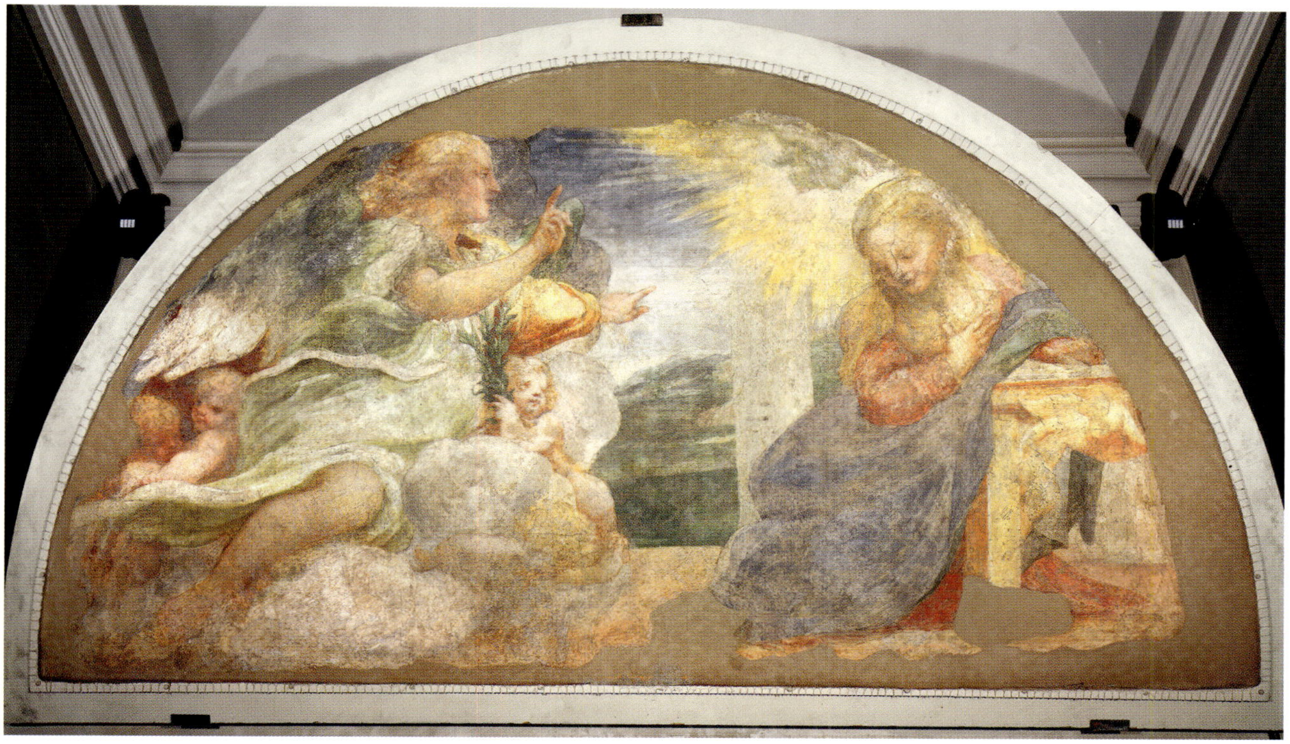

8.2 Correggio, *The Annunciation,* ca. 1525. Demounted fresco, 157 x 315 cm. Galleria nazionale di Parma. © 2013. Courtesy of the Ministero per i Beni e le Attività Culturali.

"took fright at the beauty of the cartoons," left his frescoes unfinished, and immediately fled Genoa.[10]

In one of his Royal Academy lectures, published in 1831, the Swiss painter Johann Heinrich Füssli cited Vasari and regretted that "the expression is dignified with the great name of Correggio." Doubting the authenticity of the quoted remark, Füssli considered the transmitted version a "lapse of memory."[11] He wished to disassociate Correggio from an idea of "design . . . at the point of the brush," which he connected with a basic lack of artistic quality. Füssli was not alone in offering this critique, which was directed against French rococo painting. In a conference held at the Académie royale de peinture et sculpture on 3 June 1747, Count Anne-Claude-Philippe de Caylus had tied the fixed expression "cela a été fait au bout de l'outil" (that was made at the tip of the tool)[12] to a "funeste facilité." In any case, Füssli's doubt that the statement was actually Correggio's strays from the interpretive paradigm offered by the article "Pinceau" in Watelet and Lévesque's *Dictionnaire des arts de peinture*, published in 1792. Here, Correggio's "charm" resides precisely in his use of the brush. Its author, one M. Robin,

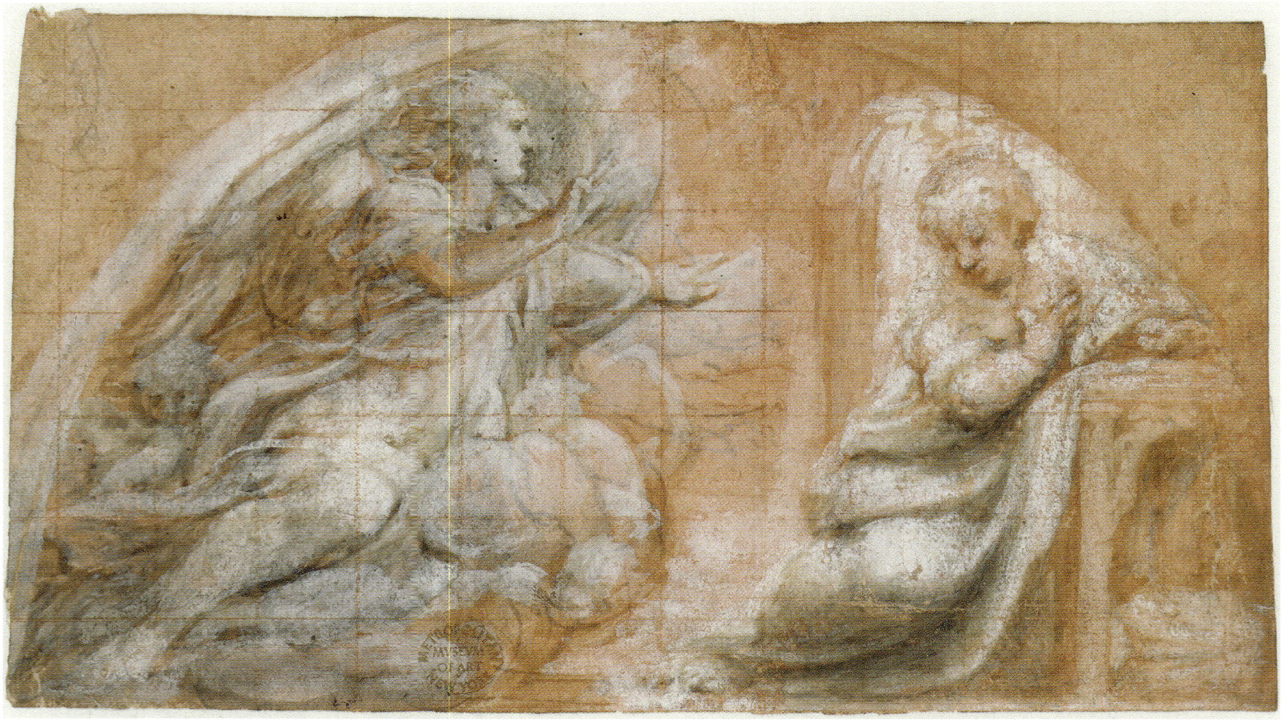

8.3 Correggio, *The Annunciation,* ca. 1522–25. Pen and brown ink, brush and gray-brown wash, highlighted with white gouache, squared in red chalk, on paper tinted with reddish wash, 9.5 x 17.2 cm. bpk, Berlin / Metropolitan Museum of Art, New York.

describes with acuity Correggio's joy in painting: "He took pleasure in applying the paint thickly and letting it flow [à empâter, à fondre], & he savoured the amusement *to lose himself & find himself again* in the color, neither making himself a slave of form nor becoming preoccupied with its choice. A *brush* so audacious, *so much in love,* so caressing, communicated to all those who viewed his works a pleasure & rapture which their author certainly knew how to enjoy."[13]

Charles-Antoine Coypel, director of the Académie royale, also tied this self-abandonment to the medium with Correggio—criticizing not his art but his example. Coypel was the painter who restored Correggio's heavily damaged *Leda*, replacing her head.[14] As is well known, the fanatically pious son of Philip of Orléans, Louis, had Leda's head cut out of the composition in 1731 and the rest of it cut into pieces.[15] According to Füssli, the *Io* suffered an even more "barbarous fate": before the "bigoted" Prince Louis tore that painting to shreds, we read, he burned Io's face[16]—a legend mostly ignored in the Correggio literature, and not without reason. A 1612 inventory for the Vienna Kunstkammer in fact lists the painting. It is very likely that the damage Füssli describes was in fact inflicted on a copy, as Cecil Gould argues. The *Io* was located, together with *Leda*, in Berlin and counts among works lost during the war. We can conclude that Prince Louis's iconoclastic act was a response to the spectator's "rapture," an emotion that, according to Füssli, was provoked by Correggio's sensual rendering of flesh and texture.

In an address to the academy's students, Coypel refers to the self-love that functions something like a defensive lie, in that "nothing is more devastating for a man born with feeling than to be forced to admit internally that he is merely mediocre in an art that suffers no mediocrity." Before Coypel juxtaposes two forms of amour propre, he warns his pupils against the complacent use of the instrument:

> You take a love for holding a pencil or paintbrush for genius, and in this point your parents deceive themselves even more than you do. Without consulting your spirit and your powers, and burning with a perilous ardor, you try to pursue the thorniest of careers. . . . If he [the misguided artist] is cold in the composition of his paintings and dry in their execution, self-love will persuade him that he has arrived at Raphael's sublime simplicity and perfection. If, to the contrary, he composes in an extravagant manner and draws incorrectly, self-love will bring him to compare himself with *Correggio*.[17]

The example of Correggio as an antipode to Raphael thus stands for a particular extravagance of praxis: an absence of orientation around any guidelines, for the sake of complete abandonment to a "love of wielding the brush."

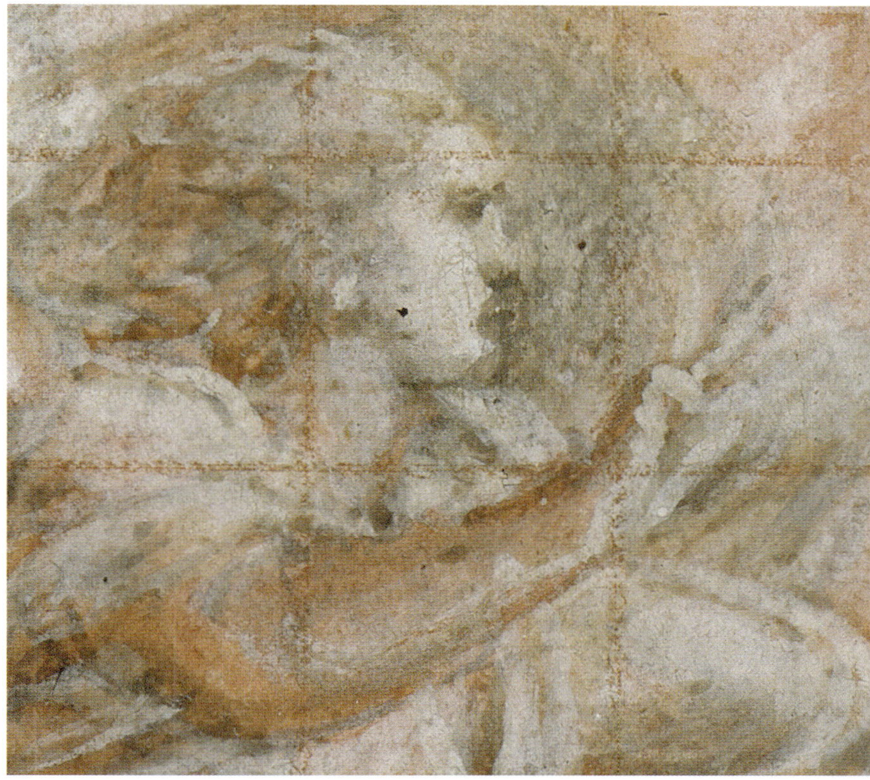

8.4 Correggio, *The Annunciation,* detail, ca. 1522–25. Pen and brown ink, brush and gray-brown wash, highlighted with white gouache, squared in red chalk, on paper tinted with reddish wash, 9.5 x 17.2 cm. bpk, Berlin / Metropolitan Museum of Art, New York.

The painter Anton Raphael Mengs also focused, critically but respectfully, on the incorrectness of the sort of drawing emerging from pure enthusiasm for one's own art. Mengs's much-quoted metaphorical description of Correggio's painting in his *Gedanken über die Schönheit und über den Geschmack in der Malerei* (*Thoughts Concerning Beauty and Taste in Painting*, published in 1762)[18] presents the scenario of a raging river that has overflowed its banks. Mengs objects here to chiaroscuro's excess, producing the flooding of the narrow boundaries of contour, as both seductive and misdirected.[19] Robin's view that Correggio, "neither making himself a slave of form nor becoming preoccupied with its choice," simply abandoned himself to feelings he could articulate through the brush's coloring undoubtedly follows Mengs's assessment. But what Robin approaches positively, in the sense of "nonchalance," sounds more problematic in Mengs's assessment that Correggio's "inventions are only produced through feeling, not contemplation; he wishes more to present his figures so that large quantities of light and shadow emerge than that they convey inner meaning—aside from a very lovely one, which his feeling often reveals."[20]

This perceived absence of contemplation, in favor of an expansion of feeling, and in the interest of formal innovation, led Mengs to qualify his judgment of Correggio as an exemplary painter: "where agreeableness can be introduced to good effect, Correggio should be imitated; but where meaning is called for, he is pernicious."[21] Later art historians have revised this view of Correggio, pointing to the extensive literary and pictorial sources on which he drew in his art. Erwin Panofsky and, more recently, both Marcin Fabiánski and Stephen Campbell have confirmed Correggio's abundant application of iconographic elements.[22] However, Fabiánski concluded that the pictorial achievement is less the illustration of a text than "a coquettish and natural representation of feelings and the senses" (intesa come vezzosa e naturale rappresentazione dei sentimenti i dei sensi), offering the viewer "pleasure full of satisfactions" and "blissful embarrassment" (lasciandoci un piacere colmo di appagamenti, e . . . un felice imbarazzo). In doing so, Fabiánski supports Mengs's assessment of Correggio. Fabiánski's choice of words speaks volumes.[23]

In 1898, in *Das Werden des Barock bei Raphael und Correggio*, Josef Strzygowski reformulated—albeit indirectly—Mengs's fundamental critique of Correggio, speaking of the painter's "naïve" and "sweetish" or "mawkish" (süßlich) self-abandonment (Sichgehenlassen). The blurring of "feeling and consciousness" in Correggio's work, Strzygowski observed, undoubtedly contributes to its particular eroticism. Of *Leda*, *Danae*, and *Io*, all reflecting the "authentic" Correggio, Strzygowski commented, "at their sight even the impartial observer feels as if moving on a thin layer of ice."[24]

The proliferation of meaning on the level of sensory impression is realized through exuberant mimesis. Such sensory density is also present in the skin, hide, and fire in Correggio's *Venus and Cupid with a Satyr*; it illustrates a "thinking on the tip of the brush" that is also apparent in the "formal tumescence" of the figures (fig. 8.5). In this painting, the overly large behind of Cupid that stretches toward the viewer cushions, as it were, Correggio's female nude. The emotional impact is kindled by the convergence of two bodies. Here, the sensual effect leapfrogs. The encroaching quality is based on the composition's paratactic dimension. Proximities with irritating significance are generated between the individual motifs and textures on the pictorial surface. The evoked reality thus functions less on the level of the sign than on that of a sensual phenomenality extending past the bodily borders, which are concealed in the *sfumato*. It is formed less in a conceptual and more in a pictorial manner, as Vasari pointed out in his praise for the artist. Vasari defines Correggio's abilities as a painter through his innovatively sensual way of depicting hair: "Where previously it was painted in a strenuous, hard, and dry way, with Correggio it seemed soft and flowing."[25] For Vasari, this epochal

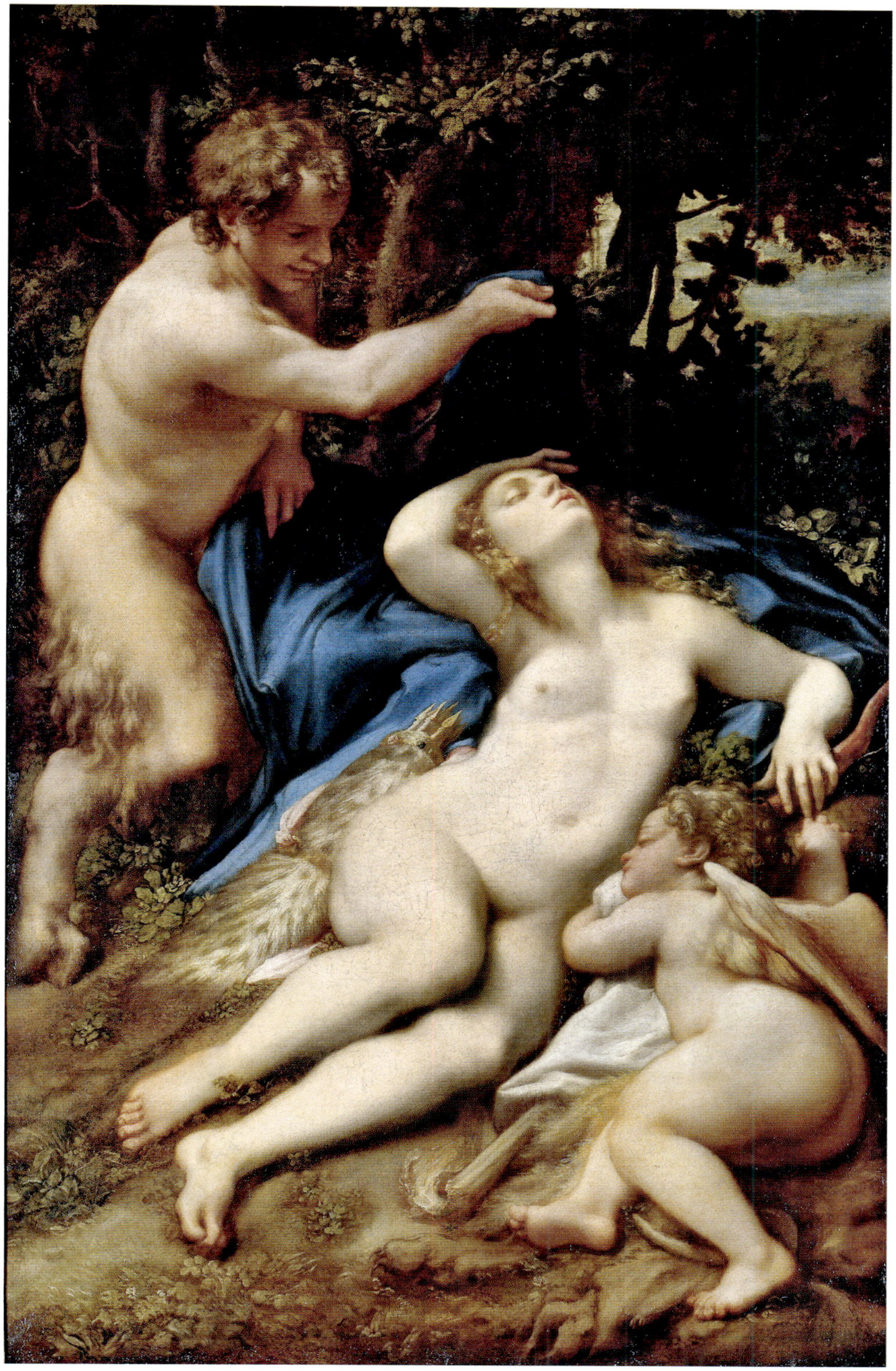

8.5 Correggio, *Venus and Cupid with a Satyr,* ca. 1528. Oil on canvas, 185.5 x 125.5 cm. Louvre, Paris. © 2013 White Images / Scala, Florence.

innovation in painting was tied to the dissolution of the hard style of the quattrocento. It was not the product of theoretical reflection preceding the act of painting and independent of it, but rather emerged from experimentation with the medium of oil paint.

In Correggio's work, "thinking on the tip of the brush" reveals itself through the mimetic force of the pictorial arrangement, which thus contributes decisively to his illustrative dynamic. Caylus did not consider Correggio a slave to form, but should he be seen as a slave to his brush?[26] The enslavement spoken of here is not the same as the supportive use of instruments criticized by Füssli, who writes, "for though neither he who anxiously forms lines without the power of embodying them, nor he who floats loosely on masses of colour, can be said to design, this being merely the slave of the brush, that of a point, yet both tools may serve alternately or indiscriminately the purposes of the real designer."[27]

Füssli points to the transgressive moment when painterly practice overruns the task of execution and creates pleasure for its own sake. The development of a praxis-based specialized skill allows a fatal shift of *ars* into *métier.*

Arent de Gelder's Handcraft

In French art theory, this approach, grounded in an instrumental application of artistic means, is captured in the expressive term *manœuvre*, which encompasses artistic praxis but not necessarily art, since it designates "handwork" (*manus-opera*) but not understanding. The unnamed author of the short article on *manœuvre* in Watelet and Lévesque's *Dictionnaire* (1792) thus describes it as denoting "the manner of forming colors and mixing their substance, using the brush, & style of the stroke. These details constitute the essentials of the vocation of painting, but the essential qualities of art are all spiritual. . . . With much intelligence, success can be achieved by bizarre means; but the singularity of the talent & not that of the process is what truly distinguishes the great artist."[28] To illustrate his basic argument, Lévesque focuses in his *Dictionnaire* article on *main* (the hand) in the artistry of the Dutch painter Cornelis Ketel, flirting as it does with the production process's sensational qualities. In his article on *main,* he criticizes Ketel's tour de force of painting without a brush as an effort, motivated by ambition, to demonstrate genius;[29] in the article on *manœuvre*, the unnamed author likewise categorizes it as a *manœuvre bizarre* through which the artist hopes to distinguish himself from others.

Despite its own strong critique of the principle of chance in artistic creation, the literature on art is full of descriptions of chance-centered malpractice aimed at producing art in new and surprising ways. An important and often cited source is Karel van Mander, who wrote that Cornelis Ketel had an "inventive, productive spirit" and appeared "to have allowed himself no rest." He felt "all sorts of impulses and occupied himself with various efforts in the realm of painting." In 1599 he began to paint with his fingers, which, we are told, "by many is held to be a ridiculous abominable urge such as sometimes happen with pregnant women who crave to eat strange, raw, or uncooked food."[30]

In order to demonstrate his uniqueness, Ketel introduced his own corporeality as an obstacle, in this way evading conventional praxis and expressing self-will in a virtuoso manner. Nicolas Galley interprets this passage as equating artistic creation with procreation, a reference to the power of divine creation at play.[31] But without a doubt this comparison is meant ironically. Van Mander underscored the general extravagance of this experimental praxis: "For many artists carry out such unusual things in order to show their skill. Some for instance shoot rifles laid on their backs or held inversely and hit what they've aimed at, not to speak of those who run along a tightrope when the earth is more suitable."[32] In ridiculing the figure of the tightrope walker, Van Mander turned firmly against the sort of notion of artistic genius proposed by Pliny through his own use of the same figure.[33] In his famous letter to Lupercus the Rhetorician, Pliny encourages him

> to treat the edge of a precipice; for a path along the heights and peaks often skirts the sheer drop below. . . . A runner risks more falls than a man who keeps to a snail's pace, but he wins praise in spite of a stumble, whereas there is no credit in walking without a fall. . . . You have seen tightrope walkers and the applause they win as they move along the length of the rope and every minute look as though they are going to fall; for it is the most unexpected and dangerous feats which win most admiration.[34]

Van Mander took up the image of the tightrope walker in order to remove the nimbus of the tragic from failure in the face of the sublime. The claim to produce what is new and unexpected is consigned to ridiculousness through the absurdity of the reflection.

In the *Dictionnaire* article on *manœuvre*, Cornelis Ketel is tied to the Rembrandt student Arent de Gelder on the basis of his own unconventional application of paint: with the thumb or palette knife. De Gelder, we are informed, inverted the brush shaft's standard function, using it to create "unique effects"; special mention is made of the almost relief-like plasticity of the paint in de Gelder's treatment of fringes and embroidery.[35] The technical information in this respect itself draws on Arnold Houbraken's *De Groote Schouburgh* (1718–19), which offers the following details: "At times, when for example he wants to paint a dress's fringe or edging, he even daubs the paint with a broad spatula onto his canvas or wood, and then scrapes out the indentation in the trimming or the threads of the fringe with his brush shaft, rejecting no means that can lead him to his goal."[36] Thus the unusual use of instruments—each and every means was appropriate—opened previously untested ways to realize a picture.

According to Houbraken, the experimental painter's joy in technique is motivated by a steady search for the new. Correspondingly, he introduces his entry on Arent de Gelder with the observation that "garments, furniture, and such things are subjected to fashion and change: and if someone precedes here, others will follow immediately. That is the same thing with painting."[37]

The removal of paint allows a coloristic interplay between the upper and lower paint layers. The crossing and interlacing of the layers creates a density of pictorial surfaces. Undoubtedly, the scraping is a "parabola," aimed at admiration in Pliny's sense. Playing with chance, risking smearing, this practice consciously deviates from the path according to which the picture's completion must hover before the spirit *before* the first stroke is completed. It thus renders the productive force of painting's medium transparent, because a clear idea of this result cannot be present during the painting's execution. The painting process delivers the picture. It determines the course things take on the canvas, revealing itself materially in

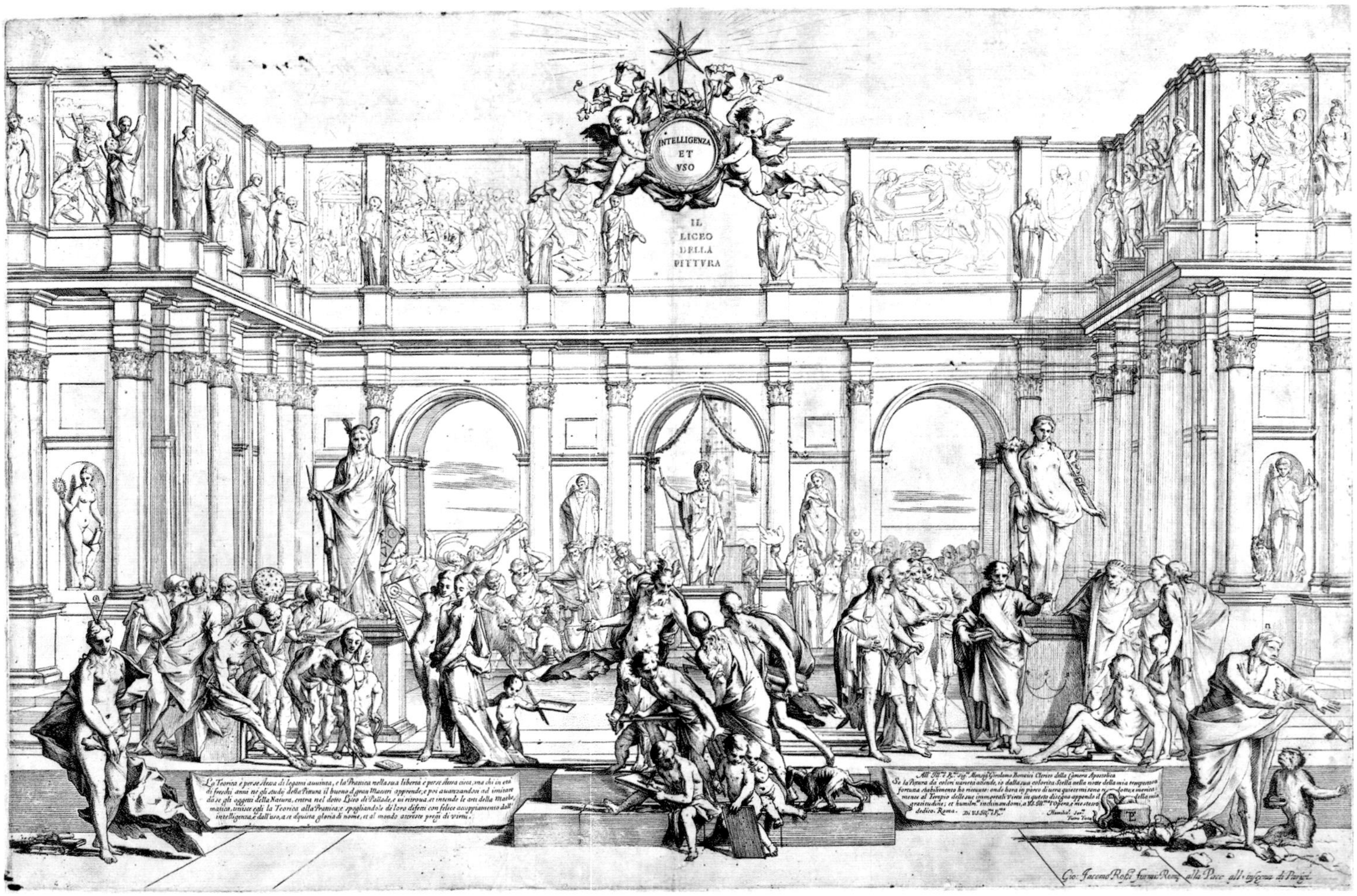

8.6 Pietro Testa, *Il Liceo della Pittura,* ca. 1637–38. Engraving, 47.9 x 73.7 cm. The British Museum. © Trustees of the British Museum.

the flow of the colors. Arent de Gelder's practice of scraping into the wet material ends where the paint dries. The praxis surrenders itself to the possibilities of a technique, possibilities that unfold only at the point of the technique's application; in the end, praxis enters realms that it does not foresee from the start but learns to master during the execution itself.

What appears to emerge arbitrarily beneath one's hand challenges the intellect in turn; hand and understanding become related in a complex way. Hand and brain work together in the development of technique, thus fusing the manual with the conceptual. Crucially, the basis of the acquired proficiency is not praxis alone but also involves "speculation and science."[38] But the role of speculation in the mastering of experimental techniques has always been called into question, for it is hard to regulate. Often, its visual results have been denigrated as products of chance that owe their existence to the space of praxis alone.

Pietro Testa's Figure of Practice

To work only from *practica* means, as the seventeenth-century German art theorist and painter Joachim von Sandrart saw it, to work "blindly *all'aventure*, without rules," the activity of painting thus being delivered to a "blind being full of uncertainty."[39] Here, a putative absence of rules in practical execution and a sense of the creative spirit's self-sufficiency meet in the form of an ignorant faith that the source of everything a painter needs can be found in oneself alone. At this point, the metaphor of thinking on the tip of the brush appears to have come dangerously close to the image of the blind man with his cane.

Praxis is in fact personified as a blind woman in an engraving titled *Il Liceo della Pittura* (created ca. 1637–38) by Sandrart's contemporary Pietro Testa (fig. 8.6).[40] Testa himself can be considered a critic of "dißegni nella stremità de'Pennelli." In his notes, he frequently speaks against ease in creating (facilità in far le cose), which he excoriates as a "mad fury" (furor matto). In this important engraving, conceptually illustrating the ideal academy of painting, we find, located on the lower right edge, across from Theoria, Practica trying with outstretched arms to find an exit from the lyceum, spread out in panorama behind her back. At its vanishing point, the sculpture of Minerva rises on a pedestal, flanked by pagan and Christian dignitaries. In the middle ground, representatives of practical philosophy populate the right side, while the figures on the left devote themselves to mathematical speculation. Directly beneath Minerva on the threshold of the stage of knowledge, a group of pupils has formed; they are directed by the personification of judgment to follow the personification of perspective. This fully coheres with a remark in Testa's sketchbook: "Those arts that take more from mathematics are the more noble because they produce more authentic reason."[41] Elevated above everything else hovers a motto held by two *putti*, "Intelligenza et Uso." The interconnection of understanding and practice is manifest in a group bent over a demonstration behind

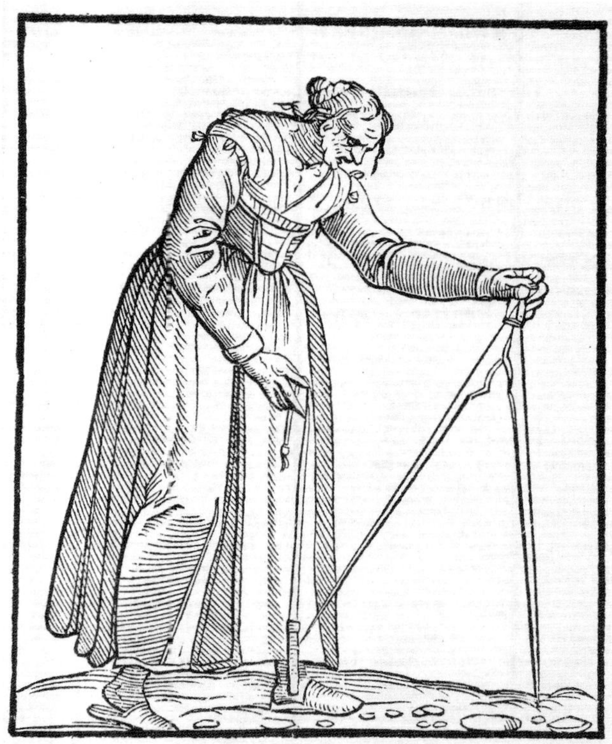

Cesare Ripa, "Practica." In Cesare Ripa, *Iconologia* (1618), ed. Piero Buscaroli, 2 vols. (Torino: Fògola, 1991), 2:285. 8.7

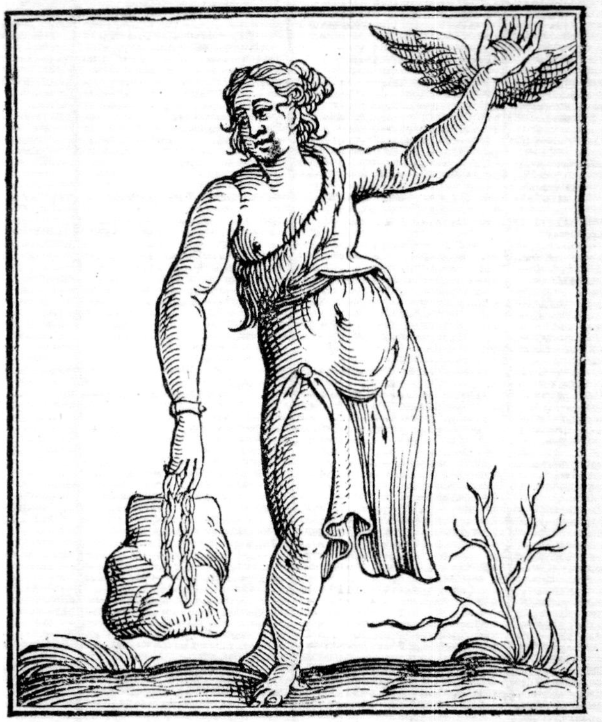

8.8 Cesare Ripa, "Povertà." In Cesare Ripa, *Iconologia* (1618), ed. Piero Buscaroli, 2 vols. (Torino: Fògola, 1991), 2:123.

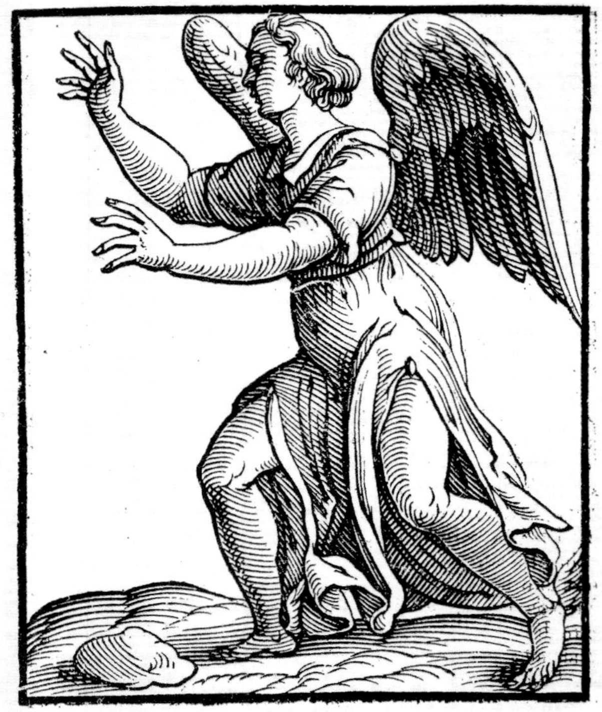

8.9 Cesare Ripa, "Volontà." In Cesare Ripa, *Iconologia* (1618), ed. Piero Buscaroli, 2 vols. (Torino: Fògola, 1991), 2:249.

Theoria, while behind Practica all attention is fixed on the raised hand of Aristotle, who is giving a lecture. The eloquent gesture—one of many references to Raphael's *School of Athens*—begets a relation with the extended hand of the old blind woman. But where the philosopher holds a stack of books in his right hand, the winged right hand of Practica holds a blind man's staff, which probes not the ground but empty space. Testa's depiction is unusual in its redefinition of practice as a result of the mixing of images offered by Cesare Ripa's *Iconologia.* On the one hand, Testa orients himself around Ripa's Practica (fig. 8.7), which is likewise depicted as an old woman. But Ripa's personification bends over a large compass, itself attached to a plumb line, rather like Testa's mathematicians. When Testa exchanges the compass for a blind man's staff and at the same time removes the figure from a fixation on the ground, then—as Elisabeth Cropper has demonstrated—he is evoking another of Ripa's personifications, the one illustrating the impoverishment of genius in the isolated execution of praxis.[42]

The personification of poverty experienced by the one "who possesses beautiful *ingenium* [alone]" (Povertà, in uno c'habbia bello ingegno) is presented in the wings on the "inactive" left hand, while the right hand is chained to a heavy stone that draws the blind figure toward the ground (fig. 8.8). The "ingenious" painter's desire, a desire that according to Ripa's explanatory text is handicapped, indeed foreclosed, by Povertà, is made visible in the winged hand extended outward. The blindness of Testa's Practica links that figure to Ripa's personification of will as itself blind (fig. 8.9): "She extends her hands forward, one more than the other, in order to hold fast to something. Blindness suits her, for in principle she sees nothing but gropes behind the sensual desires, if she is weak and ignorant, or behind reason, if she is bold and worthy."[43] Ripa himself furnishes will (Volontà) with wings (on feet and shoulders), referring to the inscription of flight (volo) within the Italian word.[44]

Where Ripa's "Povertà, in uno c'habbia bello ingegno," is internally torn and incapable of action, Testa's Practica pursues her will. In contrast to Povertà, her right hand is winged and gropes ahead, while the assisting left hand creates space. Where Ripa has the heavy stone in the left hand striving against the winged nature of the right, Testa has tossed the weight away so that it forms a stony, thorny path, which signifies proceeding without the foundation of theory. An ape—symbol of the slavish imitation of nature—shows her the way to departure from the stage of knowledge. The beggar's bowl in the ape's right paw may illustrate the scraping together of models, something no less spiritless than slavish imitation.

Where the hands of Praxis grope forward into the unknown, those of Theory are bound, her gaze lowered, while the compass's pivot rests on her parted hair. Instead of stones before her, there are books behind her, piled up in a shadow. On the base of the lyceum's stage, we can read the following explanation: "Theory by herself is chained to connections, practice by herself is blind in her freedom."[45] The distance that visibly opens up here between theoretical knowledge and practical ability is bridged by the lyceum.

Testa is depicting the long-standing expectation that guided the early modern understanding of art. The "freedom of the hand" illustrated by the wings in Testa's engraving is in itself problematic, standing opposed, in the end, to art as an activity steered by the understanding. It is already found in the passage on aesthetics that is part of Albrecht Dürer's treatise on human proportion (1528): "For the understanding must begin to grow together with practice, so that the hand can do what the will would have within the understanding. With time, the sureness of art and practice emerges from this. For these two things must exist together."[46] Dürer himself had sharp words for the blindness of a praxis removed from theory: works emerging "from a free hand's constant practice," he indicated, are "made forcefully but without reflection and according to their pleasure alone." For artists, succumbing to the freedom achieved by "practice"—a freedom releasing them from the dominance of understanding—has fatal results. "But when you have no proper [theoretical] grounding, then it is not possible to make something proper and good, and this despite having access to the greatest practice in the world through your hand's freedom. For although it seduces you, it is more a prison. Thus just as there should be no freedom without art, art remains hidden without practice. Thus the two must exist together, as said above."[47]

But the problem of art lies precisely in the mediation of its science, as Alberti explained in *De pictura* (1435): "To be sure, very often both the ambitious and the possessed are discouraged almost more by not knowing how to proceed with learning than by the effort caused by learning itself."[48] Alberti addresses the fundamental problem of artistic instruction here, especially instruction in painting: the basic perplexity involved in deciding how to begin. Leonardo da Vinci also reflected on the artist's lack of orientation. But there is a basic difference with Alberti, who describes the artist's training as a path—thus suggesting a method (we should note the meaning of *meta-hodos* here as "way through"). For Leonardo, by contrast, painting represents a sea for the crossing of which one must be properly equipped: "Those who become enamored of a praxis without science are like helmsmen who board a ship without helm or compass and are never sure where they are sailing."[49] The techniques that furnish orientation—helm and compass—provide both direction and evaluation of the accuracy of that direction in the wider context of the sea and stars. If either determining and pursuing one's goals or control over compliance is lacking, then the artist will have embarked on an unknown path. For Alberti, the path inevitably leads to despair: with the goal no longer transparent, a disorientation sets in that can be described as blindness. "While wandering around in the dark, fearful and as it were blinded," Alberti writes, these disoriented artists "test out unknown paths with the brush like a blind man with his cane, seeking despairingly for an exit."[50]

In the same year as Testa's engraving, a text appeared in which the image of the blind man also played a central role: René Descartes's *Discourse on Method* (1637).[51] Descartes used an illustration of a blind man in his section on dioptrics. Like Testa's blind woman, the man holds his staff in the air in order to measure the space before him (fig. 8.10). Rather than representing an image of pity or alarm, Descartes's blind man was meant to illustrate the process of seeing. The illustration has been repeatedly criticized in the literature on Descartes: it is, we are informed, inherently inconsistent and thus unsound to illustrate what is described in the *Dioptrique*. However valid the criticism, it is clear that for Descartes the image of the blind man was of basic significance for understanding experience and cognition.

Tied to Descartes's model of the blind man are deliberations assembled in the *Discourse on Method*. The risk of falling, which, we will recall, endows the tightrope walker with an aura of the sublime, according to Pliny, also accompanies the blind man's passage on firm earth. Descartes calls for just such uncertainty when he writes, "But like one who walks alone and in the twilight I resolved to go so slowly, and to use so much circumspection in all things, that if my advance was but very small, at least I guarded myself well from falling. I did not wish to set about the final rejection of any single opinion which might formerly have crept into my beliefs without having been introduced there by means of Reason."[52] The danger of falling provokes the intention to undertake "the conduct of my own life." Descartes even goes so far as to make the man gone astray from his path into a model: in the end, he does his best to save himself through blind determination:

> My second maxim was that of being as firm and resolute in my actions as I could be, and not to follow less faithfully opinions the most dubious, when my mind was once made up regarding them, than if these had been beyond doubt. In this I should be following the example of travellers, who, finding themselves lost in a forest, know that they ought not to wander first to

> one side and then to the other, nor, still less, to stop in one place, but understand that they should continue to walk as straight as they can in one direction, not diverging for any slight reason, even though it was possibly chance alone that first determined them in their choice. By this means if they do not go exactly where they wish, they will at least arrive somewhere at the end, where probably they will be better off than in the middle of the forest.[53]

Doubtful views and chance describe a precarious initial situation that nonetheless, and necessarily, gains orientation through execution. Where Testa maintained that the orientation on models could generate certainty, Descartes proclaims the blind man's credo: you must draw everything out of yourself. This is the basis for the ideology of creative originality. As Claus Zittel has recently argued, with this choice of terms the philosopher's *Discours* revealed itself as a method of praxis. In this respect the title, *Discours de la méthode*, is itself revealing. To his friend Mersenne, Descartes emphasized that his text had intentionally not been planned as a treatise. As Zittel has noted, the process of running or wandering around—*dis-correre*—unmistakably resonates in the term *Discours*.[54] Albeit critically, Alberti had already recognized this as a condition for discovering "unknown paths."

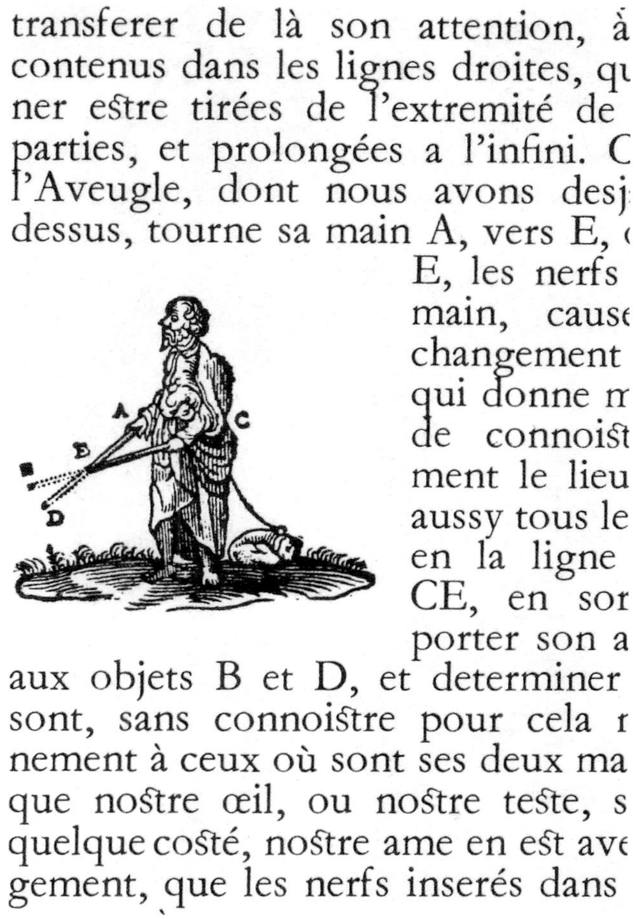
transferer de là son attention, à
contenus dans les lignes droites, qu
ner eſtre tirées de l'extremité de
parties, et prolongées a l'infini. C
l'Aveugle, dont nous avons desj
dessus, tourne sa main A, vers E,
E, les nerfs
main, cause
changement
qui donne m
de connoiſt
ment le lieu
aussy tous le
en la ligne
CE, en sor
porter son a
aux objets B et D, et determiner
sont, sans connoiſtre pour cela r
nement à ceux où sont ses deux ma
que noſtre œil, ou noſtre teſte, s
quelque coſté, noſtre ame en eſt ave
gement, que les nerfs inserés dans

8.10 René Descartes, *Discours de la méthode: Pour bien conduire sa raison, et chercher la verité dans les sciences plus la Dioptrique, les meteores et la geometrie qui sont des essais de cete* [*sic*] *methode*, ed. Jean-Robert Armogathe and Vincent Carraud (Paris: Fayard, 1986), 121.

I would like to conclude this chapter by briefly considering a painting by Joshua Reynolds, who, it seems to me, brings all these points to a head. In a surprising manner, Reynolds's self-mirroring as a young painter who at the time had not yet made his Italian journey reflects Alberti's image of a despairing, orientation-seeking artist (fig. 8.11). Light streams onto his face, its soft features showing his youth, especially in the lower portion. His left hand is raised over his eyes to protect them from the light. A strained and concentrated gaze emerges from the resulting shadow. But does it actually meet the viewer's gaze? The darkening of the eye area makes the gaze elusive. In any event, it seems perhaps slightly askew, focused

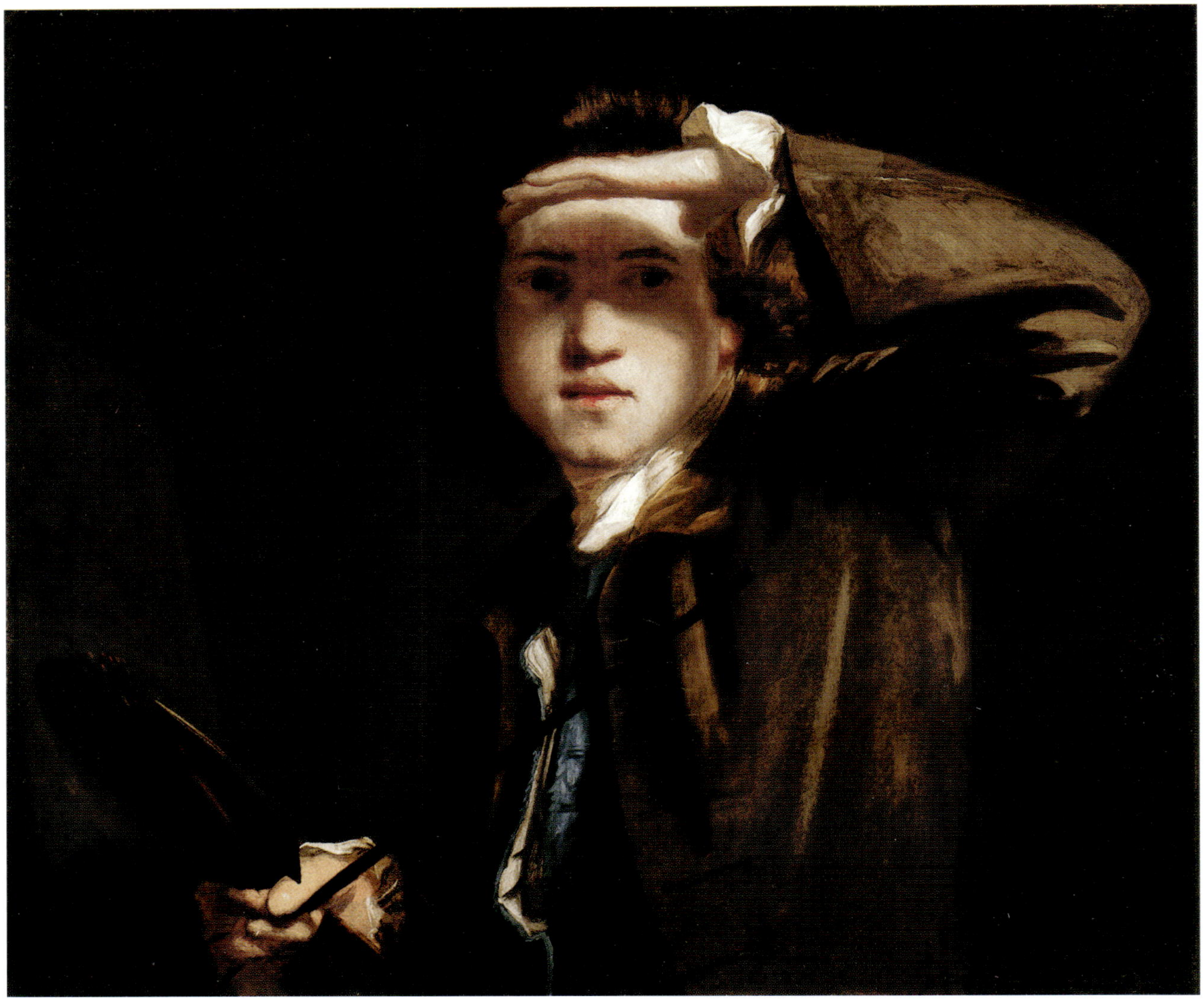

8.11 Joshua Reynolds, *Self-Portrait,* 1747–49. Oil on canvas, 63.5 x 74.3 cm. National Portrait Gallery, London. © National Portrait Gallery, London.

on nothingness—wandering blindly. What is it actually taking in? The gaze seeking a foothold is supported by an instrument—a maulstick—whose actual function is to support the painting hand. The maulstick protrudes beyond the picture's imaginary border but is spatially retrieved through an abandonment of foreshortening; planimetrically considered, it traverses several shadowy areas of the coat, their deep black appearing to swallow the stick's contours repeatedly. Its tip points in the direction of the shadowed shoulder area, which skillfully obscures the complex spatial relation between shoulder and upper arm. It seems as if a strained attention would turn to precisely this difficult place in the imaginary mirror as the only plausible object of focus.

While the right hand alone is engaged with the painting instruments—in addition to the maulstick, his palette and brush, the latter receding into darkness, thus becoming impalpable—the artist's left hand hovers over his forehead, seeming, as it were, to measure its breadth. Overlapping with the hairline, it screens the middle of his forehead. The back of the hand and upper forehead are illuminated to the same degree: *mens* and *manus* draw together. Thus, where on the painting's lower edge the right hand is preoccupied with the painter's tools, the left hand is raised to offer a gesture that in pictorial rhetoric connotes astonishment (*thauma*).[55] But the ecstatic moment of being overwhelmed in the face of witnessing a miracle is not the moment at which Reynolds is aiming. Rather, his "astonishment" is directed at his own mirror image, which—as a vision of himself—reflects the young painter's ambition. His artistic future, which appears to open itself to him in the self-reflection, is narrowly tied to the moment of the gaze's channeling. The eyes' shadowing seems to be the precondition for their encountering something, sharpening the gaze while immediately removing it from view. The positioning of the artist's left hand before his face and above his eyes, covering the high forehead, endows it with a central, eloquent role in the painting's structure: the hand dominates both eyes and understanding. The path that praxis opens for reflection is still one needing determination. Alberti's description of the blind man, using his brush like a staff to feel out "unknown paths," offers potential for the self-mirroring of a young painter. In Reynolds's self-portrait, the seemingly blind groping of praxis receives a positive turn through the painter's astonished pose, evoking an initial moment: *thaumazein*.[56] Wonder or astonishment does not stand at the start of philosophy alone.

Notes

1. "Les poetes ont leurs inspirations dans la teste ou est la nerve poëtique, et les peintres au fin bout des doigts et à la pointe scarante du pinceau." Lebrun, "Recueuil des essaies des merveilles de la peinture," 767. All translations are by Joel Golb unless otherwise noted.

2. See Horký, *Künstlerbilder—Künstlermythen*, 38.

3. "Io mi levo col pensiero più alto. . . . Io mi servo di certa idea che mi viene nella mente." Bottari and Ticozzi, *Raccolta di lettere sulla pittura*, 1:116–17.

4. Scannelli, *Microcosmo della pittura*, 9–18. Although entirely original, Scannelli's trope for the sensory process at work in acts of creation is at the same time commonplace. It has roots in antiquity and is tied to the bee analogy widespread in the early modern period. In Seneca's famous eighty-fourth letter to his pupil Lucilius, following a critical description of this analogy, we are told that "in the food we have eaten, as long as it retains its original quality and floats, in our stomachs as an undiluted mass, is a burden; but it passes into tissue and blood only when it has been changed from its original form. So it is with the food which nourishes our higher nature,—we should see to it that whatever we have absorbed should not be allowed to remain unchanged, or it will be no part of us. We must digest it; otherwise it will merely enter the memory and not the reasoning power." Seneca, *Moral Letters to Lucilius*, vol. 2, letter 84, lines 6–7. The subsequent appeal of rendering models unidentifiable, the appeal, rather, of concealing them, reveals the real target of the metaphorical description: it propagates an art of forgetting *within* the appropriation of models, in order to allow one to develop one's own new material. Thus the digestion of knowledge does not contain its elimination—no smashing of received images is necessary in the production of a reinscribable tabula rasa.

5. "E se in questi in riguardo del particolare incaminamento prevalse l'idea artificiosa fabbricata col tempo, e continuato studio; appare dall' altra parte nel Pittore da Correggio in eccesso il talento della naturale inclinatione, mediante la quale compose in breve, come da se stesso, e più tosto con modo divino, che naturale la più esquisita eccellenza della Pittura; ond' egli abhorrendo l'esercitio del

dissegnare, richiesto più volte in occorenza dell'opere de gli anticipati dissegni, nella maniera, che dicevano, esser consueti i mentoati Rafaello, e Parmegianino, era solito rispondere, *c'haveva I suoi dißegni nella stremità de' Pennelli.*" Scannelli, *Microcosmo della pittura*, 359.

6. Ekserdjian described the specific technique (mixing charcoal, red chalk, white lead, and ink) used in the preparatory drawing of the Annunciation as "pittoresque" and emphasized its innovativeness: "Questa tecnica mista è senza veri precedenti." See his essay "Correggio Disegnatore," 80.

7. See Vaccaro's entry on the *Annunciazione* in Schianchi, *Correggio*, 405.

8. Scannelli, *Microcosmo della pittura*, 359.

9. "Appare nel Pittore da Correggio in eccesso il talento della naturale inclinatione, mediante laquale compose in breve, come da se stesso, e piu tosto con modo divino, che naturale la più esquisita eccellenza della Pittura." Ibid. Scannelli's view is still shared in modern art history. See Charles Dempsey's assessment: "Correggio worked with divine grace and not a little learning, from a fundamental empirical, not fully rationalized base." Dempsey, *Annibale Carracci*, 29.

10. "Mentre, dico, che Perino stava così e non cominciava, Girolamo da Trevisi mormorava di lui, dicendo: 'Che cartoni e non cartoni! Io, io ho l'arte su la punta del pennello.' E sparlando più volte in questa o simil maniera, pervenne agli orecchi di Perino; il quale presone sdegno, sùbito fece conficcare nella volta, dove aveva ândare la stoiria dipinta, il suo cartone; e levato in molti luoghi le tavole del palco, acciò si potesse veder di sotto, aperse la sala. Il che sentendosi, corse tutta Genova a vederlo, e stupiti del gran disegno di Perino, lo celebrarono immortalmente. Andòvvi fra gli altri Girolamo da Trivisi, il quale vide quello che egli mai non pensò vedere di Perino; onde spaventato dalla bellezza sua, si partì di Genova senza chieder licenza al prencipe Doria, tornandosene in Bologna, dove egli abitava." Vasari, *Vite de più eccellenti pittori*, 5:140.

11. "It is with the same intention of emancipating your practice from an exclusive and slavish attachment to any particular tool, that you are reminded by the same authority of the proverbial expression 'Io tengo il disegno alla punta dei penelli, My design is at the point of my brush'—though I am afraid the expression is dignified with the great name of Correggio through a lapse of memory, as it appears from Vasari that it was the petulant effusion of Girolamo da Trevigi, an obscure painter, in derision of the elaborate cartoon prepared by Pierino de Vaga for his fresco-painting in the great salon of the Palace Doria at Genova." Füssli, "On Design" (lecture VII), in *Life and Writings of Henry Fuseli*, 2:309.

12. "Je ne puis cependant quitter ce sujet [his discourse on the *génie*; Su.] sans chercher à prémunir les jeunes artistes contre une erreur qui paraît considérable. C'est celle de se livrer à un faux feu, qui tient plus de la réminiscence que du génie, et de le prendre pour lui; malheur qu'une funeste facilité entretient souvent en nous, et dont on croit avoir excusé l'inconvénient en disant: *Cela a été fait au bout de l'outil.* Ce danger de l'art est d'autant plus a redouter qu'il n'est que trop commun et que l'amour-propre fournit toujours des raisons pour l'excuser." Caylus, "Comte de Caylus: Réflexions sur la peinture," 36.

13. "Le Corrège dis-je, trouva des charmes dans le maniment de la brosse; il prenoit plaisir à empâter, à fondre, & savouroit l'amusement *de se perdre & de se retrouver* dans la couleur sans se rendre esclave des formes, ni s'occuper de leur choix. Un *pinceau* si hardi, *si amoureux*, si caressant, communiqua le plaisir & l'ivresse dont son auteur savoit sûrement jouir, à tous ceux qui virent ses ouvrages." M. Robin, "Pinceau," in Lévesque and Watelet, *Dictionnaire des arts de peinture*, 5:64.

14. Füssli, "The School of Mantoua," in *Life and Writings of Henry Fuseli*, 3:386.

15. See Gould, *Paintings of Correggio*, 194–95. As an early source of the international awareness of this kind of iconoclasm, see the review in *Allgemeine Literaturzeitung* 1, no. 78 (18 March 1793): 621–22: "Mehrere trefliche Gemälde von großen Meistern, welche sich vordem in den Pariser Sammlungen befanden, sind von raubgierigen Händen des Fanatismus und der Bigotterie vertilgt, weil sie zu frey gemalt waren, Nuditäten zeigen u.f.w. So wurde z.B. die berühmte Leda zu Fontainebleau von Michel Angelo (sie ist von Aeneas Viccus gestochen) verbrannt, und eine Leda von Correggio in der Orleanschen Sammlung ebenfalls vernichtet. Das nennt man mit Recht Barbarey." The restoration of Leda's head was undertaken again.

16. "The picture of Io embraced by Jupiter, inbosomed in clouds, by a silent water in which a stag quenches his thirst, was their companion: a work to which the most lavish fame has done no justice, and beyond which no fancy ever soared. The Io shared a still more barbarous fate. Not content with mangling her like the Leda, the bigot prince burnt her head; and, were it not for the beautiful duplicate which fortune preserved in the Gallery of Vienna, we should be reduced to guess at Correggio in the fragments at Sans Souci." Füssli, "School of Mantoua," 3:386–87.

17. "Vous prenez pour génie un amour de manier le crayon et le pinceau, et vos parents, sur ce point, s'abusent encore plus que vous. Sans consulter votre esprit et vos forces, et brûlant d'une périlleuse ardeur, vous entreprenez de courir la carrière la plus épineuse. . . . Car enfin, comme vous l'avez dit, Monsieur [i.e., M. Massé] est-il rien de plus accablant, pour un homme né avec des sentimens, que d'être forcé de s'avouer intérieurement qu'il n'est que médiocre dans un art qui n'admet pas la médiocrité. L'on me dira, que ce même amour-propre, qui l'a séduit d'abord, sait aussi l'aveugler sur ses ouvrages; que, s'il est froid dans les ordonnances de ses tableaux et sec dans l'exécution, l'amour-propre lui persuade qu'il est parvenu à la simplicité sublime et à la perfection de *Raphaël*. Que, si au contraire il compose d'une manière extravagante et dessine incorrectement, l'amour-propre le porte à se comparer au *Corrège*." Quoted in Montaiglon, *Procès verbaux de l'Académie royale*, 6:205–6.

18. Originally from Bohemia, Mengs lived and worked mainly in Italy; he intensively appropriated Correggio's figural compositions—albeit under the misguided impression that his correction of

Correggio's art from a classicist's viewpoint was a key to reaching the highest level of art.

19. "Ihm waren die engen Ufer der schmalen Umrisse seiner Vorfahren viel zu ängstlich, seinen himmlichen Geist darin einzuschliessen: Da er die Vergrösserung der Theile des Lichtes und Schattens fand, so durchbrach er, wie ein geschwollener Fluß, auf diesem Wege die Ufer, und zog seine Zuschauer mit sich in das weite Meer der Gefälligkeit—Ja, er hat viele mit sich gezogen, und durch den angenehmen Gesang seiner Gratien der schmeichelnden Syrenen, an die Ufer des Irrthums geführet." Mengs, *Gedanken über die Schönheit*, 65–66.

20. Ibid., 65.

21. Ibid., 66.

22. In the conclusion of his iconographical analysis of the *Camera di San Paolo*, Panofsky writes of Correggio's "intent to display mythological knowledge and a classicizing mode of expression," adding, "but the essence of the mythical characters is obscured or at least attenuated by a whimsical preference for recondite textual sources and an almost compulsive propensity for cryptic allusion." Panofsky, *Iconography of Correggio's Camera di San Paolo*, 98. In his interpretation of the *Allegories* painted for Isabella d'Este, Campbell not only focuses on the "grotesque and monstrous juxtapositions" of individual elements no longer explicable through the pictorial tradition, but also underscores the uncertainty at work in both the titling and description of what is being depicted. See Campbell, *Cabinet of Eros*, 221–49, quotations on 231, 237.

23. Fabiánski, "Sui dipinti mitologici di Antonio Allegri," 240, 241, 243.

24. Strzygowski, *Werden des Barock bei Raphael und Correggio*, 93, 103, 104.

25. "Né si può esprimere le leggiadrissime vivacità che fece nelle opere sue Antonio da Correggio, sfilando gli innanzi a lui, ch'era difficile, tagliente e secca, ma d'una piumosità morbidi, che si scorgevano le fila nella facilità del farli, che parevano d'oro e più belli che i vivi, i quali restano vinti dai suoi coloriti." Vasari, *Vite de più eccellenti pittori*, 4:9.

26. In *Art and Anarchy*, Edgar Wind aptly formulates the idea of an inevitable dependency resulting from technique relying too heavily on its instruments: "Any instrument, it is true, brings with it the danger that it might enslave the man that it is meant to serve. Thus the pencil dominated Meissonier, Menzel, and Muirhead Bone, and their perception became as mechanical as their skill" (83). Unquestionably, Ernest Meissonier, Adolf von Menzel, and Sir Muirhead Bone were virtuoso draftsmen.

27. Füssli continues with the passage just cited; see above, note 11. Füssli, *Life and Writings of Henry Fuseli*, 2:308–9.

28. "Elle renferme la manière de faire les teintes, celle d'empâter les couleurs, le maniement du pinceau, & le style de la touche. Ces détails constituent l'essentiel du métier de la peinture, mais les qualités qui constituent l'essentiel de l'art, sont toutes spirituelles. . . . Avec beaucoup d'intelligence, on peut réussir par des moyens bizarres; mais c'est la singularité du talent, & non celle des procédés qui distingue vraiment le grand artiste." Watelet and Lévesque, *Dictionnaire des arts de peinture*, 3:379.

29. "La Hollande a produit un artiste qui peignit réellement avec la *main*. Cornille Ketel, après avoir peint pendant vingt ans, comme les autres, avec la brosse, s'avisa de quitter cet instrument de l'art, & de se servir de ses doigts au lieu de pinceaux. Pour n'avoir pas de témoins de ses premiers essais en cette manière, il commença par son portrait, & reussit. Ne trouvant pas encore ce tour de force assez singulier, il se mit à employer les doigts de sa main gauche, comme ceux de la droite, & en vint même jusqu'à peindre avec les orteils. On rapporte ce trait, moins pour le faire admirer, que comme une bizzarrerie qui ne mérite pas de trouver d'imitateurs. En effet, comme le remarque M. Descamps, peintre lui-même: 'dès qu'on peut mieux peindre avec le pinceau qu'avec les pieds & les mains pourquoi abandonner un usage plus sûr & plus facile?' . . . Il [Ketel] disoit une chose juste; c'est que tout sert d'instrument quand on a le génie. Il ajoutoit que c'étoit pour le prouver qu'il avoit quitté le pinceau; & en cela il avoit tort; car il auroit dû reconnoitre que les instrumens aident aux opérations du génie." Ibid., 3:365–66.

30. Van Mander, *Leven der Doorluchtige Nederlandsche*, 2:70; see also Galley, "Cornelis Ketel."

31. Galley, "Cornelis Ketel," 89.

32. Van Mander, *Leven der Doorluchtige Nederlandsche*, 2:72. In *De pictura*, Leon Battista Alberti declares that "the painter will never apply his brush or stylus to his work before he has clearly decided in his own mind what he is going to do and how he will do it. It is safer to remove errors within the mind than to erase them from one's work." Alberti, *De statua, De pictura*, 305. But here the technique of scraping paint away is not used to correct errors.

33. Cited in Franciscus Junius, *De pictura veterum*, 249–50.

34. Pliny, *Letters and Panegyricus*, 2:129.

35. In fact, however, scraping the brush shaft over the freshly applied paint was not de Gelder's innovation. Rather, it was an established practice that his master Rembrandt and Rembrandt's colleague Jan Lievens used for producing individual hairs or arranging light accents. Jan van Brueghel already used the practice. Although there are clear borrowings at work here from the technique of the master, Arent de Gelder developed his own particularities of practice. Ernst van de Wetering, who has analyzed them in detail, concludes that de Gelder is "a much more interesting colorist than his teacher," as the play of color in his paintings has a stronger, highly unusual autonomy. Wetering, "Verhältnis von Stil, Technik und Zufall," 28, 30.

36. Houbraken, *Groote Schouburgh*, 3:163.

37. Ibid., 3:161.

38. "Darum soll der Mahler sich dahin gewöhnen, jedesmal alle Dinge im Sinn und Verstand eigentlich zu überschlagen, bevor er Hand anlege, und seine Arbeit auf eine gute *Speculation* und Wissenschaft gruenden." Sandrart, *Teutsche Akademie*, 1.1:67.

39. "Etliche haben im gebrauch nur blind hin all'*aventure*, ohne Regeln ihrem ungegründeten Wahn und den Mutmaßungen der Augen

in allen natürlichen Dingen, (da doch solches Augengemerk dem Selbstbetrug sehr ergeben ist) nachzufolgen. . . . Ob nun wol diese beede Manieren im schwang gehen so ist doch die erste Art (wiewol sehr viele, auch von den berühmtesten Künstlern, gefunden werden, die derselben sich bedienen) nur ein blindes Wesen voll Ungewißheit deren man sich nicht untergeben soll. Es ist und bleibet auch eine lautere Unmöglichkeit, daß einig wichtiges Werk, ohne Vorwissen oder Beobachtung der wahren Regeln zu vollbringen sey: wie solte man dann, ohne diesen Behuf, in der Zeichenkunst ein vollkommener Meister können genennet werden? Also haben wir, wie sonst in allem, also auch in diesem, an die unfehlbare sichere Regeln, außer denen alles unrichtig bleibet, uns zuhalten." Ibid., 89. Similar arguments would continue to be formulated a century later, for instance, by Antonio Maria Zanetti, who viewed the "blind movement" of praxis as resulting from a propensity "to form the whole out of one's own ideas, drawing every detail from the palette or paintbrush." Zanetti, *Della pittura veneziana*, 301. See also Rosand, "Crisis of the Venetian Renaissance Tradition," 5.

40. My comments on this engraving are based on the pioneering research by Elisabeth Cropper in *The Ideal of Painting* and *Pietro Testa: Prints and Drawings*.

41. "Quelle arti che più partecipano delle mattematiche sono più nobili perché hopera[no] con più vere ragioni." Quoted in Cropper, *Ideal of Painting*, 229.

42. Ibid., 87–88; Cropper, *Pietro Testa*, 78.

43. "VOLONTÀ . . . sarà cieca, sporgendo ambedue le mani avanti una più dell'altra in atto di volersi appigliare ad una cosa. . . . La cecità le conviene, perche non vedondo per se stessa cosa alcuna, và quasi tentone dietro al senso, se è debole, & ignobile, ò dietro alla ragione, se è gagliarda, & di prezzo." Ripa, *Iconologia*, 2:249–50.

44. "Haverà l'ali alle spalle, & à piedi. . . . Si depinge con l'ali, perche si dimanda col nome di volontà." Ibid.

45. The whole text reads, "La Teoria è per se stessa di legami avvinta, e la Prattica nella sua libertà è per se stessa cieca; ma che in età di freschi anni ne' gli studij della Pittura il buono d gran Maestri apprende; e poi avanzandosi ad imitare da se gli oggetti della Natura, entra nel detto Liceo di Pallade, e vi ritrova, et intende le arti della Mathemattica, unisce egli la Teoria alla Prattica; e spogliandole de' loro difetti con felice accoppiamento dall' intelligenza, e dall'uso, a se aquista gloria di nome, et al mondo accresce pregi di virtù." Cropper, *Pietro Testa*, 76–77 (cat. no. 41).

46. "Dann der verstandt muß mit dem gebrauch anfahen zu wachsen, also das die hand kün thon, was der will im verstandt haben wil. Auß solchem wechst mit der zeyt die gewyßheit der kunst vnd des gebrauchs. Dann dise zwey müssen bei ein ander sein, dann eins on das ander sol nichtz." Dürer, *Schriftlicher Nachlass*, 3:297.

47. "Aber so du kein rechten grund hast, so ist es nit müglich, das du etwas gerechtz vnnd gutz machst, vnd ob du gleych den grösten gebrauch der welt hettest in freyheyt der hand. Dann es ist mer ein gefencknus, so sie dich verfürt. Darumb sol kein freyheit on kunst, so ist die Kunst verborgen on den gebrauch. Darumb muß es bey einander sein, wie oben gesagt." Ibid.

48. "Sed plerunque non minus studiosos quam cupidos, quod viam perdiscendae rei ignorent, magis quam discendi labor fugit." Alberti, *De statua, De pictura*, 296.

49. "Quelli che s'innamorano della pratica senza la scienza, sono come i nocchieri che entrano in naviglio senza timone o bussola, che mai hanno certezza dove si vadano." Leonardo, *Trattato di pittura*, 45n77.

50. "Si qui vero sunt pigri artifeces, hi profecto idcirco ita sunt quod lente et morose eam rem tentent quam non prius menti suae studio perspecuam effecere, dumque inter eas erroris tenebras versantur, meticulosi ac veluti obcaecati, penniculo, ut caecus bacillo, ignotas vias et exitus praetentant ac perquirunt." Alberti, *De statua, De pictura*, 304.

51. Descartes and Testa agree that geometry as an ordering system could offer clear guidelines to the understanding. See the quotation in note 41.

52. Descartes, *Discourse on Method*, 1:91.

53. Ibid., 1:96.

54. See Zittel, *Theatrum philosophicum*, 131; the letter to Marsenne was written in March 1637.

55. For the codified gesture of the *aposkopein* in the arts of antiquity, see Jucker-Scherrer, *Gestus des Aposkopein*. A prominent example for the gesture is the apostle in Raphael's *Transfiguration*, who has slid to the ground. See Preimesberger, "Tragische Motive in Raffaels 'Transfiguration.'"

56. "It is through wonder [*thauma*] that men now begin and originally began to philosophize." Aristotle *Metaphysics* A.2.982b9.

Bibliography

Alberti, Leon Battista. *De statua, De pictura, Elementa picturae*. Edited by Oskar Bätschmann and Christoph Schäublin. Darmstadt: Wissenschaftliche Buchgesellschaft, 2000.

Arent de Gelder (1645–1727): Rembrandts Meisterschüler und Nachfolger. Exh. cat. Cologne: Wallraf-Richartz-Museum, 1999.

Aristotle. *Metaphysics*. Translated by Hugh Tredennick. Cambridge: Harvard University Press, 1989.

Bottari, Giovanni, and Stefano Ticozzi. *Raccolta di lettere sulla pittura, scultura ed architettura*. Vol. 1. Milan: Giovanni Silvestri, 1822.

Campbell, Stephen. *The Cabinet of Eros: Renaissance Mythological Painting and the Studiolo of Isabella d'Este*. New Haven: Yale University Press, 2004.

Caylus, Anne-Claude-Philippe, Comte de. "Comte de Caylus: Réflexions sur la peinture." In *Conférences de l'Académie royale de peinture et de sculpture*, ed. Jacqueline Lichtenstein and Christian Michel, vol. 1, *De 1667 à 1679*, 34–41. Paris: Beaux-Arts de Paris les Éditions, 2012.

Cropper, Elisabeth. *The Ideal of Painting: Pietro Testa's Düsseldorf Notebook*. Princeton: Princeton University Press, 1984.

———, ed. *Pietro Testa: Prints and Drawings*. Exh. cat. Aldershot: Scolar Press, 1988.

Dempsey, Charles. *Annibale Carracci and the Beginning of Baroque Style*. Glückstadt: J. J. Augustin, 1977.

Descartes, René. *Discourse on the Method of Rightly Conducting the Reason and Seeking Truth in the Sciences*. In *The Philosophical Works of Descartes*, trans. Elizabeth S. Haldane and G. R. T. Ross, 2 vols., 1:79–130. Cambridge: Cambridge University Press, 1911.

Dürer, Albrecht. *Dürer: Schriftlicher Nachlass*. Edited by Hans Rupprich. 3 vols. Berlin: Deutscher Verein für Kunstwissenschaft, 1956–69.

Ekserdjian, David. "Correggio Disegnatore." In *Correggio e l'antico*, exh. cat., ed. Anna Coliva, 77–83. Milan: Federico Motta Editore, 2008.

Fabiánski, Marcin. "Sui dipinti mitologici di Antonio Allegri." In *Correggio*, exh. cat., ed. Lucia Fornari Schianchi, 239–43. Milan: Skira, 2008.

Füssli, Johann Heinrich [Henry Fuseli]. *The Life and Writings of Henry Fuseli*. Edited by John Knowles. 3 vols. London: Henry Colburn and Richard Bentley, 1831.

Galley, Nicolas. "Cornelis Ketel: A Painter Without a Brush." *Artibus et Historiae* 25, no. 49 (2004): 87–100.

Gould, Cecil. *The Paintings of Correggio*. Ithaca: Cornell University Press, 1976.

Horký, Mila. *Künstlerbilder—Künstlermythen: Graphik und Zeichnung des 16. bis 18. Jahrhunderts*. Exh. cat. Brunswick: Herzog Anton Ulrich-Museum, 2002.

Houbraken, Arnold. *De Groote Schouburgh der Nederlantsche Konstschilders en Schilderessen*. 3 vols. Maastricht: Leiter-Nypels, 1953.

Jucker-Scherrer, Ines. *Der Gestus des Aposkopein: Ein Beitrag zur Gebärdensprache der antiken Kunst*. Zurich: Juris-Verlag, 1956.

Junius, Franciscus. *De pictura veterum libri tre*. 1694. Edited and translated by Nicole Nativel. Geneva: Librairie Droz, 1996.

Lebrun, Pierre. "Recueuil des essaies des merveilles de la peinture de Pierre Lebrun, peintre (Brussels Manuscript, 1635)." In *Medieval and Renaissance Treatises on the Arts of Painting: Original Texts with English Translation*, ed. Mary Merrifield, 759–841. New York: Dover, 1999.

Leonardo da Vinci. *Trattato di pittura*. Edited by Gaetano Milanesi. Rome: Unione Cooperativa Editrice, 1890.

Mengs, Anton Raphael. *Gedanken über die Schönheit und über den Geschmack in der Malerei*. Edited by Johann Caspar Füssli. Zurich: Orell, Fueßlin, 1762.

Montaiglon, Anatole de. *Procès verbaux de l'Académie royale de peinture et de sculpture, 1648–1793*. Vol. 6. Paris: Baur, 1885.

Panofsky, Erwin. *The Iconography of Correggio's Camera di San Paolo*. London: Warburg Institute, 1961.

Petit, Béatrice, ed. *L'apothéose du geste: L'esquisse peinte au siècle de Boucher et Fragonard*. Exh. cat. Paris: Hazan, 2003.

Pliny. *Letters and Panegyricus*. Vol. 2. Translated by Betty Radice. Cambridge: Harvard University Press, 1975.

Preimesberger, Rudolf. "Tragische Motive in Raffaels 'Transfigurati-on.'" *Zeitschrift für Kunstgeschichte* 50 (1987): 89–115.

Ripa, Cesare. *Iconologia*. 1618. Edited by Piero Buscaroli. 2 vols. Torino: Fògola, 1991.

Rosand, David. "The Crisis of the Venetian Renaissance Tradition." *Arte* 11–12 (1970): 5–53.

Sandrart, Joachim von. *Teutsche Akademie der edlen Bau-, Bild- und Malerey-Künste*. 3 vols. Nördlingen: Verlag Dr. Alfons Uhl, 1994.

Scannelli, Francesco. *Il microcosmo della pittura*. 1657. Edited by Guido Giubbini. Bologna: Edizioni Labor, 1989.

Schianchi, Lucia Fornari, ed. *Correggio*. Exh. cat. Milan: Skira, 2008.

Seneca. *Moral letters to Lucilius*. Translated by Richard Mott Gummere. 3 vols. London: William Heinemann, 1919–25.

Strzygowski, Josef. *Das Werden des Barock bei Raphael und Correggio, nebst einem Anhang über Rembrandt*. Strassburg: Heitz, 1898.

van de Wetering, Ernst. "Das Verhältnis von Stil, Technik und Zufall bei Arent de Gelder: Ein Vergleich mit Rembrandt." In *Arent de Gelder (1645–1727): Rembrandts Meisterschüler und Nachfolger*, exh. cat., 18–35. Cologne: Wallraf-Richartz-Museum, 1999.

van Mander, Karel. *Het Leven der Doorluchtige Nederlandsche en eenige Hoogduitsche Schilders (1618)*. Vol. 2. Amsterdam: Steven van Esveldt, 1764.

Vasari, Giorgio. *Le vite de più eccellenti pittori, scultori e architettori: Nelle redazioni del 1550 e 1568*. Edited by Paola Barocchi. 6 vols. Florence: Sansoni Editore, 1966–87.

Watelet, Claude Henri, and Pierre-Charles Lévesque. *Dictionnaire des arts de peinture, sculpture et gravure*. 5 vols. Paris: L. F. Prault, 1792.

Wind, Edgar. *Art and Anarchy*. New York: Knopf, 1964.

Zanetti, Antonio Maria. *Della pittura veneziana e delle opera pubbliche de' veneziani maestri*. Venice, 1771.

Zittel, Claus. *Theatrum philosophicum: Descartes und die Rolle ästhetischer Formen in der Wissenschaft*. Berlin: Akademie der Wissenschaft, 2009.

9.

White Earth, or How to Cultivate Color in the Field of Painting

Still Life and Baroque Color Theory

KARIN LEONHARD

Color in Nature

From antiquity to the seventeenth century, we find an increasing awareness of an incongruity between the mixing of physical color—such as *paints, dyes, inks,* etc.—and the mixing of visible *light* emitted from differently colored light sources. The two mixtures do not follow the same rules, though from a systematic point of view they should. The difference created as many problems within the field of color theory as it did for the relationship between color theory and practice: how can we develop a concept of color mixture when we cannot base it on a correspondence between the optical *mixing of light color* and the physical *mixing of pigments?*

Early microscopy certainly changed the notion of color: now a finely grained microstructure lay beneath the visible world. As Galileo once put it, "Any material, be it stone, or wood, or metal," when studied attentively, and best with a microscope, will display all its "colors divided into minute particles."[1] Later in the seventeenth century, Newton would state that "the opakest Bodies, if their parts be subtily divided, . . . become perfectly transparent,"[2] an argument that gained its momentum from a revival of Lucretius. At the same time that such radical changes were occurring in theories of vision and matter, artists and scientists began to share an interest in colors as paint or natural phenomena. Not only was there a disciplinary interpenetration of theories about natural and artificial color and color mixing, along with an exchange of practical techniques and experiences, but a significant reconceptualization of the boundary emerged between artificial and natural color itself. It prompted the idea of letting natural color coincide with the ratio of pigment mixture, an idea inspired by a renewed interest in atomism and a corpuscular microstructure of the visible world. Such an interest nurtured an explanatory model that equalized divergent color concepts and, in the end, completely revolutionized color theory.

This chapter tackles the questions of *when and why color stopped being an inherent quality of bodies and became the product of surface textures interacting with light*. A great deal of my focus is on the reconstruction of a peripatetic tradition in color explanation and some historical steps that led out of it, but in addition to looking into some written sources, I am concerned with the genre of still life painting as yet another important home of color discourse. I argue that it was precisely the fascination with processes of color generation, mixture, and change in natural bodies that prepared the ground for discussions between still life painters, on the one hand, and natural philosophers and scientists, on the other; and, furthermore, that this fascination acted as a driving force behind the development of still life painting as a genre. In examining these issues, I proceed in three steps, beginning with an explanation of ancient ideas about an originally "white earth" or "white world" that successively gained color by some sort of natural dyeing process; this includes a close reading of one of the principal ancient texts on chromatic change in nature and art, the Pseudo-Aristotelian *De coloribus*. The focus then shifts to the genre of still life painting of the early modern period, still under the historical presumption that color is an inherent property of an object. But although still life painters obviously dwell on processes of color generation and change in objects such as fruit and flowers, they also reflect on the problem of how to render luminous color in paint. The third and final part of the chapter contains an analysis of developments within color theories of the second half of the seventeenth century, especially when colors were examined under the microscope. This analysis does not claim to give more than a brief insight into the complex relationship between form and matter in baroque thinking, but it is intended to provide a basis for my discussion of when and why color stopped being seen as an inherent quality of bodies. The ancient concept of a "white world" was inextricably bound to the doctrine of the four elements. When we therefore end with a seemingly similar concept of a "colorless world" in the thought of early modern natural scientists, we have to be very much aware of the crossovers and differences between the two beliefs: *color is now no longer an inherent quality of bodies, and where it once had been white earth, we now find a crystalline structure* as an explanatory model of how color comes into the world. Yet these two views do have something in common: they both contemplate the fact that because of differences in color, the properties of matter become visible. The moment a color changes, then, is of the greatest interest. Let us look at some paintings.

The flower bouquet (fig. 9.1) displays an abundance of color: two sunflowers, one darker, one brighter, are surrounded by pink roses and long-stemmed red carnations. Iris and morning glory add spots of blue color, and

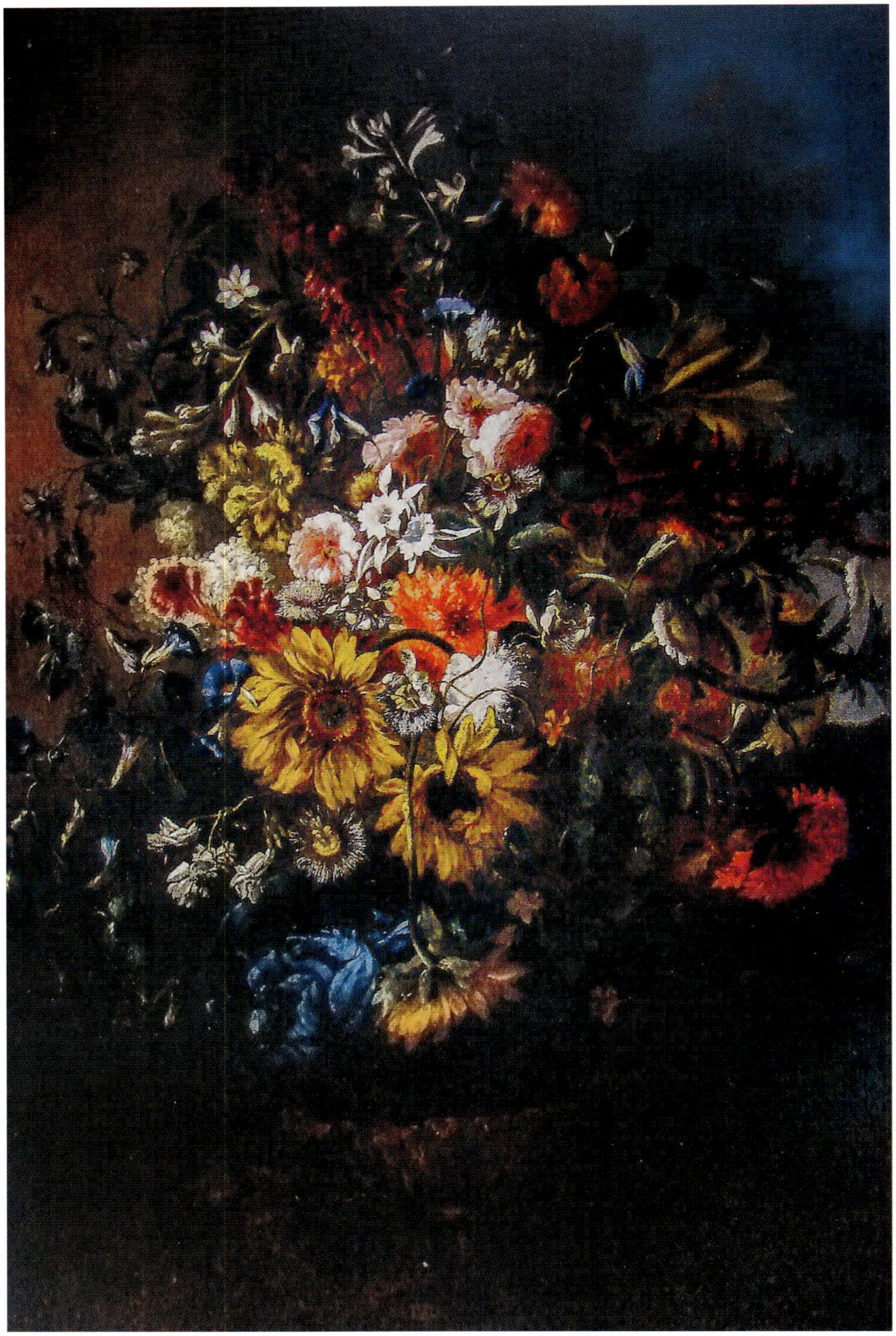

9.1 Karel van Vogelaer, known as Carlo de' Fiori, *Metal Vase with Flowers,* 1680s. Oil on canvas, 152 x 109 cm. Private collection.

there are rich blossoms of peonies and the exotic passion flower, lilies, columbine, anemones, and white bunches of the guelder rose. Deep down, obscured in shadow, we can just make out a spray of humble violets. Somehow, we cannot shake the feeling that the selection of flowers makes sense, that the arrangement has been put together rather consciously. In addition, the energy that flows from the scene is arresting and manages to draw us in, to make us take part in an aesthetic exchange.

Karel van Vogelaer is a major exponent of late baroque still life painting. Born in Maastricht, he settled in Rome in 1675, where he established himself as a painter of large-sized flower pieces in the manner of Mario Nuzzi and Paolo Porpora. His *Metal Vase with Flowers*, painted in the 1680s, is a fine example of his work, recognizable in the tight clusters of brushwork and the glowing tints of color in which the flowers emerge from the dark background of the picture plane. This is one example for our discussion of some possible implications of baroque concepts of color and color change in still life painting. Others include works of art in the Bimbi circle as well as of Bimbi and Nuzzi themselves. However, our discussion is not confined to Italian still life painting alone. As much as it was stimulated by Flemish flower pieces of the first half of the seventeenth century, such as the Brueghel school, from the 1650s onward, the Italian *natura morta* reciprocally attracted attention from Dutch artists such as Willem van Aelst, Matthias Withoos, and Karel van Vogelaer, among others.

Van Vogelaer's flower piece displays the full spectrum of the colors of the rainbow. The sky is immersed in blue and red light that seems so dense and material that it truly expresses the sensation of paint. Closer to the ground, the colors darken and then form a substantial base for the metal vase. It is significant that the painting works its way from one color to another and from one gradation of tone to another. Most of its brightness is concentrated in the center, and as a consequence the bouquet bends forward and curves the pictorial space. This manner of chromatic spatial modeling is known as "houding" in Dutch art theory, a term that was much propagated by Van Hoogstraten and de Lairesse, who both devote chapters of their books on painting to this subject.[3] But in contrast to color perspective that creates a sense of depth by opening up an illusionary view, here it is the forceful power of color that moves out to meet the eye. Not only can the color's substance be experienced visually, but it also hints at possible correspondences (and incongruities) between seventeenth-century concepts of color in natural philosophy and in painting. They derive from the peripatetic assumption that any color arises from the blending of black and white, or transparency and opacity.

WHITE EARTH

In the history of color theory, it is impossible to ignore how the influence of antiquity, especially of Aristotle, reached well into early modern times. Besides the peripatetic tradition, we find traces of Empedocles, Democritus, Plato, Avicenna, and Alhazen in the writings of late sixteenth- and early seventeenth-century artists and natural philosophers. The Greek text *De coloribus*, however, was first ascribed to Aristotle and then to Theophrastus and Strato, but was probably created by an anonymous member of the peripatetic school.[4] It was well known during the Middle Ages and studied well into the early modern period; we know that there are two medieval Latin translations, known as *Translatio vetus* and *Translatio vulgata*, and a version by Caelius Calcagnini of Ferrara, dating from the first half of the sixteenth century, which was widely read and hugely influential among artists, natural philosophers, and natural historians.[5] *De coloribus* starts with an enigmatic passage: "Those colours are simple which belong to the elements, fire, air, water and earth. For air and water are naturally white in themselves, while fire and the sun are golden. *The earth is also naturally white, but seems coloured because it is dyed*" (Ps.-Arist. *De. col.* 791a1–5).

How are we to understand this passage? It is necessary to start with the explanation of ancient ideas about an originally "white earth" or "white world" that successively gained color by some sort of natural dyeing process. It is through Theophrastus's account and critique of the views of other writers regarding sense perception (his *De sensibus*) that some of the ancient thoughts on how color is produced and mixed have been passed down to us. Democritus, we learn in *De sensibus*, claimed that "red, when mixed with white, gives almost a 'pure' pale or yellow green and not a black" (78). And, just as cryptically: "Leek green is of crimson and deep woad blue, or of pale green and a crimson tinge" (77).[6] Ancient Greek descriptions of the mixing of colors have puzzled historians of philosophy and science from the outset. In 2001, however, in a commentary on a Dutch edition of the peripatetic treatise *De coloribus*, clues emerged that led to a better understanding of Greek color terms and the concept of blending. Until then, there had been confusion over whether the colors that result from mixing, especially as described by Democritus and Plato, were to be understood as painter's pigments or as phenomena in nature.[7] This recent commentary, in turn, corresponds to a thesis that was advanced a century ago in Walther Kranz's paper "Die ältesten Farbenlehren der Griechen." Democritus's mixtures, Kranz argued, should not be understood as standard paint mixtures at all. In support of this, Kranz quoted a passage from the *De coloribus* by Pseudo-Aristotle that had been mistranslated by one word. The new translation reads: "But we must make our investigation into these things not by mixing these colours as painters do, but . . . by considering the mixing of colours *in nature*" (instead of "investigating the true nature of colour blends") (792b16–20). Kranz had already understood the passage in this sense, and he used the shift of meaning for his own argument: not to interpret ancient color blends in the way painters mixed their color, but as chromatic phenomena that are evident in nature owing to the interaction of light and matter. Kranz had tried to decode Greek color terms accordingly but did not get beyond first trials. For Ferwerda and Struycken, the doctrine of the Pseudo-Aristotelian *De coloribus* also provides the key to understanding Democritus's or Plato's puzzling views on color mixture, and therefore provides insight into a broad ancient tradition of how to understand and explain color blending.[8]

For a better understanding, we must reconstruct a number of steps in the history of Greek color theory. Parmenides opens the discussion by introducing two elements—fire (light) and darkness (night)—as so-called basic colors and associates them with white and black. Any other colors result from the interaction of these two elements, which again coincide with two pairs of opposites: hot and cold, and dry and wet (Parm. frags. B 8.55–59, B 9 DK).[9] Later developments show that color theory increasingly draws upon direct observation and experience of the natural world and tries to tie them back to basic pairs of terms. Democritus's color theory, for instance, falls into line with the opposing set of primary colors "white" (leukon) and "black" (melan) as proposed by Parmenides, but Democritus extends it by one more pair, that of "red" (erythron) and "pale green" or "yellow-green" (chlōron) because of a need to link color to the material elements from which everything originates. However, we note that for Democritus, the elements are not colored themselves, but their chromatic appearance results from different combinations of otherwise colorless atoms (Democr. frags. B 123, A 49, B 125 DK; Thphr. *De sens.* 50–55).

It was Aristotle who in his *Meteorologica* and *On Sense and Sensible Objects* crucially extended the color palette. According to him, three stages of color blending may be distinguished, the first consisting of the well-known opposites of white and black. They are the only primary colors. Aristotle viewed Democritus's attempt to assume a larger number of primaries with some suspicion (as did Theophrastus in *De sens.* 82) and thus rejected it as a possible concept. The opacity and color of objects on a second level, Aristotle argued, arises from different mixtures of invisibly small particles of white and black (Arist. *De sens.* 439b19–26). The purest or most intense colors, such as light yellow, crimson, purple, dark green, and blue, result from a whole-number mixture ratio of light and dark similar to the whole-number divisions of a vibrating string that produces musical notes or harmonics. Together with black and white, Aristotle thereby obtained a range of seven colors, with red and purple as the "most attractive" or harmonious (439b). Further colors result from a blending of secondary colors and are therefore located at a tertiary level. Color brightness, according to the mixture ratio, derives from numerical proportions of white and black and can only be expressed in terms of "more" or "less." As a consequence, quantitative aspects of color must at the same time be read in terms of quality.

The peripatetic text on colors, *De coloribus*, together with Aristotle's *Meteorologica* and *De sensu et sensibilibus* and Theophrastus's influential *De sensibus* form the main corpus of ancient color works, which continued to be influential well into the seventeenth century. Though tampered with in some passages, *De coloribus* nevertheless carefully delivers a description of the generation of colors and color change attributable to the interaction of the four elements. But it has one special feature: in contrast to Democritus, only three of the four elements are, in their unmixed state, described as "white," while fire is referred to as "gold" or "yellow" (791a1–5). The author then explains that a certain moistness or humidity, set in parallel to the color black in combination with fire, accounts for the dyeing of the white elements. These take on color only when (dry) brightness mixes with (wet) darkness. Thus the colorless or white matter is transformed into chromaticity because of the dyeing agency of (yellow) heat, on the one hand, and (black) humidity, on the other.

Once again, there are cosmological implications. In its primordial state the physical world merely possesses a potential visibility that must be realized by opposing pairs of light and darkness, or dryness and wetness, and from then onward must be activated over and over again by the same principle. John Gage rightly notes that in the technical literature in general, from the peripatetic *De coloribus* of the fourth century B.C. (795b10, 797a5) to the *Onomastikon* of Julius Pollux, compiled in the second century A.D. (1.49), treatises laid particular emphasis on color change,[10] meaning *photochemical* developments in nature (e.g., Thphr. *De sens.* 74, 81). If they dealt with chromatic effects of light, then in the sense that light (understood as the action or heat of the sun) acts on material substances that as a consequence are warmed up, and take on and change color, the tones depend on their specific properties and the strength of the sunlight. Such a biochemical concept of color forms an important part of ancient discussions on the processes of generation and corruption in nature, such as the change of color in growing plants, ripening fruit, or fading flowers, the different tones in animal fur, and the graying of men's hair in old age, for example. Ancient descriptions of color blending were for the most part, as we will see, based on precise visual observations.

PURPLE AND LEEK GREEN

The color purple, for instance, emerges when the day fades into night, when the light weakens and mingles with darkness, which is a result of the blending of white and dark. On the painter's palette, such a mixture would have resulted in a totally different color: a hue of gray. For Aristotle, however, as much as for Plato or Democritus, colors are part of the eternal transformation of the elements, outcomes of the processes of generation and corruption in nature. Accordingly, a red or purple sunset occurs when the day is replaced by the night, and the same color appears, conversely, when a black coal or iron is put into fire and made to glow, because "black things when burned always change to the color red" (Ps.-Arist. *De col.* 792a12–13). Such diverging observations of a sunset in an evening sky or iron in a blacksmith's forge are closely related because they share the same pattern of "white/yellow heat meeting black coldness." In fact, purple is elevated to the most significant color (after white/yellow and black) precisely because of its intermediary position between light and darkness, its mutual saturation of heat and moistness, and it therefore seems to be the most balanced and powerful color of all: "Purple is gay and bright whenever the rays of the sun are a weak mixture of white and shady. Consequently at the hours of sunrise and sunset the air seems to have a purple tint, the sun being at its rising or setting" (792a14–16).

It may be too early to show how such notions passed into sixteenth- and seventeenth-century art literature, and their significance may only become evident in due course. But in some of his well-known passages on landscape painting, Karel van Mander, who was well aware of the tradition of *De coloribus*, which he quotes in his *Schilder-boeck* (1604), refers to the rising and setting of the sun with the careful attention of someone who conceives purple not only as a pigment color but also as a dyeing agent in natural processes, such as the transition from day to night and night to day.[11]

Rules of color change must be deduced from the behavior of light in materials or the atmosphere. Sunset and sunrise, the glowing coal and the heated iron, are not the only instances described in *De coloribus* of how the color purple is "mixed." The ripening of grapes is another classic example: "The colour of dark wine occurs when sunlight rays are mixed with what is pure black and what is glittering, like the berries of the grape; for their colour is said to be wine-dark at the moment of ripening; for, when they are growing black, red changes to purple" (792b6–12). Flowers and fruits make especially rewarding subjects for the investigation of color change and color variation because they show different stages of maturity and ripeness in the same crop or bouquet. It is easy to observe, for example, how a fruit takes on one color as it loses another: "For in many of them owing to the fact that the ripening of all the colours does not take place at once, but some form earlier and others later, they change from one to the other, as in grapes and dates. For some of these become red at first, but as the black is formed in them they turn to wine-dark; but at last they become purplish, when the red colour is mixed with a large quantity of pure black. For those colours which are formed later, when they prevail, cause the previous colours to change" (795b23–33).

A similar passage adds an explanation of general color change in plants: "The primary colour of all plants is green; for shoots and leaves and fruit are all pale green. . . . As the moist part blackens, the green becomes very dark, and of the colour of a leek. Consequently the older shoots are much blacker than the young ones; the latter are yellower because the moisture in them has not yet turned black" (794b20–795a4). This passage is most illuminating. It sheds light not only on the etymology of the color term *prasinon* (leek-green) (in contrast to *chlōron*, pale or yellow green) but also on Democritus's puzzling statement that "red, when mixed with white, gives almost a 'pure' pale or yellow green and not a black," while "leek green is of crimson and deep woad blue, or of pale green and a crimson tinge" (Thphr. *De sens.* 78, 77). For Democritus, the color red was associated with the element of fire or with heat. In his description of the color terms "pale green" or "leek green," he had in mind the image of seedlings or stalks of leek, which are white in color when they lie beneath the ground but turn to yellow-green when they are warmed by sunlight and grow out of the earth.

Another addition of "red" (or "heat") causes the color to change to darker green. The reason for this is easily given: as the internal disposition of warm and cold, dry and wet, is changed under the influence of the heat of sunlight, the plant's sap coalesces and slowly darkens the plant's color (red + white = yellow-green; yellow-green + red = dark green).[12] In fact, when summarizing Democritus's mixture of *chlōron*, Theophrastus gives this explanation himself, as it also "accounts for the fact that plants at first are of such a green before there is a heating and dispersion" (*De sens.* 78). Many other passages in ancient writings on color can be interpreted in this way,[13] but this is an opportunity for a fresh look at the coloring in paintings, especially in still life and landscape painting of the early modern period. In fact, there is reason to believe that the development of still life painting as a genre was inspired not least by a profound interest in the complex relationship between vision, color, and light. We must thus examine the powerful tradition of peripatetic thinking, which continued well into the seventeenth century, and acknowledge the increasing appropriation of Aristotelian texts that deal with visual phenomena. Contemporaries who made or admired still lifes understood painting through color theories in the tradition of *De coloribus*, *De meteorologica*, and *De sensu et sensibilibus*.

Joris van Son's *Still Life with Grapes*, for instance, is set on a rock before a garden view and displays not only bunches of white and red grapes in fine tinges of colors and various degrees of ripeness but also a spray of lushly autumn-colored grape leaves (which in *De coloribus* are extensively discussed under the category "natural color change") and some species of peaches and pears, again in different stages of maturity (fig. 9.2). The painter has put a great deal of emphasis on color gradation and color change and has even tried to evoke the fruit's sweetness by showing us how the grape seeds float in the juicy flesh that basks in the sun. The interaction between moistness and sunlight makes a fair motif for fruit still life painting, and one might wonder whether the fire (which has, for no apparent reason, started in the background and darkens the sky with its smoke) refers to the idea of the elementary mixture of fire and air, as we read, for example, in Aristotle's *Meteorologica*. There, the diffusion of air and smoke is responsible for the color purple (the air, for Aristotle, being white and the smoke dark).[14]

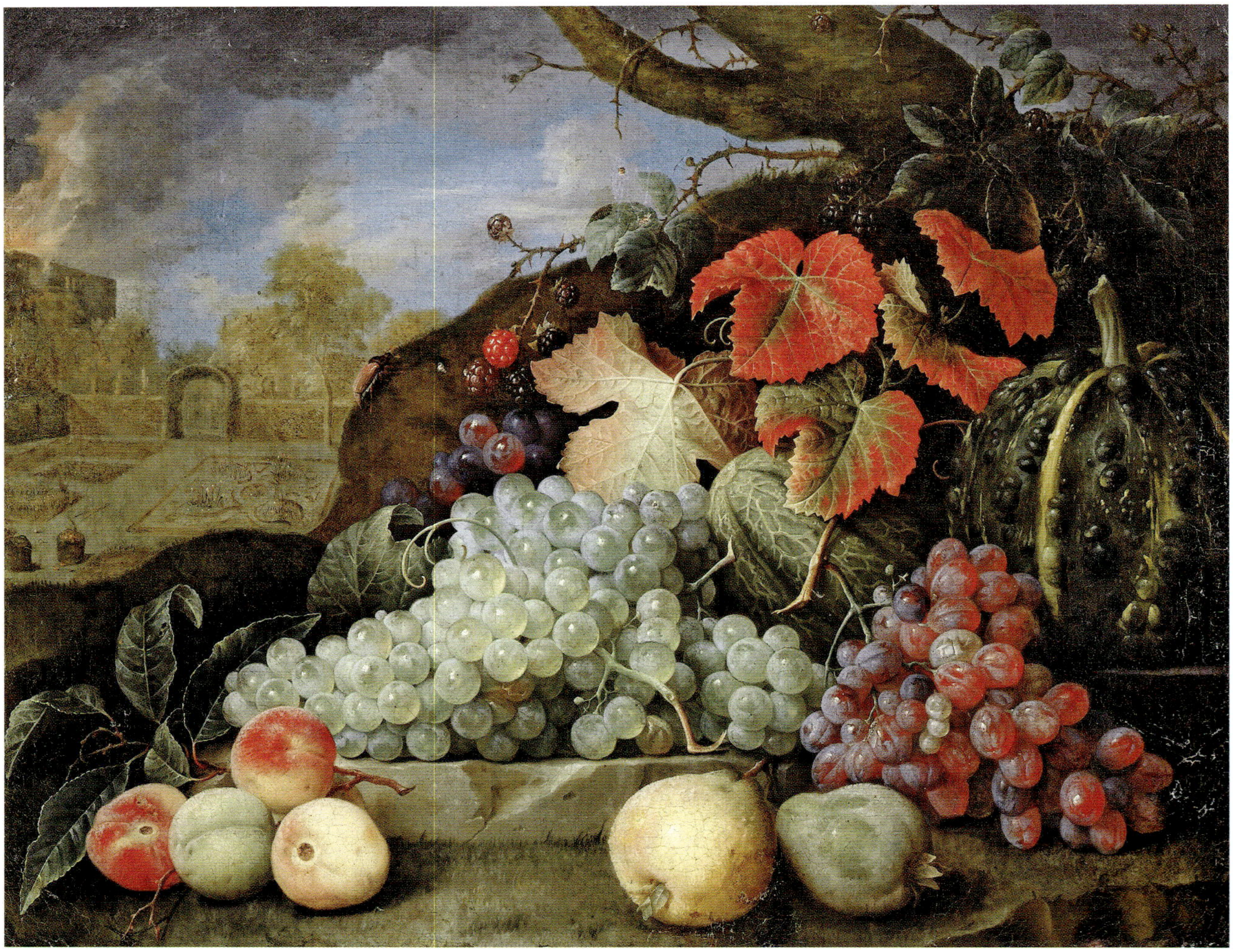

9.2 Joris van Son, *Still Life with Grapes, Melons, Peaches, and Pears in Front of a Formal Garden,* 1660s. Oil on canvas, 52.6 x 67.6 cm. Private collection.

Now, it is perhaps not a matter of chance that many examples of seventeenth-century still life painting show an accord between colors of the sky and colors of the depicted fruit. In Bartolomeo Bimbi's *Still Life with Plums,* for instance, painted for Cosimo III de' Medici as part of his program to assemble "tutte le sorti di frutte, d'agrumi, d'uve e di fiori" in his garden and villa la Topaia, a fruit basket has toppled over and all kinds of plums are spilled on the ground (fig. 9.3). The various tints of color—purple, lilac, mauve, magenta—make a splendid sight as the fruits lie close together, resembling a huge bunch of grapes. But they also perfectly match the colors of what

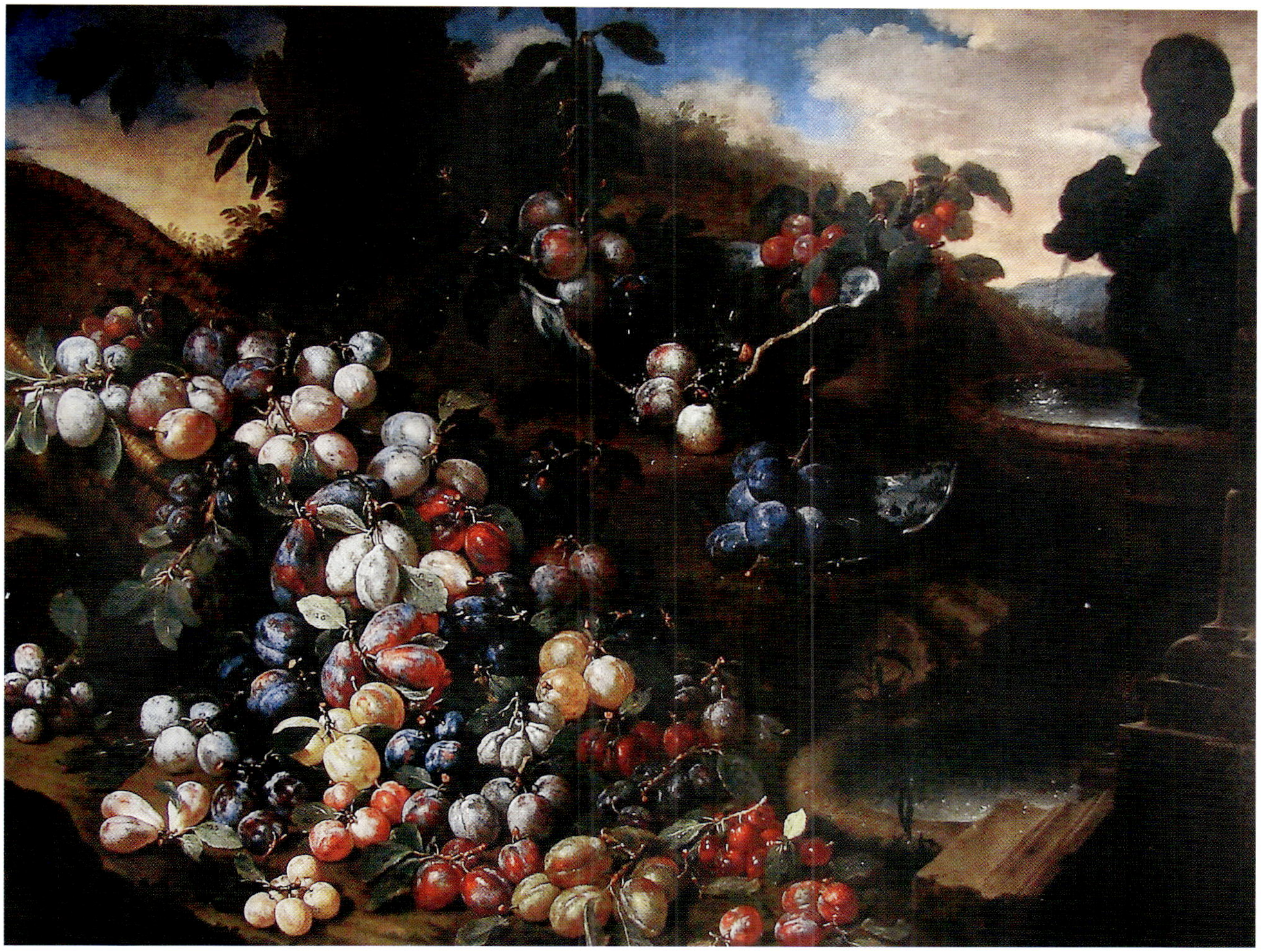

9.3 Bartolomeo Bimbi, *Still Life with Plums*, 1699. Oil on canvas, 174 x 228 cm. Villa Medicea di Poggio a Caiano, Museo della natura morta, inv. Castello 1910, no. 595. Courtesy of the Ministero per i Beni e le Attività Culturali.

seems to be an evening sky. Purple clouds gather on the horizon, where we glimpse the last rays of sunlight. Otherwise, the sky is of a particularly intense blue. The colors fade slightly toward the horizon, while the dark outlines of a figure on a fountain, by contrast, rise up into the sky, creating a sense of tranquility and solemnity. The saturated colors and the glimmer of light in the water basins evoke humidity, fertility, and the rich abundance of an aristocratic country estate.

In addition to documenting the numerous varieties of fruit species in the Medici gardens and throughout the grand duchy, the painting's richness lies in the great variety of color hues and shades with which the fruits are at once bound together and differentiated. This is why the sky's tinges of purple and blue interact strongly with the chromatics of the low and earthbound part of the painting, linking the accurate depiction of botanical species with a more general view of nature's generation by the elements. In this way, the painting combines a scientific mode of description with pictorial or even painterly values, and also makes us see how the particular and the universal intersect in nature's operations. Color richness and variation, if nothing else, are linked to the cycle of seasons and culminate in autumn, because "then all the fruits become ripe; and as the moisture in them is also warmed by the sun and the heat of the atmosphere, each combines to take the colours from the juices, just as those which are dyed from the flowers. So they are coloured little by little, and most of all those which are turned towards the sun and the warmth. So that the colours change in accordance with the seasons. This is obvious; those of a green hue all change as they grow ripe to their natural colour. For they are all white, black, grey, yellow, blackish, dark, dull-coloured red, wine-dark and saffron and exhibit almost all the difference of colour. . . . One can see this in many cases; for some fruits, as has already been said, exhibit many variations at the time of ripening" (Ps.-Arist. *De col.* 795a22–34, 796a18–20).

Chromatic interactions between microcosm and macrocosm, according to the elements and changes of season, similar to those featured prominently in *De coloribus* can be found in many fruit still life paintings of the period. In Nicola van Houbraken's *Fruits and Vegetables in a Landscape Setting*, for example, which dates from 1717, when it was bought by Cosimo III and sent to the Medici villa of Castello, we recognize what might be common root vegetables and herbaceous plants (fig. 9.4). Oranges are placed in a copper dish, and they, along with the red grapes and pumpkins, suggest they were painted in late autumn or early winter. The stone plate, however, and the glazed pot that so marvelously forms the painting's slightly shifted center, together with the winter vegetables and fruits, are painted in a range of dull, burnished tones of green and brown, lit up by intensely luminous touches of white.[15] But it is not only that the colors of the fruits and vegetables match the bleak landscape setting and winter/night sky, but rather that they are blended into each other: for example, the grapes and pumpkins that are piled up in the background fade off into the sky owing to the shared color palette and the elusive brushwork, which makes the leaves resemble shreds of clouds.

But what do these observations really teach us about the relation between color and objects? It is obvious that the dry, cold earth displayed in Van Houbraken's painting in various forms (as stone, soil, clay) alludes to other subterranean material such as copper or roots. A closer look reveals a turnip still buried in the ground, like a footnote to the main theme of winter's chthonian dominion. The season is dry and cold, a natural reason for the painting's tenebrous tones and lack of chromaticity; because of the season, the picture lacks light: "In the case of those in which the moisture is not mixed with the rays of the sun, their white colour persists, unless it grows black by lasting a long time and drying up first. Consequently in all plants the parts above the earth are green at first, but beneath the earth stalks and roots are white" (Ps.-Arist. *De col.* 795a10–15; see also 794a25–28). The color has literally drained out of the earth, just as it has out of the painter's palette: a striking feature, because the light hits the bowed shape of the so-called cardoon, or artichoke thistle, in the foreground.

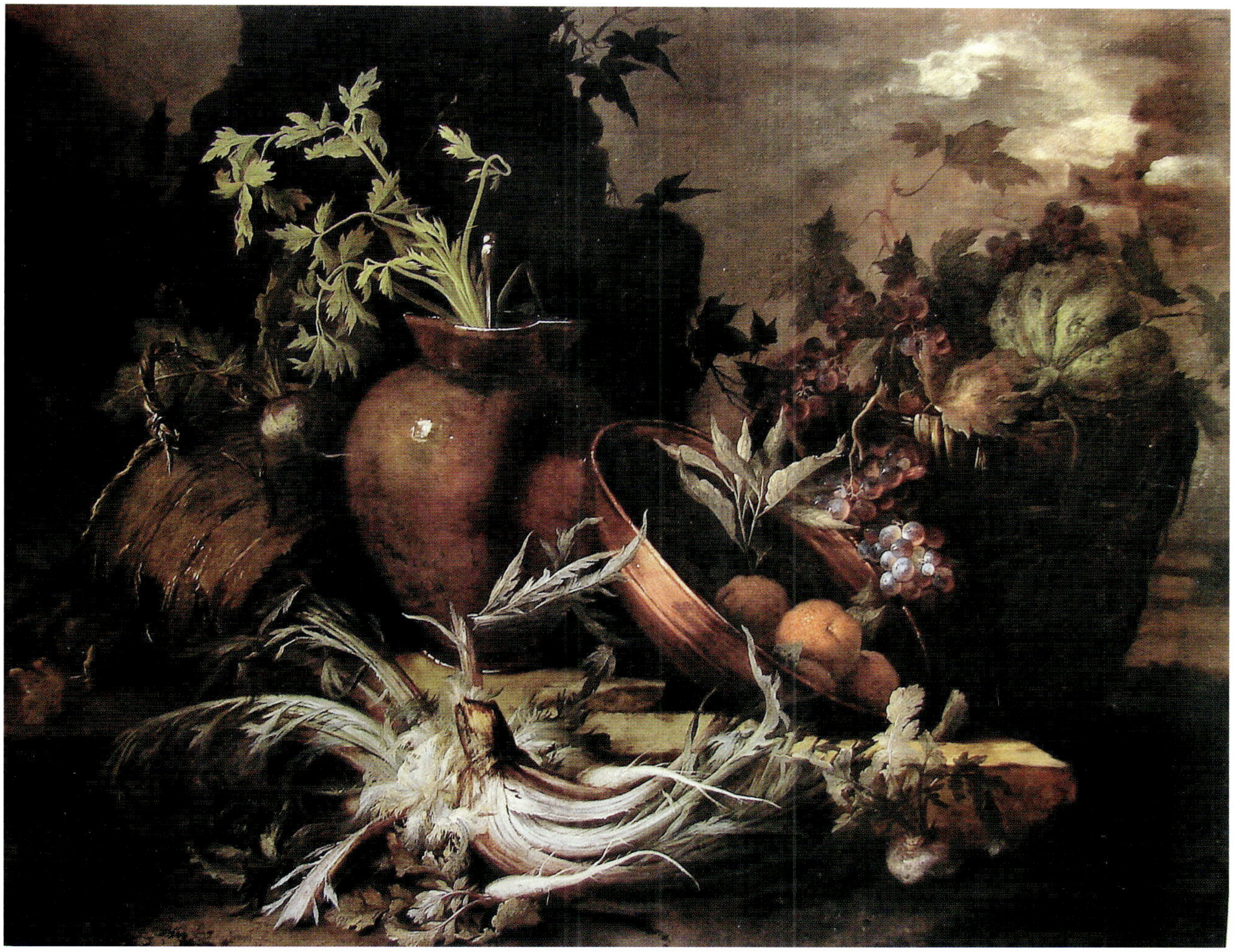

Nicola van Houbraken, *Fruits and Vegetables in a Landscape Setting,* 1717. Oil on canvas, 115 x 160 cm. Villa Medicea di Poggio a Caiano, Museo della natura morta, inv. 1890, no. 5137. Courtesy of the Ministero per i Beni e le Attività Culturali. **9.4**

Grown as a vegetable and blanched by being wrapped or packed in earth, the cardoon is also valued for its striking silvery thistlelike foliage. It is now pulled out of the ground so that we can see its bleached stems, which fade only slightly into hues of pale green, while the leaves look like they have been dipped into green color or dyed.[16] The traditional explanation for such a natural color change would have been that "shoots, again, if they are below the earth are white, but if the earth is removed from around them they all become green right from the first, as has been said before, because the moisture which passes down into them through the shoots has this colour naturally" (795a15–19). Van Houbraken's painting therefore dwells to some extent on the subject of earth and winter light, and on how to cultivate color in the field of painting.

Van Houbraken was of Flemish origin, though he was born in Sicily, where his family had settled after accompanying Anthony van Dyck on his journey to Italy. From 1674 until his death in 1723, he lived in Tuscany, and his works were collected by the Medici circle. He is also known for successfully blending the Flemish tradition of still life and flower painting, including its contemporary expressions developed in Rome, with the practice of the Italian still life, thus achieving his own personal style.[17] Karel van Vogelaer, in turn, had come from the Dutch province of Limburg, settling in 1675 in Rome, where he soon incorporated features of the Roman *natura morta* into his flower pieces, especially those of Mario Nuzzi (or de' Fiori), who was among the first Italians to specialize in still life painting of flower arrangements and was an influence on Van Vogelaer, with whom we began. If we begin to see baroque still life painting, at least in Rome and Florence from the 1650s on, as part of an investigation into the natural conditions of color change, which is closely related to questions of primary and secondary colors and color mixture in art, then our context and points of reference become much larger. In this way, *color in nature* and *color in art* are as closely connected as they are kept apart, because of the prevalence of natural over artificial color production during this time.

The view that no mixed color in art could match the luminescent color mixtures in nature was once seminally expressed in Aristotle's *De meteorologica* (372a). Alexander of Aphrodisias summed up this view in his commentary, writing that the "colors of the rainbow can neither be procured nor imitated by painters."[18] Such statements opened up the competition between nature and art and also became a driving force behind the development of still life painting as a genre (that is, as I have argued, as the home of color discourse). When Plutarch, for instance, called pigment mixture a "deflowering" of color in terms of its force, saturation, and luminosity, he coined a phrase that endured but was much opposed in Dutch art theory around 1600—for example, when Karel van Mander confidently referred in his *Schilder-boeck* to the painter's ability to reproduce the rainbow's colors; or when Philips Angel, with the same air of confidence, claimed that painting was superior to sculpture in the *paragone*, in that painters can imitate "the rainbow, rain, thunder, lightning, clouds, light, reflections and such things,"[19] while sculptors cannot. Still life painters in particular strove to make their pigments compete with the colors in nature, whether by working within contrasting or complementary color hues in order to increase their luminosity, or by the use of subtle gradations that depicted every stage between the "flowering" and "deflowering" of colors. Within baroque art literature, depending on the different degrees of saturation and luminosity, these tonal stages are described as either weak or strong.

In Van Vogelaer's *Metal Vase with Flowers*, for example, the whole range of rainbow colors described by Aristotle and Alexander—yellow, crimson, purple, dark green, and blue—are "painted" on the canvas. It is precisely because of the association of the flower bouquet with the spectrum of the rainbow that the painting achieves its climax, as the heavenly colors are now re-created with pigments. The flower piece simultaneously demonstrates its materiality and transforms itself into a piece of art: nature

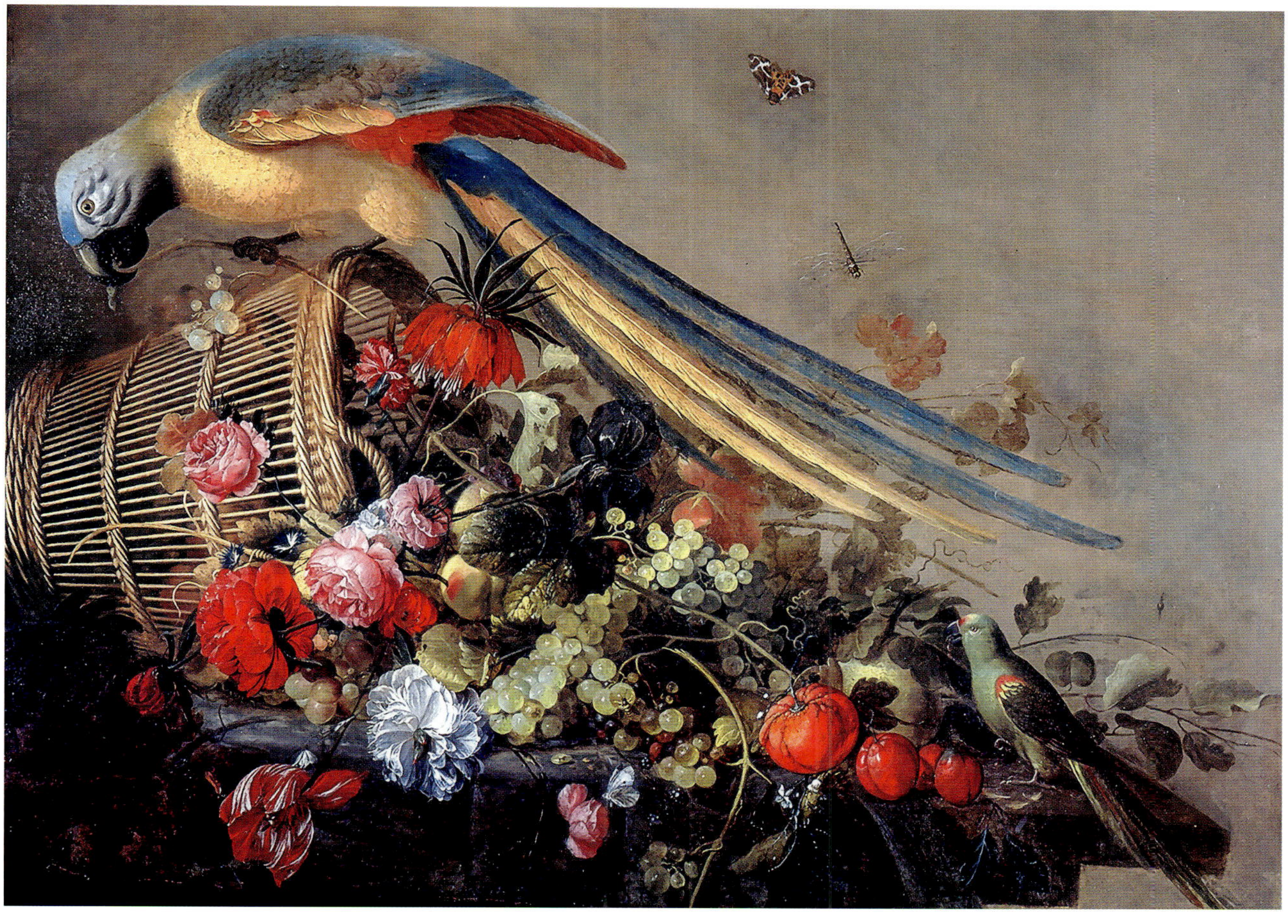

Cornelis de Heem, *Still Life with Parrot and Flower Basket,* ca. 1670. Oil on wood, 85 x 119.5 cm. Private collection (courtesy of the Galerie De Jonckheere). **9.5**

turned into a spectacle, a brilliant, almost dazzling feast for the eyes. The sky is divided into blue and red (or cold and warm, wet and fiery) zones of color according to the baroque system of the elements. As if they were powerful agents of its visible presence, two sunflowers form the center of the bouquet. They seem to brighten it up and visibly move the flowers toward the eye, while other parts of the painting fade into denser brushwork and darker patches of color that still seem wet with paint. The flowers appear to be literally soaked in or saturated with color. Cornelis de Heem's *Still Life with Parrot and Flower Basket*, by contrast, encircles the subject of mimesis both on the level of motif (the parrot signifies the imitation of nature) and on a formal level—in the *paragone* between color in art and color in nature (fig. 9.5). Perched on a basket that has tipped over with the weight of flowers, a huge parrot

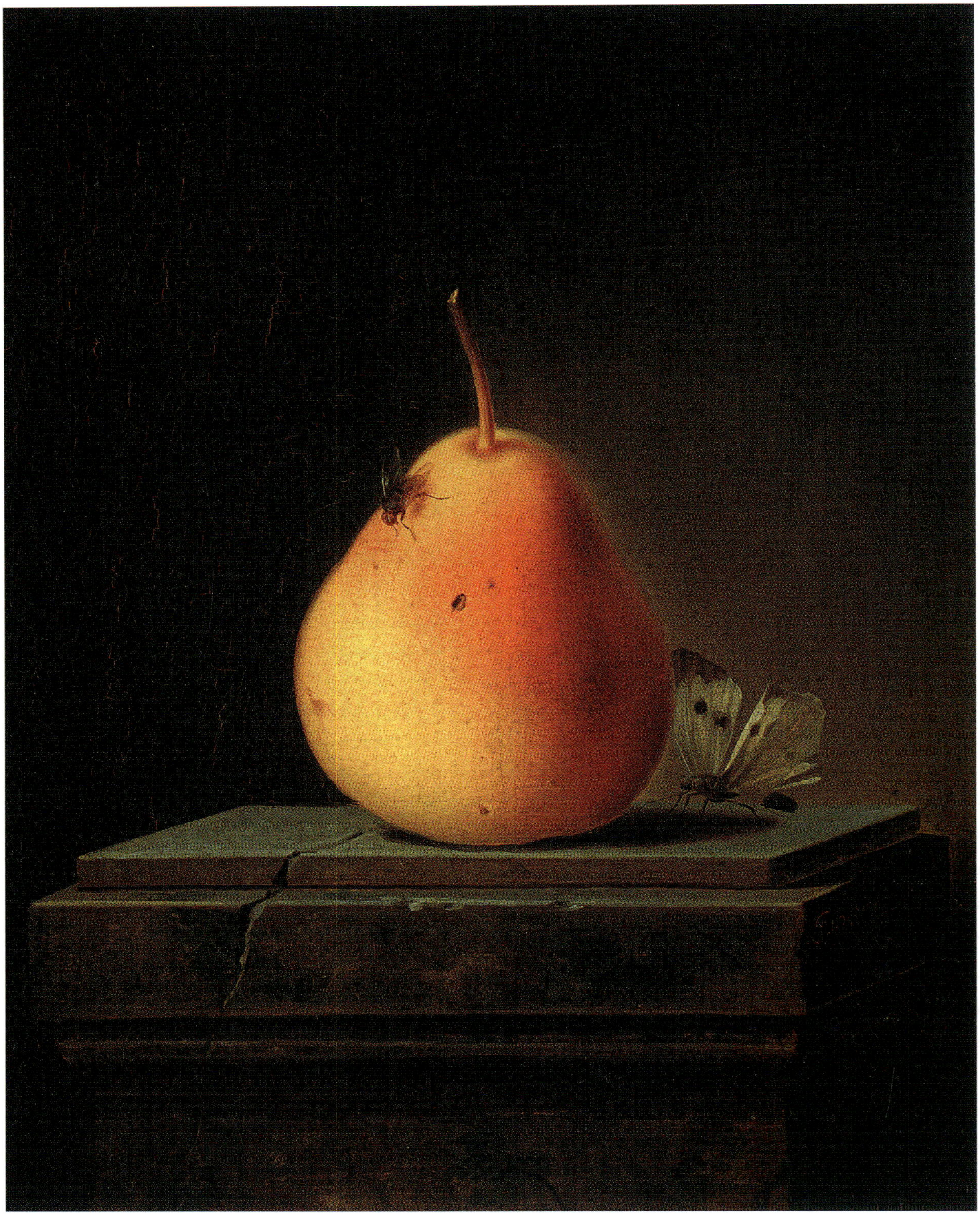

spreads its plumage, dazzling in the light. Our attention, however, is directed not only to the way the tail feathers are drawn in loose brushstrokes that echo the gesture of the artist's hand, but also to the fact that they stretch across the canvas like the stripes of a rainbow, each color lit up with different degrees of intensity. So powerful is the association that it unites such different aspects of color discussion as the materiality of paint, the coloration of the natural object, and color sensation itself (in the eye of the beholder). Still life painting reflects on the differences between pigment, natural object, and visual perception as much as on their similarities and interdependence. It is therefore crucial to understand that, for painters, the competition between natural and artificial colors pertains not only to the variety of color hues that must be produced but also to their varying degrees of saturation, brightness, and luminosity.

Color in Art

AUSTERUS—FLORIDUS

Of course, the artists of antiquity also knew how to mix paint. Nevertheless, as John Gage has shown, ancient writers were concerned with the problem of mixture, because the process produces change and therefore putrefaction.[20] Plutarch, for example, states in his *De defectu oraculorum* that "pigments ground together lose their own color in the process," as "mixing produces conflict, conflict produces change and putrefaction is a kind of change" (Plut. *Moralia* 5.47); he also remarks that this was the reason why painters called a blending of colors a "deflowering" (*phthorax*) (a term Aristotle used for "passing away")—a concept of particular resonance for still life painting. Obviously, it is not flowers alone that flourish and then shrivel but color itself, depending on its application in a pure or mixed form.[21] The crucial question of whether a compound color could be "de-compounded" and dissolved back into pure colors occupied the minds of many ancient writers, and the opposition between blended and pure colors may also have been reflected in the two genera of color (the *austeri colores* and the *floridi colores*) that were used as classificatory terms in treatises on painting since antiquity. Pliny gives the most extensive description: "Now some colors are 'austere' and others are 'florid.' They derive either from nature or from mixture. The florid colors . . . are *minium* [red lead, vermillion], *Armenium* [blue, ultramarine, from azurite], *cinnabaris* [a blood color, see *Hist. nat.* 33.115], *chrysocolla* [copper green], *Indicum* [indigo], *purpurissum* [purple]; the other colors are austere" (Pliny *Hist. nat.* 35.30).[22] It has been argued that "the *floridi colores* simply signified colors which were bright, while the *austeri colores* signified those which were more somber (possibly as a result of having been blended with other colors)."[23] In addition, the oppositional terms reveal a close resemblance to familiar patterns in rhetoric tradition, as the term *austerus* was etymologically connected to the Greek word αυστηροζ, which meant to be "dried up," bitter, or harsh, while a "florid" style may have been associated with elegance, fluidity, and "sweetness."

Other senses were thought to operate as well. Because colors were believed to either flower or "deflower," they consequently tended toward bitterness or sweetness.[24] Aristotle discusses color vision together with other sense perceptions, especially smell and taste, and, furthermore,

9.6 Justus Juncker, *Pear and Insects,* ca. 1765. Oil on wood, 25.8 x 21.5 cm. Städel Museum, Frankfurt am Main / Artothek.

sees taste and color as analogous with respect to their mixing from two primaries (Arist. *De sens.* 442a12–21). Subsequently, Aristotle classes "white with yellow, as rich with sweet," while he compares the color black to a bitter or harsh taste. In fact, the analogies can be extended to both primary and secondary colors (442a17–25). Correspondingly, we also acquire a definition of odors by arranging them according to the species of flavor they resemble. Sharing the same mixture ratio, they become interrelated. There are also certain odors, Aristotle continues, that are pleasant, and to illustrate the distinction, he chooses the smell of flowers, as this is always pleasant. A similar view is held by Plato, who also states that we enjoy the smell of flowers as a pure pleasure (*Philebus* 51a–e).

Flavor and odor, though not representative as such, nevertheless form a genuine part of still life painting's iconography. Together with the imagery of flowering and deflowering, or generation and corruption of color, they make still life a genre that reflects deeply on sense perception. Perhaps Justus Juncker's *Pear and Insects*, in a mildly ironic gesture, sums up this tradition by taking it to extremes (fig. 9.6). The pear's maturity, so splendidly conveyed through mellow and glowing colors that they at once evoke the sweetness and lushness of fruit, contrasts with the cold stone plinth on which the fruit is placed with a great deal of consideration. With its sensory qualities transformed into painting, it has become its own monument.

LAYERS OF COLOR

A passage in *De coloribus* suggests how close coloring in nature comes to coloring in painting. Apart from the more general idea of flowering and ripening dealt with in the text, it explains how nature itself lays colors on the (white) object world, like a painter who works over the canvas with layers of colors and glazes of tints: "The same thing happens with fruits. In their case it is easy to see that the colour of the fruit changes, as the colours are laid on it afterwards. For the fruit of the pomegranate and the petals of the rose are white to begin with, but at last as the juices in them get tinted by ripening, they become shaded off and change again to the colour of sea-purple and red" (796a19–26). The pomegranate, especially, is a favorite example used to demonstrate the different stages in color change (799a10–15). The varied plumage of birds such as the peacock and the pigeon is discussed in similar terms, for at first it is all black but "later on the plumage takes on all these varied hues" (799b9–15).

In Bartolomeo Ligozzi's *Flowers, Fruits, and a Parrot*, we meet the well-known formula of a dark, clouded sky with glimpses of sunlight (fig. 9.7). The composition portrays fruits on different levels and a glass vase with flowers and a parrot, all painted against a landscape background with a fountain, based on a formula common to Roman paintings of the time, particularly in the works of Mario de' Fiori. The landscape setting, too—a sky veined with clouds—is another element of Roman origin, later developed in Florence in the paintings of Bartolomeo Bimbi.[25] The painting was commissioned by the grand duchess Vittoria della Rovere in 1688 for the apartment in the villa of her son, Cardinal Francesco Maria, a passionate collector of still lifes. The painter himself, grandson to the more famous Jacopo, had stayed in Rome during the 1670s. His selection of flowers and fruits displays a bright range of colors. It is crowned by a parrot sitting on a pair of pomegranates so ripe that their skins have burst open, revealing the flesh of the fruit. Their carefully shaded tones and hues finish with a slight tinge of golden color that indicates their softness and maturity. The fruits gleam in the light, and the seeds, still tightly contained within the skin, are painted in the brightest red. The parrot itself joins the lively colorful effect as its plumage dazzles in the light, with shots of green, yellow, and red enriching its feathers, giving it an iridescent glow.

This effect, however, comes close to that of the changeable, or *cangiante* (iridescent), fabrics that were made of silk or lustrous man-made filament yarns, with the warp and filling dyed in the contrasting colors so fashionable in the Renaissance. Viewed from different angles, one or another color predominates, giving the painting its changeable aspect. Within color theory and its crucial subdivisions of "real" and "apparent" colors, the changing colors of bird feathers—which we see in the parrot's plumage, the peacock's tail, and, most often, the pigeon's neck—were placed decisively in the latter category. This was because they derived from the reflection of light on reflective surfaces (more precisely, the spectral interference in the light), meaning that the optical coloration looks different from the "real" pigmentation and therefore produces a visually striking effect. Apparent colors also varied with conditions of illumination and different viewing angles and thus resembled the ephemeral rainbow. True colors, by contrast, occurred substantially in the natural world, as a result of the object's elementary mix. This classification changed, at the latest, around the end of the seventeenth century.

One cannot overestimate the significance of the idea that the variety of colors does not depend on the variety of natural bodies or substances alone. The peripatetic concept of color vision identified a problem that was already acknowledged in ancient times and keenly discussed throughout the Middle Ages.[26] It was only in the seventeenth century, however, that this problem received a new level of scrutiny. How are "apparent" colors generated?

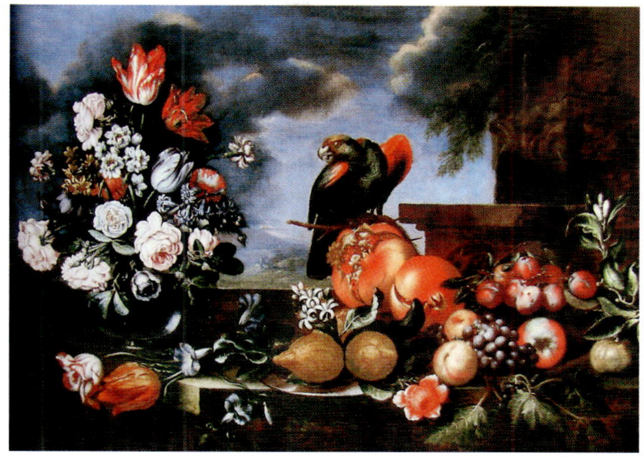

Bartolomeo Ligozzi, *Flowers, Fruits, and a Parrot,* 1699. Oil on canvas, 106.7 x 73.7 cm. Villa Medicea di Poggio a Caiano, Museo della natura morta, inv. Poggio Imperiale 1860, no. 1774. Courtesy of the Ministero per i Beni e le Attività Culturali. **9.7**

The iridescence of the rainbow or the *colori cangeanti* of silk fabrics or birds' feathers cannot be explained without optical models of light, at some point also introducing the spectator and his specific viewpoint into the explanatory model. In fact, the question was closely connected to the difficulty in explaining perspective or subjective vision in general, because the position, quantity, and distance of objects in relation to the beholder cannot be explained by substantialism. Medieval and early modern writers on optics therefore worked on the differentiation between real and apparent colors, and on the reflection and refraction of light, thus gradually breaking away from peripatetic essentialism.

Color Change—Changing Color Concepts

The epistemic shift from substantialism to subjective perception is one of the most thrilling developments in seventeenth-century color theory. Changing fabrics, birds' feathers, the rainbow, the alteration of color in vegetables and fruit: the examples stay the same, but we move on to a revised concept of how color is produced and modified. Robert Boyle picks up the tradition of *De coloribus* when he refers to natural color change in processes of growth and decay, but he no longer approves of the idea that a variety of colors depends on a variety of substances. Because of the correspondence of Boyle's theory to motives in contemporary still life painting, he deserves to be quoted at length:

> 'Tis often thought that a diversity of colours constantly argues an equal diversity in the nature of bodies wherein they reside: but I cannot wholly give into this opinion. For not to mention the taffities, the blue and golden feathers on the neck of the pigeons, and several water-fowl, natural and artificial rainbows, &c., the colours whereof philosophers call not real, but apparent; we see that the contiguous feathers in various birds are some of them red, others white, blue, yellow, &c., and that in several parts of the same feather there is frequently the greatest disparity of colours: so in tulips, july-flowers, and other vegetables; even several parts of the leaf are frequently found of different dyes; tho' no difference is observed in their other properties.[27]

Such are the first lines of his *Experiments and Considerations Touching Colours*, written in 1664. Boyle acknowledges that there is in bodies "a certain disposition of the superficial particles, whereby it sends the light reflected, or refracted to our eyes thus and thus altered," "disposition" in this case meaning the arrangement of corpuscles. One might have the impression that Boyle thought that colors were the result of textures and thus reducible to the mechanical properties of the external object. However, Boyle makes clear that it is the light itself that "produces the sensation of color": "But before we descend to a more particular consideration of our subject, 'tis proper to observe, that colours may be regarded either as a quality residing in bodies to modify light after a particular manner, or else as light itself so modified as to strike upon the organ of sight, and cause the sensation we call colour; and that this latter is the more proper acceptation of the word colour."[28] Thus it becomes clear that colors are not present in the objects but are the result of the interaction of modified light and the eye, which means that all color is apparent color. Boyle had thought of physical-chemical processes of the "composition" and "decomposition" of pigments when referring to color mixture, demonstrating a great interest in atomistic material theories.

Indeed, Democritus had never been completely forgotten, and he was given a new lease on life during the Scientific Revolution. For him, colors, like any sensory quality, are not present in the objects themselves, and knowledge of them is not knowledge of the objects themselves. Democritus was the first to address the problem with which every logically thinking atomist was to be confronted: that sensory qualities can only be considered "secondary qualities" of the corporeal world.

We know that although Aristotle had made flavor, odor, and color analogous, he decisively rejected the idea, expressed by Democritus and others, that all of these sensations could be related and even reduced to touch. The idea that the two basic sets of black and white, bitter and sweet, could be complemented by the opposites of rough and smooth never seemed right in peripatetic thinking (Arist. *De sens.* 442a30–b12). For Democritus, by contrast, it was essential to understand that color and flavor were not, as such, inherent in bodies, for they result from effects caused merely by the sizes and shapes of atoms and their position within compounds—for example, white color and sweet taste were made of smooth and mobile atoms that had no edge and did not cast much shadow, whereas the atoms of black objects were rough, jagged, and uneven. To be precise, "'bitter' is composed of small, smooth, round atomic figures whose surfaces moreover are furnished with hooks; as a consequence bitter is sticky and viscous" (Thphr. *De sens.* 66), while "what is smooth is white; since what neither is rough nor casts shadows nor is hard to penetrate,—all such substances are brilliant" (73–75).

Boyle picks up the analogy between color tone and surface structure when he tells the story of "a blind man who distinguished colours by the touch." This story was reported to him by John Finch, a Cambridge-educated anatomist who was appointed British ambassador to Florence. During a stay in Maastricht, Finch was introduced to a blind man who could allegedly tell apart the color of dyed ribbons; the touch of his fingers allowed him to sense the difference in the degree of asperity in the textile.[29] The idea that it was possible to feel color was supported by the corpuscular model of matter prominent in scientific circles in the second half of the seventeenth century. One of the truisms of that era, however, also shows how these discussions were accompanied or even promoted by advanced methods of producing lenses and optical instruments. The tip of a blind man's finger and a microscope lens form a close alliance, for the senses of both touch and sight, at their most acute, will arrive at the same conclusion. Color becomes the result of light meeting and interacting with a rigid surface structure, the light being analogous to the blind man's finger. "Yet I am apt to suspect," Boyle concludes, "that if we had perfect microscopes, we might discern in the physical surfaces of bodies, both a great many latent inequalities, and the particular size, figure, and situation of the extremely little bodies that cause them; and perhaps might perceive, among other varieties, how those little protuberances and cavities interrupt and dilate the light."[30]

CORPUSCLES AND PIGMENTS

A similar interest in the structure of compound matter was evident at the court of Rudolph II in Prague. It had certainly been at the center of discussions on color, and Anselmus de Boodt's influential book on gems, for example, is of particular interest, as de Boodt speculated about the existence of a distinct atomic structure in minerals. Diaphanous bodies, he reasoned, are bodies in which the atoms are so tightly compounded that one cannot discern their single particles. But one could also conclude from reading de Boodt that crystal has no real color in itself, for when it is ground to fine powder it becomes diaphanous, and that this may apply to any gemstone when sufficiently ground. De Boodt then focused on the gemstones' optical properties that depend on their crystalline structure and morphologies, which result in (apparent) color effects: "In perfectly transparent or perspicuous gems or stones, colours have their originall from the refraction or reflection of the light, which diversity of reflection in Crystalls, is caused by the multiplicitie of their superficies." In his lapidary, de Boodt famously presented a three-color primary system as the basis for color mixing that derived from the use of gems as pigments: "The *primarii colores*, or especiall colours which arise not from commixtion of other colours are these: White, black, blue or skie colour, yellow, red. . . . Almost all kinds of colours are caused by the various mixtures of these colours. But in their own originals they are not mixt with others."[31] Subsequently, he described white and black as adding shades to color, not generating it in the first place. Here, two lines of our inquiry meet, as ground crystal obviously intermingles with color pigment, and the efforts to explain coloration in nature combine with efforts to explain it in art.

By the beginning of the seventeenth century, then, the quest to provide a modern foundation for painting culminated in the discovery that color mixing was governed by a simple law: that all colors may be made from just three, the painter's primary colors, together with white and black.[32] In fact, most human color perceptions can be generated by a mixture of the three primary colors yellow, red, and blue, a discovery that became quite important for color reproduction in print media at the beginning of the eighteenth century. No mixture of colors, though, can produce a fully pure color perceived as identical to a spectral color. But the idea of letting natural color coincide with the ratio of pigments in pigment mixtures completely revolutionized color theory. It was because of the coincidence of the explanatory models that ancient color terms became more and more difficult to understand, given that the two domains were clearly separated in the past. For the same reason, and at the same time, we can observe a "modern" tendency to explain color mixing after the model of painting.

Scientists now paid attention to art. Robert Boyle analyzed the blended colors on the painter's palette by examining them under his microscope. He even apologized for his disciplinary encroachment "to meddle with the mixing of Pigments," as this was "no inconsiderable part of the Painter's Art" and not within his scope as a scientist. But pigment mixture became a paradigm of Boyle's corpuscular color concept, and he therefore used it as a model in distancing himself from peripatetic substantialism. Boyle then proved scientifically what painters already knew through doing: "And with an Excellent *Microscope*, where the *Naked* Eye did see but a Green powder, the *Assisted* Eye as we noted above, could discern particular Granules, some of them of a Blew, and some of them of a Yellow colour, which Corpuscles we had beforehand caus'd to be exquisitely mix'd to compound the Green."[33] In 1664 Boyle confirmed that the painter's palette could reproduce all of the tones and hues seen in nature through a combination of these primary colors. Pigment mixture may lack the luminosity of light, but with the addition of black or white to make shades and tints it can mirror any chromatic gradation, which means that the colors of art can be regarded as parallel to the colors of nature.

> There are but few Simple and Primary Colours (if I may so call them) from whose various compositions all the rest do as it were Result. For though Painters can imitate the Hues (though not always the Splendor) of those almost Numberless differing colours that are to be met with in the Works of Nature, and of Art, I have not yet found, that to exhibit this strange Variety they need imploy any more than White and Black, and Red, and Blew, and Yellow; these five, variously Compounded, and (if I may so speak) Decompounded, being sufficient to exhibit a Variety and Number of Colours, such, as those that are altogether Strangers to the Painter's Palletts, can hardly imagine.[34]

We find similar remarks in the private notes of one of Boyle's contemporaries, the physician and geologist Martin Lister, a fellow of the Royal Society who was also looking keenly at paintings when reflecting upon color: "That wil[l] looke blew by day, looks Green by night: the reason is, for that the light of a candle is yellow, yellow superindulged to blew makes a green, as is known in painting. The great secret of Veralst [*sic*] that Dutch flower painter, was not to mix colours, but to paint dark first, & then to wash over with a light e.g. if he paints a green Tree."[35] Color in nature and art partakes of the same visual regime—that is, in the diffusion of light into apparent and ephemeral color. By this time, contemporaries would have noted the delight Verelst must have felt in painting the optical effects of reflected lights and shiny glints of luminosity. His flowers shine like an apparition, immortalizing an otherwise fleeting moment (fig. 9.8). There is no longer a desire to touch and hold, but a slight prickle in the tip of his finger, which led Samuel Pepys to note in his diary that this was "the finest thing I ever saw in my life, the drops of dew hanging on the leaves, so as I was forced again and again to put my finger to it to feel, whether my eyes were deceived or not."[36]

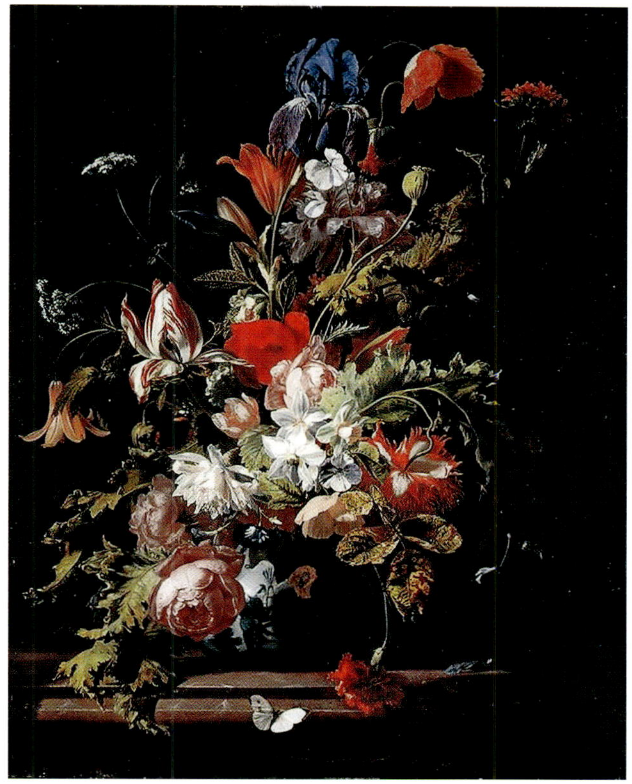

9.8 Simon Pietersz Verelst, *Flower Still Life,* 1680s. Oil on canvas, 80 x 66 cm. Private collection.

A CRYSTALLINE WORLD

Particle theory developed over time, of course, and color is no longer thought to be inherent in natural bodies. With Robert Boyle, Robert Hooke, Antoni van Leeuwenhoek, and other microscopists and natural scientists of the seventeenth century, the scope of visibility was widely explored afresh. However, it was owing not only to the new visible world discovered by the microscope but also to the revival of ancient Greek atomism that peripatetic substantialism was discarded and the solid object world converted into a vibrating particle universe. The seventeenth century witnessed the growth of a range of experimental sciences, a development of considerable epistemological significance. Nevertheless, the early atomistic approaches of Pierre Gassendi, Isaac Beeckman, Kenelm Digby, and René Descartes are not completely detached from the idea of a primordial "white earth"; they only explain it in a different way. Their originality lies in the capacity to put an old idea to new use, and, as a consequence, the definition of color and color vision became the proof stone for scientific discussions of the era.

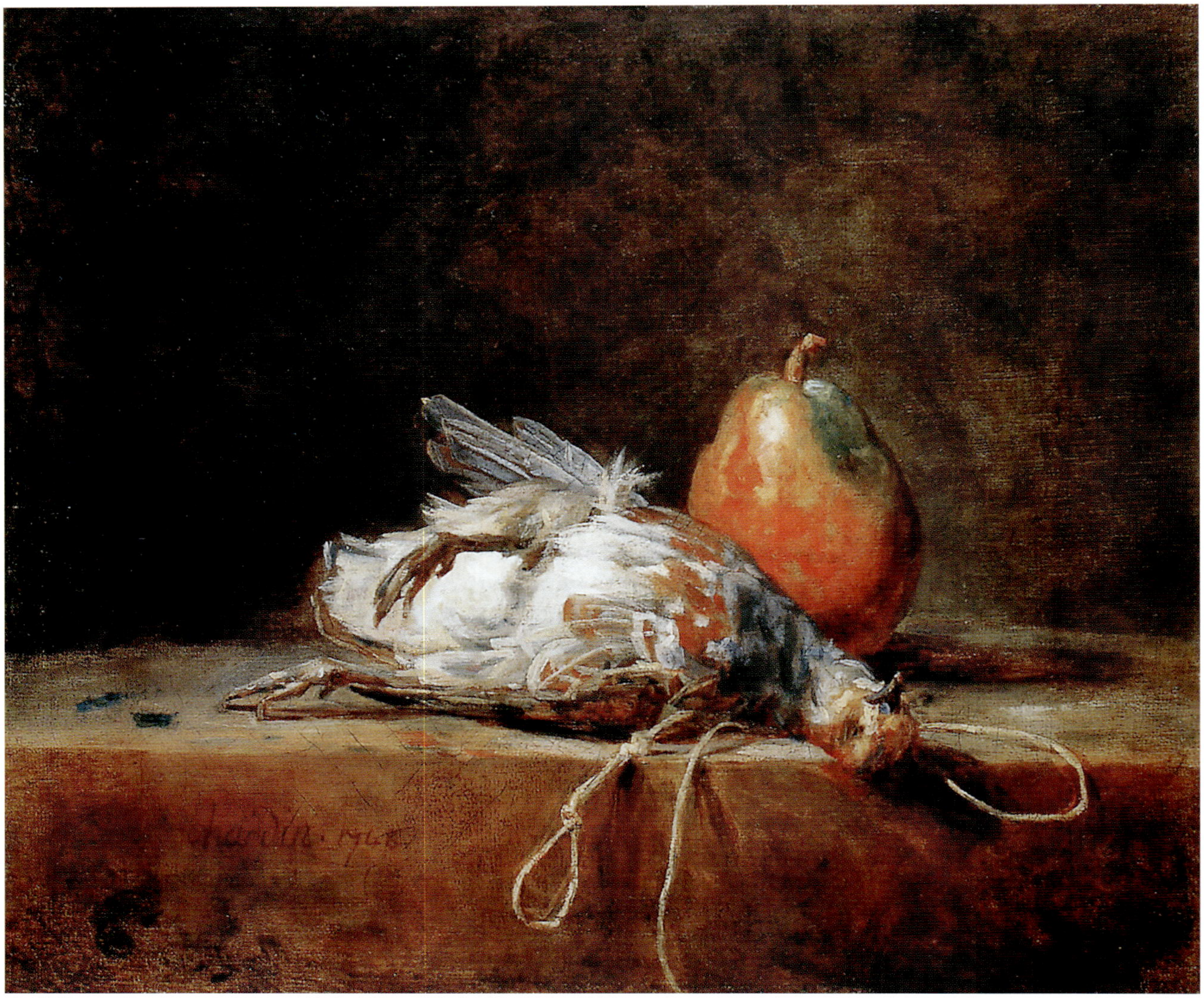

9.9 Jean-Baptiste-Siméon Chardin, *Dead Partridge with Pear on a Stone Ledge*, 1748. Oil on canvas, 39 x 46.5 cm. Städel Museum, Frankfurt am Main / Artothek.

When looking through a microscope, color does indeed seep away. Physical matter becomes diaphanous, just as the ancients believed that the elements of earth, water, and air were white. Under the microscope, the colors of flower petals, leaves, soil, even blood cells are washed out and turned into clusters of transparent globuli or highly porous network structures. Early microscopists were deeply impressed by the constancy of geometrical shape, right down to the particle level, and interpreted it as the result of an internal force by which the corpuscles were formed into a special configuration. The leaf of an apple tree or a red blood corpuscle seen through Van Leeuwenhoek's single-lens microscope revealed hardly any color at all. Van Leeuwenhoek, when using his single lens, described how difficult it was to distinguish the discrete blood units from their surrounding plasma, and observed that they were not at all silhouetted against their background. It is of similar significance that when viewed through a microscope, not even the earth itself is clothed in color. Granulation can be found practically everywhere, and there is a geometrically regular arrangement of crystalline particles that absorb, reflect, and refract the white light, thus creating the sensation of color only in the eye of the beholder.

There may have been painters in the first half of the eighteenth century who were interested in the experience of the optical nature of light and its interaction with paint and pigment. The velvet skin of Chardin's fruits, for instance, is no longer warmed and colored by the sun but is gently grazed by the artificial light of the artist's studio (fig. 9.9). And is it even skin, or just paint? There is only one kind of matter or particle universe in which the objects exist, and the rhymes of their shape and form are reinforced by the subtle interchange of color and reflection among them, offering an infinite series of correspondences.[37] Chardin's interest in tonality enables him to enhance his paintings with a soft diffusion of light that appears to fill the air around the objects he paints. The light, just like his brush, touches, passes over, and reveals the surfaces of the objects, their velvety substance now expressing pigment as well. Already with Willem van Aelst or Simon Verelst, color has slid into the register of powdery and iridescent paint, which was so characteristic of still life paintings of the late seventeenth century. Art can surely not be explained by science alone, but we need to learn much more about the historical discourses on sense perception, color, and pigments if we are to understand how author and playwright Louis-Sébastien Mercier, in his "Epître à un ami" (1784), could say:

Pray, what gives the sun its gracious warmth
The star at night its solitary beauty
The flowers, their hue, the fruits, this gift
The silent hedgerows, their captivating affinity?
They are your senses, friend; kings of nature
Godly creators of sensuality so pure.
The eye paints the blue that colors the skies
The ear forms the tones of melodious cries
The universe, devoid of their sweet potent magic
Would be but chaos: no color, no fabric.
It is you who composes brushstrokes of the scene:
The world is a blank canvas, and you hold the brush.[38]

Notes

1. Galileo, *Saggiatore*, 105.
2. Newton, *Opticks*, 2.3.54.
3. See Taylor, "Concept of Houding."
4. I use the following translations in this chapter: Loveday and Forster's translation of Pseudo-Aristotle, *De coloribus* (abbreviated Ps.-Arist. *De. col.*); W. S. Hett's translation of Aristotle, *De sensu et sensibilibus*; H. D. P. Lee's translation of Aristotle, *De meteorologica;* and G. M. Stratton's translation of Theophrastus, *De sensibus.* All other translations are my own, unless otherwise noted.
5. See Gottschalk, "*De Coloribus* and Its Authors," 59.

6. See Struycken, "Colour Mixtures According to Democritus and Plato," 287, 291. See also Stratton, *Theophrastus and the Greek Physiological Psychology*.

7. Struycken, "Enkele Oudgriekse denkbeelden over kleur."

8. See ibid.

9. See Lloyd, "Hot and the Cold."

10. Gage, "Colour in History," 106.

11. "This is what the painters should seriously study. . . . At sunset we can see different things tinged by red colour, as with the earthen ground, the stones, the tiles and also the faces of people: and when the rays of the sun reach them, or at least cause a mild reflection, they assume a luminous red, fiery and glowing colour." Van Mander, *Schilder-boeck*, chap. 7, "Den Grondt der edel vrij Schilder-const," fol. 29r–v.

12. See Struycken, "Enkele Oudgriekse denkbeelden over kleur."

13. See ibid. for numerous examples.

14. Aristotle *De meteorologica* 342a34–342b12. See also Theophrastus *Libellus de igne* 75.

15. See Casciù and Chiarini, *Natura morta*, 32.

16. On the cardoon, see Theophrastus *Historia plantarum* 6.4.10–11; Pliny *Historia naturalis* 19.152–53; Athenaeus *Deipnosophistae* 70a–71c; Palladius *Opus agriculturae* 11.11.1. See also Dalby, *Food in the Ancient World*, 28.

17. Casciù and Chiarini, *Natura morta*, 32.

18. Alexander of Aphrodisias *Commentaria in Aristotelem Graeca* 3.2. For analysis, see Gage, "Locus Classicus," 9.

19. "Maer om ons geding tot een eynde te brenghen, soo willen wy de Beeld-houwers den laetsten doodt-slach geven: wy seggen dan, dat de Schilder-Konst veel al-gemeener is, om dat sy de Natuyre veel over-vloedelijcker weet na te bootsen: want boven dien dat sy aff-beelt alderley Dieren, als, Vogelen, Vissen, VVormen, Vliegh-en, Spinnen, Ruspen, soo kanse ons oock verthoonen alderhande Metalen; onderscheydende de selve, als Goudt, Silver, Metael, Koper, Tin, Loodt, en wat des meer is. Men kann door haer uyt beelden den Regen-Boogh, Regen, Donder, Blixem, VVolcken, Waesem, Licht, Weerschijn, en dierghelijcke dinghen meer, als, 't opgaen vande Zonne: de Morgenstondt: het dalen van de Zonne: den Avondt: de Nacht-verlichtende Maen, met haer Leydtslieden, de Sterren: de Spiegelinghe in 't Water: het Hayr van Mensche: het schuymbecken der Paerden &c. van welcke de Beeldt-houwers gheen van ellen weten na te botsen." Angel, *Lof der schilderkunst*, 25–26.

20. Gage, "Colour in History," 118–19.

21. Ibid.

22. Quoted in Pollitt, *Ancient View of Greek Art*, 322.

23. Ibid., 324.

24. See Fehrenbach, "Calor nativus—Color vitale."

25. Casciù and Chiarini, *Natura morta*, 46.

26. But intermediary steps must be considered. Already in *De coloribus*, for example, the opposition between true and apparent colors is softened because of the different perceptual and media-based conditions of vision. It is, though, not invalid in principle; it is only confused in reality (793b13–30). We very rarely see pure color: color is almost always blended, whether by a physical mix in which color has material existence, by a perceptual mix, or, usually, by a combination of both. According to traditional peripatetic physics, bodies alter one another by acting through their elemental qualities of moisture, dryness, heat, and cold. Thus alteration means replacing one quality with another, or one form with another, and color change can to some extent be explained by processes of heating, cooling, moistening, or drying. But it is also recognized that the chromatic variety does not derive only from the elementary disposition of substance. Color is not necessarily the property of natural objects; it is also the result of some kind of optical mixing of several species of color that overlap one another, such as can be found, for example, in the perception of "water, clouds and birds' plumage. For these owing to their smoothness and the rays that fall upon them, mixed in various ways, produce various colours" (793b8–12).

27. Boyle, *Experiments and Considerations Touching Colours*, 1.

28. Ibid., 2.

29. "To discern the colour of the ribbon, he places it betwixt his thumb and his fore-finger, but his most exquisite perception is in his thumb, and much better in the right than in the left. . . . Then the doctor desired to know what kind of difference he found in colours by his touch. To which the blind man reply'd, that all the differences he observed, was a greater or less degree of asperity; for, says he, black feels like the point of needles, or some harsh sand, while red feels rather smooth." Ibid., 11.

30. Ibid., 10.

31. Nicols paraphrasing de Boodt in *Lapidary*, 10–14, quotations on 12 and 14.

32. "The notion that white and black are different from the chromatic colors and could not generate them, was perhaps the most radical fruit of Renaissance revisions of the concept of color. Since antiquity these contraries had dominated the realm of color, and now they were cast out and placed with the grays." Shapiro, "Artists' Colors and Newton's Colors," 601.

33. Boyle, *Experiments and Considerations Touching Colours*, 41.

34. Ibid., 220.

35. MS Lister 39, Bodleian Library, Oxford, fol. 88v.

36. Pepys, *Diary of Samuel Pepys*, 9:514–15.

37. Cf. Hughes, *Nothing If Not Critical*, 75–76.

38. Mercier, "Epître à un ami," in *Mon bonnet de nuit*, 1:220.

Bibliography

Alexander of Aphrodisias. *Commentaria in Aristotelem Graeca*. Edited by Michael Hayduck. Berlin, 1899.
Angel, Philip. *Lof der schilderkunst*. Leiden, 1642.
Aristotle. *De sensu et sensibilibus (On Sense and Sensible Objects)*. Translated by W. S. Hett. In Aristotle, *On the Soul, Parva naturalia, On Breath*. Cambridge: Harvard University Press, 1936.
———. *Meteorologica*. Translated by H. D. P. Lee. Cambridge: Harvard University Press, 1951.
Boyle, Robert. *Experiments and Considerations Touching Colours*. London, 1664.
Casciù, Stefano, and Marco Chiarini. *Natura morta: Still-Life Painting and the Medici Collections*. Florence: Contemporanea Progetti, 2006.
Dalby, Andrew. *Food in the Ancient World*. London: Routledge, 2003.
Fehrenbach, Frank. "Calor nativus—Color vitale: Prolegomena zu einer Ästhetik des 'Lebendigen Bildes' in der frühen Neuzeit." In *Visuelle Topoi: Erfindung und tradiertes Wissen in den Künsten der italienischen Renaissance*, ed. Ulrich. Pfisterer and Max Seidel, 151–70. Berlin: Deutscher Kunstverlag, 2003.
Gage, John. "Colour in History: Relative and Absolute." *Art History* 1 (1978): 104–30.
———. "A *locus classicus* of Colour Theory: The Fortunes of Apelles." *Journal of the Warburg and Courtauld Institutes* 44 (1981): 1–26.
Galileo Galilei. *Il saggiatore*. Rome, 1625.
Gottschalk, H. B. "The *De coloribus* and Its Authors." *Hermes* 92 (1964): 59–85.
Hughes, Robert. *Nothing If Not Critical: Selected Essays on Art and Artists*. London: Harvill Press, 2001.
Kranz, Walther. "Die ältesten Farbenlehren der Griechen." *Hermes* 47 (1912): 126–40.
Kuehni, Rolf G. "Cangiante: A Fabric and a Coloristic Device in the Art of the Renaissance." *Color Research and Application* 21, no. 5 (1996): 326–30.
Leibniz, G. W. "Betrachtungen über die Erkenntnis, die Wahrheit und die Ideen." In Leibniz, *Philosophische Schriften*, ed. Hans Heinz Holz, 1:32–47. Darmstadt: Wissenschaftliche Buchgesellschaft, 1985.
Lloyd, G. E. R. "The Hot and the Cold, the Dry and the Wet in Greek Philosophy." *Journal of Hellenic Studies* 84 (1964): 92–106.
Mercier, Louis-Sébastien. *Mon bonnet de nuit*. Vol. 1. London: G. G. & J. Robinson, 1798.
Newton, Isaac. *Opticks*. London, 1704.
Nicols, Thomas. *A Lapidary, or The History of Pretious Stones*. Cambridge: Thomas Buck, 1652.
Pepys, Samuel. *The Diary of Samuel Pepys*. Edited by Robert Latham and William Matthews. Vol. 9. London: Harper Collins, 1976.
Pollitt, J. J. *The Ancient View of Greek Art: Criticism, History, and Terminology*. New Haven: Yale University Press, 1974.
Pseudo-Aristotle. *De coloribus (On Colours)*. Translated by T. Loveday and E. S. Forster. Oxford: Clarendon Press, 1913.
Shapiro, Alan. "Artists' Colors and Newton's Colors." *Isis* 85 (1994): 600–627.
Stratton, G. M. *Theophrastus and the Greek Physiological Psychology Before Aristotle*. London: Allen & Unwin, 1917.
Struycken, P. "Colour Mixtures According to Democritus and Plato." *Mnemosyne*, 4th ser., 56 (2003): 273–305.
———. "Enkele Oudgriekse denkbeelden over kleur." In *Aristoteles' over kleuren*, ed. R. Ferwerda and P. Struycken, 97–157. Budel: Damon, 2001.
Taylor, Paul. "The Concept of Houding in Dutch Art Theory." *Journal of the Warburg and Courtauld Institutes* 55 (1992): 210–32.
Theophrastus. *De sensibus*. Translated by G. M. Stratton. In Stratton, *Theophrastus and the Greek Physiological Psychology Before Aristotle*, 15–64. London: Allen & Unwin, 1917.
van Mander, Karel. *Het Schilder-boeck*. Haarlem: Paschier van Wesbusch, 1604.
Zaccolini, Matteo. *De colori*. MS Ashburn 1212 (ca. 1620). Florence, Biblioteca Medicea Laurenziana.

10.

Counterfeit Chimeras

Early Modern Theories of the Imagination and the Work of Art

CLAUDIA SWAN

Imaginations are for the most part false.

—ARISTOTLE *DE ANIMA* 3.3

"Many Strange Forms of Monstrosities and Demons"

The German painter and printmaker Martin Schongauer's vivid engraving *The Tribulations of Saint Anthony* (ca. 1470–75) famously features a pack of nine hideous demons whose bodies are conjugated of bestial parts (fig. 10.1). The fantastic creatures who beset the serene saint comprise an array of naturalistic animal parts: tentacles, spines, wings, fur, horns, fins, hooves, and scales have been recombined in bodies of creatures that form a concentric team of monstrous apparitions. The rendering of the ascetic saint under siege has recently been called "a milestone in the representation of demonic corporeality."[1] Since the time of its creation, it has also been recognized as an artistic feat worthy of emulation. Schongauer's ability to render the hybrid and unstable qualities of these demons by reconfiguring identifiable features—his ability to create naturalistic phantasms—occasioned the print's early renown and wide impact. In the later fifteenth century and the early decades of the sixteenth, it spawned a number of copies and variations—from prints by the Netherlandish artists Monogrammist FVB and Israel van Meckenem and a convoluted woodcut by Lucas Cranach, to Mathias Grunewald's gruesome version in a panel of the *Isenheim Altarpiece*, painted for the hospital associated with the Monastery of St. Anthony in Schongauer's hometown of Colmar (figs. 10.2, 10.3).[2] Like other prints by Schongauer, *The Tribulations of Saint Anthony* gained renown in Italy, too. It even, it is said, inspired the first painting by Michelangelo Buonarroti. Giorgio Vasari and Ascanio Condivi, in their respective biographies of Michelangelo, both describe his re-creation of Schongauer's nightmarish evocation.[3] The relevant passage of Vasari's *Vite de più eccellenti pittori, scultori e architettori* concerns the development of Michelangelo's divine gifts. Vasari writes that, through the practice of art, the extraordinary talents of the young Michelangelo became more readily evident,

> as began to be made clearly manifest in the copy that he executed of a printed sheet by the German Martino, which gave him a very great name. For there had come to Florence at that time a scene by the above-named Martino, of the Devils beating S. Anthony, engraved on copper, and Michelagnolo copied it with the pen in such a manner that it could not be detected, and then painted the same sheet in colours, going at times, in order to counterfeit certain strange forms of devils, to buy fishes that had scales bizarre in colouring; and in that work he showed so much ability, that he acquired thereby credit and fame.[4]

Condivi, too, emphasizes the artist's recourse to the natural entities from which Schongauer's hybrid demons are constructed; Condivi writes of the print that Michelangelo

> copied it on a wooden panel; and, having been provided . . . with paints and brushes he composed it in such a way and with such differentiations that it . . . aroused wonder in anyone who saw it. . . . In making this little picture, since it contained, besides the image of the saint, many strange forms of monstrosities and demons, Michelangelo worked with such diligence that he would not apply color to any part without first consulting nature. Thus he would go off to the fish market, where he observed the shape and the coloring of the fins of the fish, the color of the eyes and every other part, and he would render them in his painting, so that by bringing it to that perfection of which he was capable, from that time he excited the admiration of the world.[5]

Translation of printed to painted forms would necessarily require articulation in color, and for that the observation

This article has been a long time in the making and has benefited from the generous advice, expertise, and support of a number of friends and colleagues—especially Stuart Clark, Lorraine Daston, Frank Fehrenbach, Jessica Keating, Sarah McPhee, Marco Ruffini, Jon Swan, John Tresch, and Fernando Vidal. For their invaluable feedback, I thank the audiences at the Penn Humanities Forum at the University of Pennsylvania, the Netherlands Institute for Advanced Study, Northwestern University, and the Art Institute of Chicago, where I have presented versions of this text. For her support and especially for making it possible for me to participate in the initial workshop meeting of Vision and Its Instruments in the Early Modern Period, in Florence, I remain deeply grateful to Alina Payne. For gamely engaging in a very long-term conversation about the matters at hand, many warm thanks to Peter Parshall. All monstrous errors and misconceptions are my own.

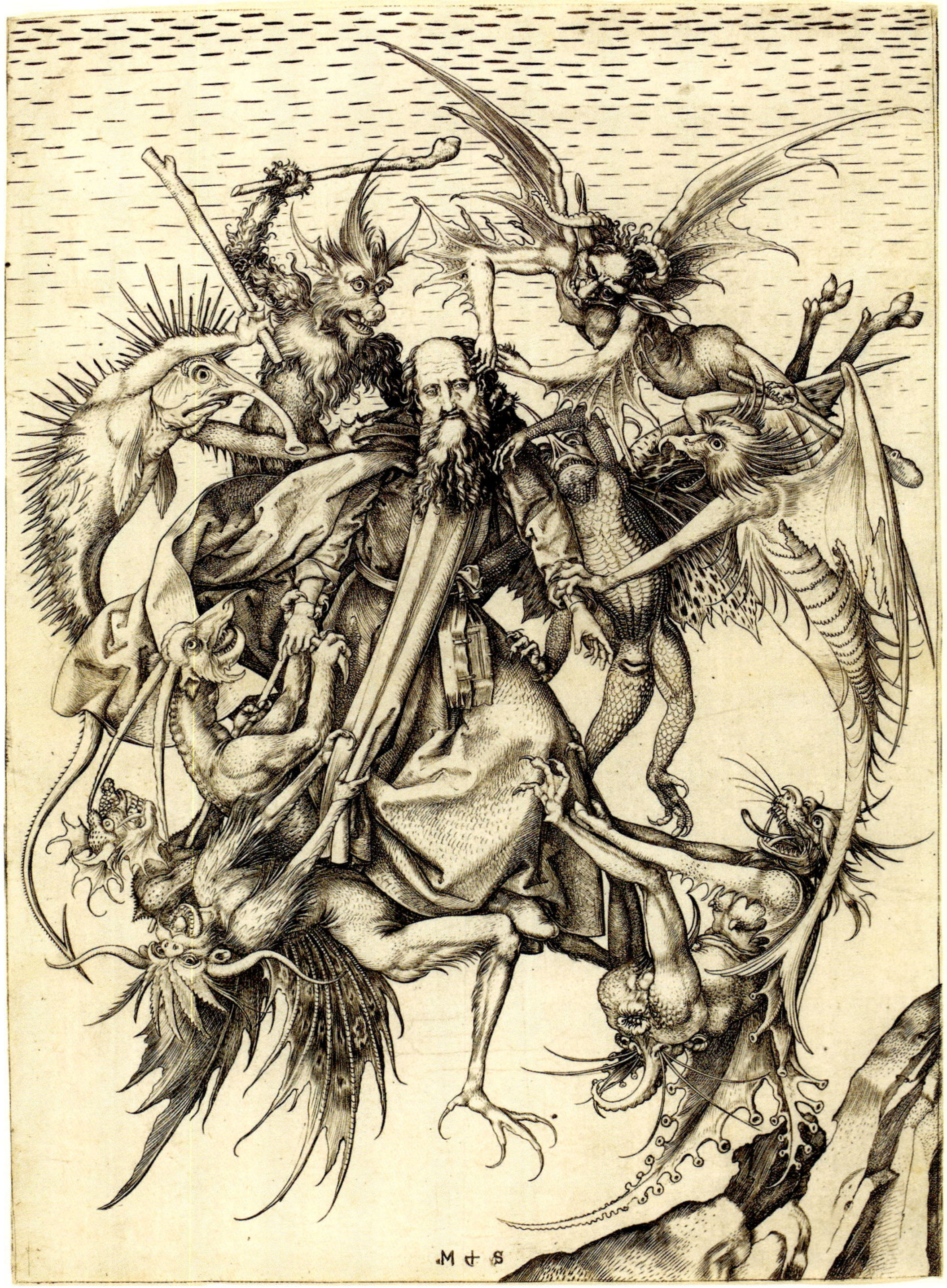

10.1 Martin Schongauer, *The Tribulations of Saint Anthony,* ca. 1470–75. Engraving on laid paper, 31.5 x 23.2 cm (12 3/8 x 9 1/8 in.). National Gallery of Art, Washington, D.C., Rosenwald Collection 1943.3.57. Courtesy of the National Gallery of Art, Washington, D.C.

of actual specimens was surely useful. In fact, Condivi introduces this work as Michelangelo's first painting on the heels of citing the young artist's ardent desire to use colors. That the naturalism of these hybrid fictive creatures was so crucial is as significant, in my view, as Vasari's use of the verb *contrafare* to describe the process by which Michelangelo imitated Schongauer's creations: Vasari writes that Michelangelo went to the fish market "per contrafare alcune strane forme di Diavoli."[6] Whether the small panel painted in tempera and oil, recently acquired by the Kimbell Art Museum in Fort Worth, Texas, is indeed Michelangelo's first painting (fig. 10.4), it illustrates the rendering, in a faithful copy, of Schongauer's evocation of demonic phantasms—and thus embodies the sort of imitative exercise for which Michelangelo acquired "credit and fame."

Michelangelo's early imitative exercise is at least as novel as it is complex. In both Vasari's and Condivi's *Lives* of Michelangelo, the young artist is said to gain sudden prominence on account of his imitating nature and/or other works of art so closely that his copies surpass their sources; both biographies subsequently describe similar works, all indiscernible as replicas. Moreover, imitation seems to have encompassed process as well as product. While Michelangelo could legitimately imitate Schongauer's composition by copying it ("la ritrasse di penna, di maniera che non era conosciuta"), in order to *contrafare* the demons, Vasari writes, he returned to the real to reengage it, to produce his own hybrids—mimicking the process by which products of the imagination originate in what we actually perceive. Michelangelo's rendering of the demons incorporated "scales bizarre in colouring," which the Italian artist observed in the marketplace and transformed from slippery object into subject matter of another artist's conjuring. They were conjured from experience—as, in turn, were Schongauer's—and transformed into wondrous products by a process of recombination. The hybrid demons animating Schongauer's engraving are so many figures of

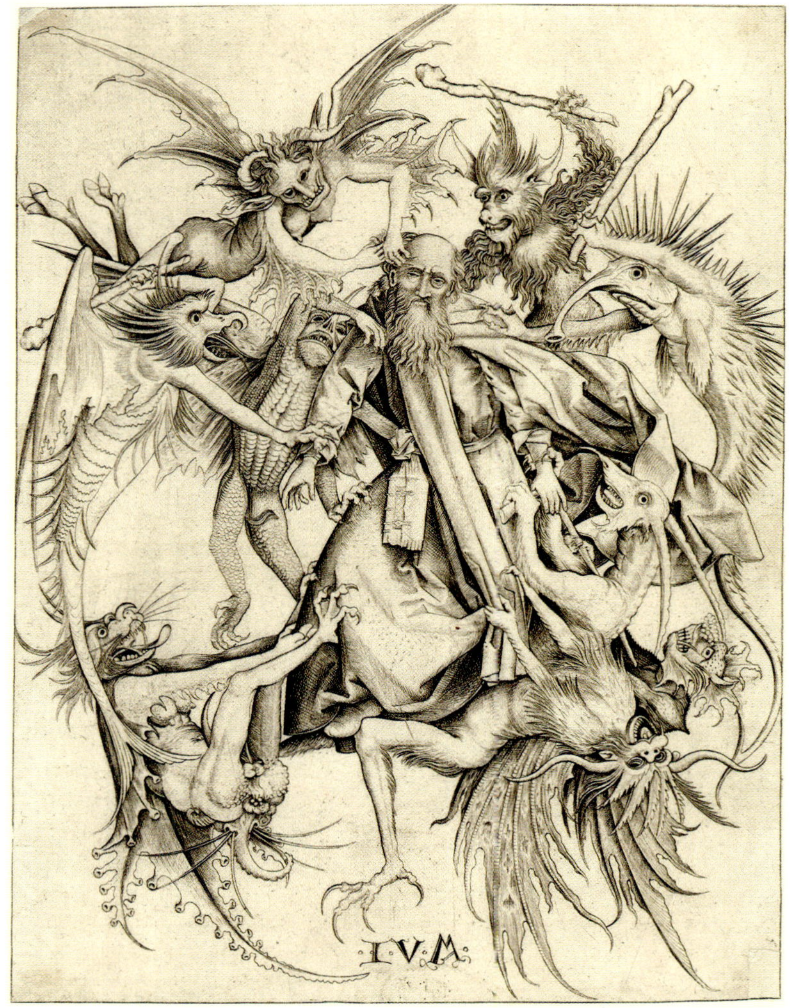

10.2 Israel van Meckenem, *The Tribulations of St. Anthony*, ca. 1480. Engraving, 28.4 x 22.2 cm. The British Museum 1845,0809.343 AN59870. © Trustees of the British Museum.

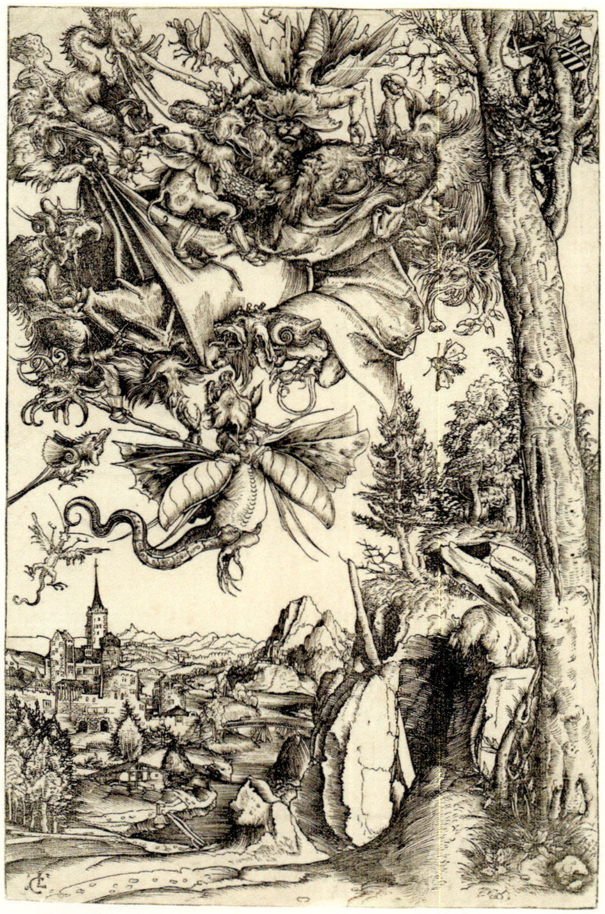

10.3 Lucas Cranach, *The Temptation of St. Anthony,* 1506. Woodcut, 40.4 x 27 cm. The British Museum 1895,0122.255 AN48707. © Trustees of the British Museum.

the imagination or, as this chapter aims to show, represent combinatory *phantasia.*

Real chimeras, imaginary creatures consisting of naturalistic parts, figure broadly in early modern art literature and practice, as they do in a range of literary, philosophical, and even medical contexts of the era. In his defense of painting as a liberal art, worthy of elevation to the status of a theory, Cennino Cennini (ca. 1370–ca. 1440) wrote that painting requires "imagination [fantasia] and skill of hand," and deemed painters capable of discovering "things not seen, hiding themselves under the shadow of natural objects, and . . . [of] fix[ing] them with the hand, presenting to plain sight what does not actually exist." He invoked the power to create real fictions—real chimeras—when he argued that the artist be able to produce precisely the sort of chimera Horace had proscribed: "the painter is given freedom," Cennini asserted, "to compose a figure, standing, seated, half-man, half-horse, as he pleases, according to his imagination."[7] The fearsome power of the imagination to recombine sensory data—in the form of images—and to produce new, previously nonexistent entities was experienced by dreamers and artists alike. Insofar as they test the limits of credibility, represent the maker's skill at conjuring illusions, and evince the power and the vulnerability of the imagination, real chimeras occupy a fertile, troubled intersection between perception and judgment.

What, for early modern artists, was at stake in generating an image, a semblance, of an imaginary entity—a convincing fiction, or "real chimera"? And how do naturalistic illusions and accounts of such illusory constructs engage radical epistemological matters? In a broad sense, this chapter tackles one of the most familiar tropes of early modern representation and the discourses surrounding it—namely, the deceptiveness of images, which make the absent present, which seem real yet are not. This theme is present throughout the following pages, and addressed directly by way of an analysis of instances where images were deemed "counterfeits." Current use of the term "counterfeit" to signal a fake—a work whose purpose is to deceive—does not capture its early modern connotations,

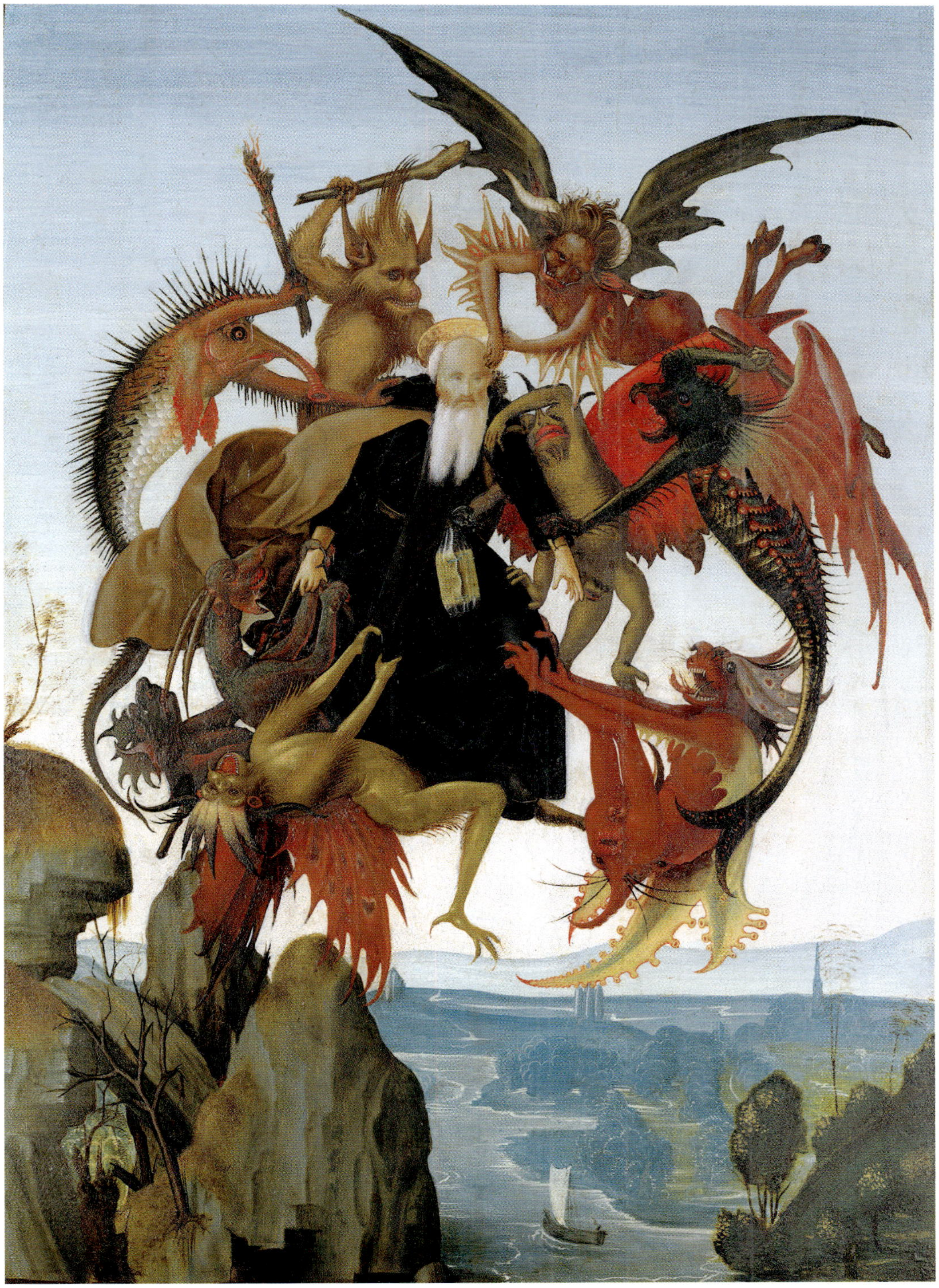

10.4 Michelangelo, *The Torment of Saint Anthony,* ca. 1487–88. Tempera and oil on panel, 18 ½ x 13 ¾ in. (47 x 34.9 cm). AP 2009.01. Kimbell Art Museum, Fort Worth, Texas.

where to "counterfeit" meant to portray, and usually to do so in ways considered trustworthy (portraits were often designated, in inventories and in art literature alike, "counterfeits").[8] Although the term encompasses both significations, I propose that it was used in the early modern period to denote verisimilitude without connoting truthfulness. Where Vasari, in the *Life* of Michelangelo, uses the term *contrafare*, or counterfeit, he raises a key question concerning the epistemological threat posed by what were also masterly fictions. Artistic interest in producing natural-seeming imaginary animals is key because it engages the possibility that something unreal or imaginary could be entirely naturalistic in appearance and truthful in effect—and widely appreciated by observers. Visual paradoxes were often highly intentional and sought after, even when, or precisely because, they challenged the limits of knowledge and experience.[9]

Real Chimeras

The anecdote about the young Michelangelo translating Schongauer's real chimeras into his own wondrous work by way of studiously recording natural forms illustrates one way to make a false image, and represents a broader set of practices and concerns. It records a process cultivated and documented by Leonardo da Vinci as well. In a vivid passage in his notebooks, Leonardo offers a paradoxical recipe for making a real chimera. If, he writes, "you wish to make one of your imaginary animals appear natural—let us suppose it to be a dragon—take for its head that of a mastiff or setter, for its eyes those of a cat, for its ears those of a porcupine, for its nose that of a greyhound, with the eyebrows of a lion, the temples of an old cock, and the neck of a water tortoise."[10] A pen-and-ink drawing of a fierce composite animal in the Royal Collection, UK, offers evidence that Leonardo followed his own recipe, producing a credible preternatural beast by recombining features of a variety of animals—in this instance, the nose of a dog, the horns of a steer, the teeth and beard of a lion, the claws of a bird of prey (fig. 10.5).[11]

Two further works by Leonardo were produced in a similar way. The first is a very early painting he devised on a misshapen piece of fig wood in the form of a small shield ("una rotella," or buckler) that his father passed along to him. The painting does not survive, but Vasari's account of its production is gripping. Leonardo prepared the support and then set to thinking about "what he could paint upon it, that might be able to terrify all who should come upon it, producing the same effect as once did the head of Medusa."

> For this purpose, then, Leonardo carried to a room of his own into which no one entered save himself alone, lizards great and small, crickets, serpents, butterflies, grasshoppers, bats, and other strange kinds of suchlike animals, out of the number of which, variously put together, he formed a great ugly creature, most horrible and terrifying, which emitted a poisonous breath and turned the air to flame; and he made it coming out of a dark and jagged rock, belching forth venom from its open throat, fire from its eyes, and smoke from its nostrils, in so strange a fashion that it appeared altogether a monstrous and horrible thing; and so long did he labour over making it, that the stench of the dead animals in that room was past bearing, but Leonardo did not notice it, so great was the love that he bore towards art.[12]

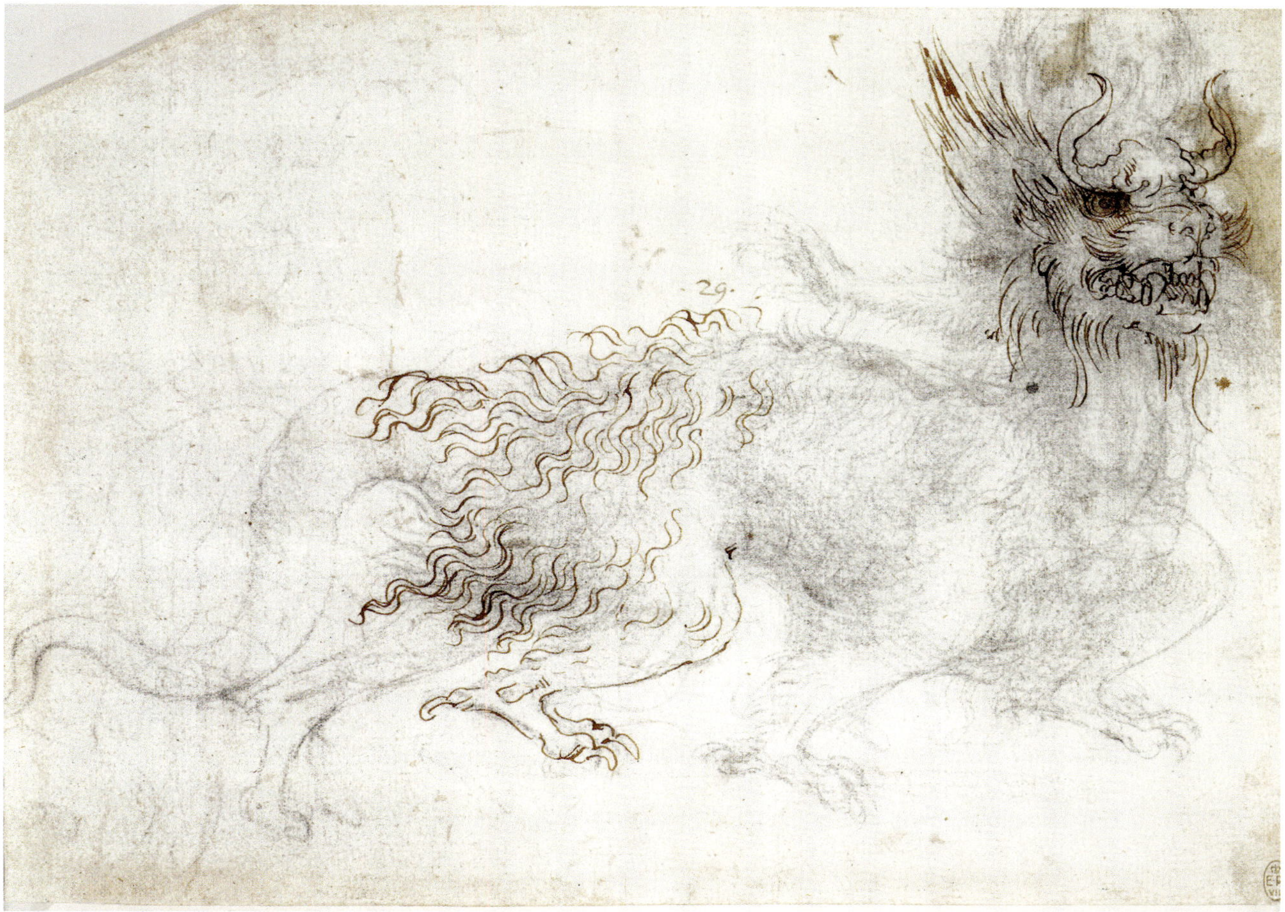

10.5 Leonardo da Vinci, *A Design for a Costume of an Imaginary Beast,* ca. 1517. Black chalk and pen and ink on paper, 18.8 x 27.1 cm. Royal Collection Trust, UK (RL 12369). Royal Collection Trust / © Her Majesty Queen Elizabeth II 2013.

By combining different naturalistic elements, Leonardo composed a "monstrous and horrible thing." Vasari's elides the production of an actual chimera with a painted one when he writes that Leonardo "formed a great ugly creature, most horrible and terrifying," "out of the number of [reptiles and insects and other creatures] variously put together."[13] The creation of this monster is described as if it occurred without the use of paint—and at the same time attests to Leonardo's love for art. Later in his biography of Leonardo, Vasari offers an account of Leonardo's production of a real chimera in three dimensions, such as is evinced in the account of the *rotella* painting. Vasari writes that in Rome, at a time when Leonardo was greatly concerned with philosophy and intensely involved in

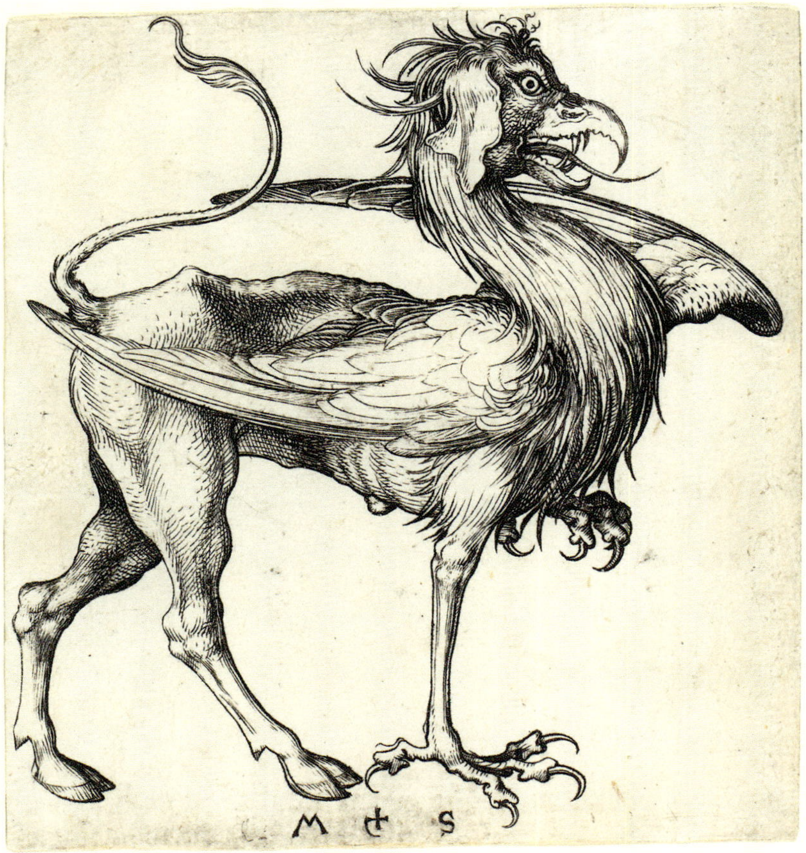

10.6 Martin Schongauer, *The Griffin,* ca. 1480–90. Engraving, 10.1 x 9.7 cm (4 x 3 13/16 in.). National Gallery of Art, Washington, D.C., Rosenwald Collection 1951.10.121. Courtesy of the National Gallery of Art, Washington, D.C.

alchemy, he "shaped animals very thin and full of wind, and, by blowing into them, made them fly through the air. . . . On the back of a most bizarre lizard, found by the vine-dresser of the Belvedere, he fixed, with a mixture of quicksilver, wings composed of scales stripped from other lizards, which, as it walked, quivered with motion; and having given it eyes, horns, and beard, taming it, and keeping it in a box, he made all his friends, to whom he showed it, fly for fear."[14] The principle of combination was nothing short of a cliché in art literature of the era, most of which harked back to "the prime topos of ideal imitation in the Renaissance," the "immortal anecdote quoted *ad nauseam* in Renaissance writing" of Zeuxis and the maids of Croton.[15] Like the description of the *rotella* painting, the anecdote of the flying lizard-chimera from which Leonardo's friends flew in fear suggests that, in a topsy-turvy version of the effect of Zeuxis's Helen (or Venus), the naturalism of hybrid monsters is measured by responses to them. The legend of the image that resulted from Zeuxis's combination of various features of several women is one of beauty and ideal form, whereas the real chimeras that Leonardo composed represent another locus of meaning—the imagination. Albrecht Dürer refers to such works in his "Aesthetic Excursus" as "dreamworks." On the limits of human creativity, bounded by the forms of nature, Dürer wrote, "Duch hůt sich ein yedlicher, das er nichtz vnmüglichs mach, das die natur nit leyden kün. Es wer dann sach, das einer traumwerck wolt machen, inn solchem mag einer allerley creatur vnder einander mischen."[16] While the impossible exceeds what nature offers, it is possible to create a "Traumwerck" by mixing together all sorts of creatures.

Interest in forging credible composites of existing animal parts is manifest in a range of works produced around the turn of the sixteenth century. In addition to his *Tribulations of Saint Anthony*, Schongauer produced a hieratic and imposing engraving of a chimera, which occasioned several copies, each marked proudly with the monogram of the derivative artist (figs. 10.6, 10.7). Schongauer's invention consists of a fantastic creature whose hind legs are those of a goat or donkey and whose front legs conclude in fierce talons, whose wings are exaggerated protrusions from the hairy neck of a furry animal, and whose head is its own microcosmic composite of beak, tongue, ears, horns, eyes, and fur of various beasts. Usually identified as a griffin, this creature contains a wider variety of constituent parts than the mythical creature said to consist of the body of a lion, the wings of a dragon, and the head of an eagle. Even allowing for confusion and diversity among monstrous confections, Schongauer's beast, like Leonardo's, features, and conjures delight in, a staggering multiplicity of component parts. Israel van Meckenem, who also copied *The Tribulations of Saint Anthony*, copied the chimera in an engraving he marked prominently with a decorative form of his monogram (in both cases the original composition is reproduced in reverse) (fig. 10.7). A third copy, an etching of the same animal set in a landscape and bearing a plaque on which Schongauer's monogram has been incised, further attests to the interest taken in this image and in remastering it. Whereas it is unsurprising that works by a groundbreaking printmaker occasioned copies, the fact that the chimera, a fictive hybrid of naturalistic parts should—like the demons haunting Anthony—have inspired exact copies is enormously compelling.

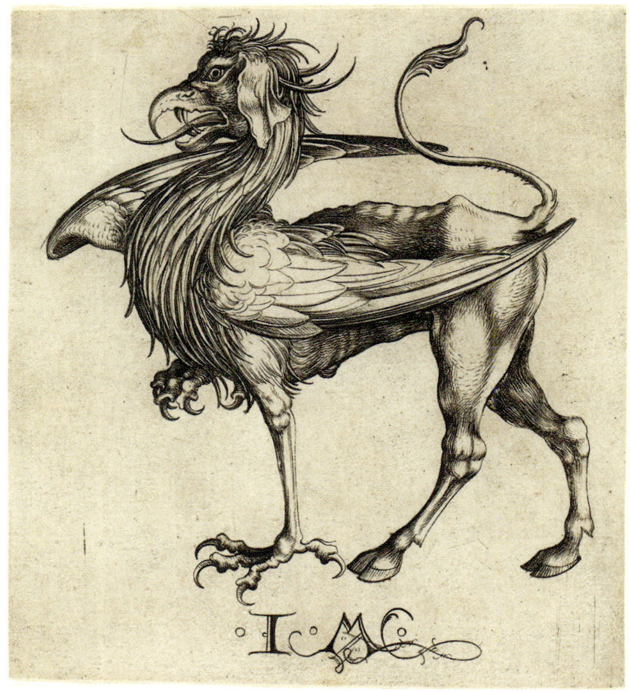

10.7 Israel van Meckenem, *The Griffin*, ca. 1490–1500. Engraving, 11.1 x 9.1 cm. National Gallery of Art, Washington, D.C., Rosenwald Collection 1946.21.141. Courtesy of the National Gallery of Art, Washington, D.C.

The production of new entities conjured from experience and rendered visible by art was a primary aspiration among early modern painters and sculptors. Dürer wrote, for example, that an artist's mind is a storehouse of internal images, from which he can produce an infinite array of previously unknown, previously unconceived things: "jr gemüt voller bildnuß . . . der wirdet durch die krafft, die Gott dem menschen geben hat, alle tag vil newer gestalt der menschen vnd andrer creature auß zu giessen vnd zu machen haben, das man for nit gesehen noch ein ander gedacht het."[17] Here, Dürer attributes the ability to unleash new forms to divinely bestowed power, but in some cases hybrid creatures evoked darker, more terrifying powers—recall the "Traumwerck" cited above, for example. As in his *Griffin*, the extraordinary demons Schongauer conjured in *The Tribulations of Saint Anthony* conform to Leonardo's strategy for composing a naturalistic imaginary creature—they comprise variously the scales, talons, horns, hooves, pelts, beaks, tentacles, and wings of actual animals. In the case of the demons, the monstrous admixtures of disparate parts add up to an array of agents of brutality whose attacks are nonetheless insufficient to disquiet Saint Anthony or compromise his ability to remain aloft: he knows them to be false images. These fantastic creatures are (merely) figures of the imagination—the saint's tormented imagination and the artist's fertile imagination alike. The fear they instill is also a fear of the power of illusion. Any early modern thinker or artist worth his salt would have understood that as in these images or in Leonardo's prescription, so too in dreams we experience combinatory images that, by virtue of the naturalism of their constituent parts, appear to be real, although they are not.

From Chimeras to the *Leviathan*

The dominant paradigm concerning the imagination, which held from Plato through the seventeenth century, posited that the imagination operates very much the way that artists do—for better or for worse. Early modern considerations and representations of chimeras directly evince the workings of the imagination and suggest their familiarity and significance. The imagination, designated "the single mediator between the incorporeal soul and the corporeal human body,"[18] was cultivated and examined in the early modern era by physicians, philosophers, theologians, and artists alike, and was valued in new ways for its ability to recombine the data (images) provided by sensory experience. In the words of Francis Bacon, the imagination, "not being bound by any law and necessity of nature or matter, may join things which are never found together in nature and separate things which in nature are never found apart."[19] Thus dreams, monsters, or works of art.

In the course of the early modern period, numerous artists took recourse to, engaged, and effectively transformed an extensive and venerable history of considerations on *phantasia*.[20] Within the history of art, the reappraisal of the combinatory powers of the imagination effected what has been described as a shift from "mimesis to *fantasia*" and from "the image of the artist as *imitatore* towards an image as *inventore*."[21] Such works as I have discussed, by Schongauer and Leonardo and Michelangelo and others, represent the workings of *phantasia*, or imagination, in ways that reflect long-standing, principally Aristotelian conceptions of internal images of nonexistent entities and posit new ways of thinking about the status of such images.

For a very long time, the relationship between sight, the external sense of vision, and imagination, construed as an internal sense, structured relationships between seeing and believing. Early modern medico-philosophical and artistic theories of the workings of the imagination depended, often explicitly, on the classical heritage. For all of the variability and complexity of the Aristotelian conception of imagination, it held fast—in a more or less consistent way—over centuries, as did a conception of the imagination as equal parts reliable and prone to falsity. In Platonic usage, the term *phantasia* refers to appearances in general and mental representation in particular; *phantasmata* are construed either as images produced by the realm of the senses for the purposes of the mind, or as visions, or as both.[22] In general, Platonic philosophy distinguishes between two kinds of imitation: imitative likenesses represent forms and are *eikastic*; paintings of such forms, however, belong to the realm of *phantasmata*. In other words, *phantastic* images are not anchored in actual forms, making them illusory and, to Platonic thinking, untrustworthy.

The image-based (in Platonic terms, illusory) nature of *phantasia* is also fundamental to Aristotelian formulations, which predominate well into the early modern period.[23] Recall Aristotle's hugely influential proclamation in *De anima* that "the soul does not think without images [phantasmata]."[24] Thomas Aquinas reiterated and amplified Aristotle's claims, and also distinguished between two roles *phantasia* (or its Latin synonym, *imaginatio*) could play.[25] On the one hand, it "serves as a kind of storehouse for forms received through the senses," which is to say that it is a receptacle for the images it produces out of sensory perception.[26] On the other hand, *phantasia* actively combines records of sensory perception and, in the case of dreams (or chimera), for example, generates mental images of nonexistent entities. Sensory data, transferred into images the mind could process, moved through the ventricles of the mind toward storage in the memory. When memory was activated, or when one dreamed, these images were brought out of storage, as it were, and perceived again by the mind's eye. In general, imagination was counted as one of the internal senses—along with common sense, reason, cognition, judgment, and memory—fitted out to process data supplied to the mind by the external senses. Whether identified as *imaginatio* or *phantasia* (and sometimes both were present), what is denoted is the faculty of mind responsible for producing the images on which the workings of the mind depend.

The workings of the imagination are variously represented over time, as are the moral dimensions of this faculty of mind and mode of perception. Book 3 of Aristotle's *De anima* explores, broadly, the relationship, vis-à-vis the human soul, between senses and knowledge, perception and thought. It opens with a passage that suggests that the faculty of the imagination is a tiebreaker of sorts, a referee in matters of truth and falsehood. Aristotle clearly distinguishes perception and thinking from each other, in contrast to "ancients" such as Empedocles, by way of the problem of error. Whereas the two faculties or capacities of mind (perception and thought) had been construed as identical, Aristotle tests their respective abilities on the forge of error, proposing imagination as a third term that is capable of reconciling thinking and perceiving even as imagination comes into its own as a faculty of mind. "Imagination," he posits, "is different from either perceiving or discursive thinking, though it is not found without sensation, or judgment without it." Much of the rest of book 3 is devoted to assessing just what the imagination is, and how it operates with regard to perception, belief, understanding, and thinking. In Aristotle's view, we may choose to engage imagination or not—in contradistinction to believing. We must believe things to be either true or false, whereas, "in the case of the imagination we are just as if we saw the terrible or encouraging things in a picture." The imagination is the domain of representation, and representation is not actual. Moreover, whereas perception cannot be false—we can be blind or numb, but what we perceive is not measured in units of truth or falsehood—imagination is inclined to be untrue. While knowledge and intellect are always true and perception cannot be false, beliefs can be false or true; but "imaginations are for the most part false." It is therefore, nonetheless and all the more, a worthy subject of philosophical deliberation. Imagination, on Aristotle's account, comes and goes, and is neither a potentiality, like sight, nor an activity, like seeing. Imagination is "thought to be a kind of movement . . . a movement taking place as a result of actual sense-perception."

The steadfastness of the relationship between seeing and believing and the claims made for it are well illustrated in Plutarch's *Lives*, and in particular by a conversation in the *Life* of the treacherous senator Brutus between him and Cassius. Brutus calls on Cassius, Plutarch writes, to explain a vision or, in John Dryden's translation, a "wonderful sign" he has experienced. "One night . . . [Brutus] was very late all alone in his tent, with a dim light burning by him . . . and reasoning about something with himself and very thoughtful, he fancied some one came in, and, looking up towards the door, he saw a terrible and strange appearance of an unnatural and frightful body standing by him without speaking." Plutarch recounts that Brutus calls on Cassius to help explain what he has seen. The response follows Cassius's self-proclaimed Epicurean beliefs. "It is the opinion of our sect," he explains,

> that not all that we feel or see is real and true; but that the sense is a most slippery and deceitful thing, and the mind yet more quick and subtle to put the sense in motion and affect it with every kind of change upon no real occasion of fact; just as an impression is made upon wax; and the soul of man, which has in itself both what imprints, and what is imprinted on [it], may most easily, by its own operations, produce and assume every variety of shape and figure. This is evident from the sudden changes of our dreams; in which the imaginative principle, once started by any trifling matter, goes through a whole series of most diverse emotions and appearances. It is its nature to be ever in motion, and its motion is fantasy or conception.[27]

The proximity of claims for the veridicality and the mutability of what we believe we see inflects this passage deeply. Just as the soul of man bears the imprints or seals of perception and at the same time is capable of imprinting or producing the "most diverse emotions and appearances," so too the motion that is fantasy or conception figures and disfigures.

Aristotelian precepts such as those recounted in Plutarch's tale (and articulated in Dryden's early modern usage) continued to inform conceptions of the workings of the mind, and of the imagination in particular, well into the early modern period, and animated a variety of media and genres. The views of the English philosopher Thomas Hobbes on the imagination are rarely cited in accounts of seventeenth-century English political theory; and they are virtually absent from the art-historical record, although it is worthy of mention that Hobbes wrote three treatises on optics, a subject that informed much of his natural philosophy.[28] The second chapter of his greatest work, *Leviathan* (1651), is devoted to the imagination, construed principally as an image-producing faculty of mind and prone to significant malfunction.[29] For Hobbes, imagination is the (distinctly visual) processing mechanism that allows sense perception ultimately to be filtered to memory. Hobbes also speaks of "simple" and "compounded" imagination; the former involves perception of a thing once seen, and the latter, the generation of fictitious, combined things, such as centaurs—or chimeras.[30]

Hobbes's primary concern is the resistance imagination is capable of mounting toward civil obedience, and the integrity of the commonwealth. And precisely because his aim is to align man and his faculties with the workings of the state, Hobbes devotes considerable attention to the pernicious manipulations of the imagination, as well as to common misunderstandings of its function. These misunderstandings are acute in the distinction between dreams and perception; giving credence to what is produced by a malfunctioning imagination is dangerous and potentially subversive. "From this ignorance of how to distinguish dreams, and other strong fancies, from vision and sense, did arise the greatest part of the religion of the Gentiles in time past, that worshipped satyrs, fawns, nymphs, and the like; and now-a-days the opinion that rude people have of fairies, ghosts, and goblins, and of the power of witches."[31] Control over the faculties hinges on control over imagination, and faulty attribution of agency amounts to a threat to social order in *Leviathan*. The handmaiden of sense perception, imagination has the power to lead the faculties (and with them the political subject) astray. It is also the case, in Hobbes's view, that the imagination accommodated the most deceptive phantasms of all—demonic illusions.[32]

Interior Temptation

The imagination is, then, the locus of false images, the verisimilitude of which is powerful and potentially dangerous. As Cassius explains to Brutus, "It is [the imagination's] nature to be ever in motion, and its motion is fantasy or conception." Universally required to enable the mind to process sensory data, its combinatory powers and ability to transform sensory data made the imagination a fraught and crucial faculty of mind. One of the more dire instances, in early modern thought and discourse, of the transformation of sensory perception by imaginative powers was demonology. The effects of the devil, as rendered by early modern artists, compellingly instantiate the workings of the imagination. They are but images—inherently false and strikingly verisimilar. Indeed, such pictures, artistic renderings of figures of the imagination, offer a commentary on the nature of images both internal and external.

Discussion of mental images permeates demonological literature of the early modern period. The profoundly influential late fifteenth-century *Malleus Maleficarum* (*Hammer of Witches*) explains, for example, that "the apparitions that come in dreams to sleepers proceed from the ideas retained in the repository of the mind, through a natural local motion caused by a flow of blood to the first and inmost seal of their faculties of perception." This is a universal physiological operation, but it is also possible, according to many authors, that this motion be caused by devils, even in people who are awake. "For in these [faculties] also the devils can stir up and excite the inner perceptions and humors so that ideas retained in the repositories of their minds are drawn out and made apparent to the faculties of fancy and the imagination, so that such men imagine these things to be true. And this is called interior temptation."[33] In other words, the devil and demons beset the operations of the imagination, and their effects are mental apparitions, images. (Saint Anthony's tormentors are only and entirely illusory.) That descriptions of the faculty of the imagination offered in demonological literature resonate with Renaissance theories of artistic creativity is fundamental. Ideas about how the devil impressed illusions on, or by his artifice created images in, the mind are entirely congruent with artistic meditations on cultivating that very same realm of the fantastic.

Early modern demonological literature holds consistently that sorcery is a trick of temperament, which amounts to harboring images of unnatural or preternatural events and effects in one's imagination. Often, the trick is staged by the devil himself, who creates those images. From the later sixteenth century onward, skeptics argued that those possessed by the devil were the pathetic instruments of their own deluded imaginations. Their weakened imaginations were particularly susceptible to the machinations of the devil. In his *De praestigiis daemonum*, first printed in 1563 and published in its final, revised form in 1583, the Rhineland physician Johannes Weyer (Jan Wier) writes of how the devil dements his victims by "work[ing] his way into [their] thoughts and weav[ing] the fabric thereof, varying it with wondrous art and skill by repeating the past, adding the present, and sprinkling in admonitions of the future . . . causing them to imagine that they see present before them the things which he has imposed upon them."[34] He goes on to explain the "artistry" of the devil and the impressionability of women who harbored these sorts of images: "With subtlety and inimitable cunning this spirit mocks and deludes these instruments of his who incline toward his promptings, these poor feeble-minded, bewitched, and idle women, whom he so maddens that . . . they falsely believe . . . that they themselves have done all the things that he puts into their imagination. . . . And yet, all of these things are known to them only through phantasms or dreams."[35] Weyer's interpretation of the devil's machinations is consistent with many of his contemporaries' views, as are questions about the reality of witchcraft and the extent to which many or all of its effects could be ascribed to the powers of the imagination.

The nexus of demonic illusion and visual art has been the subject of a number of recent studies by historians and art historians. Stuart Clark's *Vanities of the Eye*, a broad-ranging study of early modern European conceptions of vision and perception, draws extensively on the myriad ways in which discourses on demonology and witchcraft evince and affect conceptions of vision. Early modern relations between seeing and believing are fraught by and with demonological effects and other such illusions. Michael Cole has explored the close ties between the literature of magic and the demonic, on the one hand, and the cultivation of painting as an art, on the other. "The idea of the demonic," Cole writes, "cuts across not only the early modern literature of magic and witchcraft but also that of art, making possible a kind of refraction that illuminates both."[36] Cole focuses primarily on the agents of illusion rather than the status of illusionistic images. Allowing that the literature of the period supports the notion that "demonic illusions exist both as things and as the likenesses of things," Cole nonetheless is primarily interested in the painter, for he "provides a function for the demonic imagination and the figure of the demon [who] provides the artist with a conception of medium."[37] In a powerful study of the emergence of the figure of the witch in European art circa 1500 and the secularization or disenchantment of art, Christopher Wood identifies reverberating strains of thought about the power of illusion in magic and art alike: "Magic was *only ever* a way of describing the art of allure, indeed only ever a way of accounting for art in general, the various technologies of illusion and fictionalization."

The case of demonic apparitions exacerbated concerns about illusionistic art and the power of fiction. Witchcraft or magic called into question the reliability of visual perception, just as mimetic art challenged the boundaries of the real; and the two combined powerfully. "Art," Wood writes, "ungrounded and in perpetual motion, suspends the question of the reality of the witch by provisionally occupying the same kind of 'space between' that witchcraft had claimed for itself."[38] In the case of apparitions, beholders were encouraged to foster "discernment," a specialized mode of judgment concerned to assess what in or of an apparition was real. "False imaginations" were believed in all too many cases to be real, and to tragic ends: in Clark's view, the real tragedy of Macbeth lies in his miserable failure as a discerner of spirits.[39] Taking the false to be real was also a threat in the domain of art, one broadly defined in the early modern era by the term "counterfeit."

Counterfeit Temptations

Is it surprising, in the context of our considerations of illusionistic naturalism, that Vasari refers to the process by which Michelangelo copies Schongauer's hybrid demons as "counterfeit"? As Peter Parshall writes, in his path-breaking study of the early modern uses of the term *contrafactum*, "*Conterfeit* had a special place in the Renaissance lexicon of terms having to do with identity and representation. Most striking is its insistent claim to truth, implicitly or explicitly a truth based upon the testimony of direct witness. Although at first *conterfeit* was most often employed for portraits, it was quickly extended to include other subjects that for one reason or another seemed to ask for a special order of legitimacy."[40] Vasari's invocation of this recording mode is complex: while Michelangelo is said to have *portrayed* Schongauer's engraving ("la ritrasse di penna"), he *counterfeited* the forms of the demons ("per contrafare alcune strane forme di Diavoli"); and the passage that immediately follows Vasari's description of the painting of Saint Anthony describes, again using the verb *contrafare*, how Michelangelo imitated various Old Master drawings and passed them off as the originals. "He also counterfeited sheets by the hands of various old masters, making them so similar that they could not be detected, for, tinting them and giving them the appearance of age with smoke and various other materials, he made them so dark that they looked old, and, when compared to the originals, one could not be distinguished from the other."[41] Identity—between new and old, copy and original—is very much at issue here, though in this passage the term "counterfeit" is used to describe the production of illegitimate, if masterly, copies. This amounts to a compelling inversion of the use of the term "counterfeit" and its "insistent claim to truth," as reconstructed by Parshall.

Vasari's account of the forged drawings opens with the verb *contrafare* itself, which resonates with the previous passage, in which he describes Michelangelo's initial (pen-and-ink) copy of Schongauer's print as indiscernible from the original, made "in a manner that was not known" (di maniera che non era conosciuta). Just as Michelangelo went to the fish market to *contrafare* the demons in color, so too he "counterfeited other sheets in the hands of various Old Masters." In his description of these drawings, Vasari emphasizes that the real could not be distinguished from the fake: they were "so similar that one could not recognize them" (che non si conoscevano), and comparing them to the originals, "one could not distinguish one from the other" (non si conosceva l'una dall'altra). Here, Vasari uses forms of the verb *conoscere* twice in succession and both times in the negative, to attest to Michelangelo's ability to dupe viewers, echoing its occurrence in the description of the pen copy of the Schongauer print. These images were "counterfeits," forged in such a way that the originals and their copies were equivalent.

In the subsequent pages, which contain the famous account of Michelangelo's forgery of a carved marble faun for Lorenzo de' Medici, Vasari relies consistently on the verb *contrafare* to describe the artistic process: "he set out to counterfeit with a piece of marble a head that was there of an ancient faun," and "Michelangelo succeeded in counterfeiting it so well that it astonished il Magnifico [Lorenzo de' Medici]."[42] This is followed by a description of an early relief sculpture in which Michelangelo "counterfeited" the manner of Donatello, a work Leonardo himself donated to Lorenzo the Magnificent. If, on the one hand, to counterfeit is to imitate so effectively as to replicate an original to the point of evincing wonder, counterfeit works are, on the other hand, forgeries. They are in both cases convincing and in both cases deceptive. Likewise, it requires judgment (or knowledge) to distinguish between the original and a copy, or true and false. While Vasari does not seem troubled by the falsity of the "counterfeit," his use of the term does raise the issue of the discernment of distinctions between forgery and original, or image and reality.

In its early modern uses, the term "counterfeit" does not irrevocably signal truth; implied falsehood is integral to conceptions of images to which "counterfeit" pertains. Where the term is adduced in early modern theory and practice more generally, counterfeit seems to refer to sense impressions made by external stimuli, and the word comes very close to signifying something on the order of *phantasmata,* or images. Timothy Bright, the author of a seminal treatise on melancholy published in 1586, wrote of "those thinges which are sensible, and are as it were the counterfettes of ou[t]ward creatures."[43] Here, Bright's counterfeits are like the portraits or verifiable, trustworthy visual records that are commonly referred to as "counterfeits," per Parshall. It is, as we have seen, a trope of demonological literature of the sixteenth and seventeenth centuries that the devil is the consummate artist or imitator or illusionist. And, indeed, he was also deemed a master of counterfeits and counterfeiting. The Swiss Reformation physician and theologian Thomas Erastus writes of the devil that he can "deceive by means of counterfeits, and . . . substitute for things themselves, semblances and empty illusions."[44] The minor classicist Meric Casaubon (1599–1671) wrote in the later seventeenth century that the devil can "so imitate and counterfeit that we shall find it a very hard task, to distinguish between the reality of that which he cannot, and the resemblance, which he doth offer unto our eyes." Casaubon continues, "He may cast before our eyes such shapes of those things, which he cannot create; or so work upon our phancy, that it shall create them unto us so vigorously, so seemingly, that he may attain his ends by those counterfeits, as effectually, perchance, as if all were in good earnest, what it appears to our deluded eyes."[45]

Erastus's invocation of the term builds on the common usage, where "counterfeit" or its vernacular counterparts refer to "the image as a statement of fact," but it twists this conception, calling attention to the ability of an image to deceive.[46] Elsewhere in his text, Bright, too, writes of false or illusory images, calling them counterfeits. The melancholic humor "annoyeth the heart and passing up to the braine, counterfetteth terrible objectes to the fantasie . . . [causing] it without external occasion, to forge monstrous fictions." When the brain "fancieth not according to trueth," Bright explains further, it disturbs the mental faculties, which then "forgeth disguised shapes . . . [and] conterfet goblins." The melancholic, he writes, is tormented by "those conterfet goblins, which the braine dispossessed of right discerning, fayneth unto the heart." Deprived of discretion, the melancholic mind is rendered incapable of sorting out the true from the false, or counterfeit.[47]

Chimeras such as Leonardo prescribed, drew, constructed with the aid of wax, or painted, or such as Schongauer forged and Michelangelo copied, challenged the reliability of vision and of naturalistic images in ways that, beyond inspiring in the viewer fear (of the real) and awe (at the skill), also engaged judgment and discernment in signal ways. In some cases, the works may have been produced with the express intention of provoking an encounter with the actuality from which illusion takes its form. Consider, in this regard, the conclusion of Vasari's account of Leonardo's buckler painting, the early work in which he conjured from "lizards great and small, crickets, serpents, butterflies, grasshoppers, bats, and other strange kinds of suchlike animals" a painting he wished to have the effect of the Medusa. Vasari writes that when Leonardo had completed the work, he called his father to his room.

> Ser Piero having . . . knocked at the door, Leonardo opened to him, telling him to wait a little; and, having gone back into the room, he adjusted the buckler in a good light on the easel, and put to the window, in order to make a soft light, and then he bade him come in to see it. Ser Piero, at the first glance, taken by surprise, gave a sudden start, not thinking that that was the buckler, nor merely painted the form that he saw upon it, and, falling back a step, Leonardo checked him, saying, "This work serves the end for which it was made; take it, then, and carry it away, since this is the effect that it was meant to produce."[48]

Leonardo's first painting was a real chimera, insofar as it recombined real forms to produce a hybrid so naturalistic that it evinced terror in the beholder. In Leonardo's view, it served its purpose by evincing (or counterfeiting) the effect of an actual monster, not "merely painted." Incidentally, the immediately subsequent works described in Vasari's *Life* tend to share the attribute of being hypernaturalistic, or more alive than life itself. In particular, Leonardo is said to have "counterfeited" a vase of flowers and to have painted the head of Medusa. About the still life, which was painted into a painting of the Virgin, Vasari writes that Leonardo "counterfeited a glass vase full of water, containing some flowers, in which, besides its marvelous naturalness, he had imitated the dewdrops in the flowers, so that it seemed more real than the reality." The account of the Medusa, whose success would depend on its high degree of illusionism, directly and doubly invokes the imagination: "The fancy came to him [vennegli fantasia] to paint a picture in oils of the head of a Medusa, with the head attired with a coil of snakes, the most strange and extravagant invention that could ever be imagined [la più strana e stravagante invenzione che si possa immaginare mai]."[49]

The judgment elicited by "counterfeit" pictures hinges on whether we can trust our eyes, and in this sense these images pose the same epistemological problems as the imagination. Where artists translated imaginative entities into physical images, or where they mimicked the recombinatory imaginative process by making real chimeras, for example, they plied the boundaries between the real and the imaginary, the illusory and the reliable. As the early modern physician Julius Caesar Scaliger (1484–1558) pointed out, "wise men loved Homer's inventions . . . and kings loved fools. . . . The human mind, being infinite by nature, loves new things, especially those, 'like pictures of monsters,' that 'surpass the common boundaries of truth.' The wise man praises pictures, 'even though he hardly fails to realize that they are feigned.'"[50] Scaliger goes on to suggest that the ability to create a convincing simulacrum, and thereby to fool a viewer, pertains to the quasi-divine capacity of the artist. Scaliger's text was printed half a century after the artists featured in this chapter produced their real chimeras and other truth-defying pictures. Universal statements about the pleasure of illusion do not clarify why, in the decades around 1500, the constellation of artists discussed, both north and south of the Alps (and others too—neither Hieronymus Bosch nor Dosso Dossi should go unmentioned), so productively cultivated the arena of imaginative production and at the same time so purposefully plied the boundary between the true and the false. It has been argued that this was a moment at which the powers of the imagination came to be newly esteemed and defined.[51] Questions about the potential veridicality and reliability of vision persisted in a variety of discourses throughout the century. Bright, for example, was one of many authors whose conception of eyesight and its deformations responded to and informed contemporary estimations of the powers of the imagination. Alongside the new interest in the imagination, the sixteenth century also witnessed the emergence of powerful forms of naturalism in representation; and both contributed to a new urgency on the part of the beholder's judgment. Where the imagination came to be seen as a motile, fertile, active locus of novelty and nightmare rather than, or in addition to, a processing mechanism nestled among the mental faculties, it also gave rise to specters, and thus to questions about whether to see is to believe. The value accorded images as counterfeits of the real, as simulacra of sense perception, had much the same effect. Saint Anthony, beset by hybrid demons whose torments he summoned the inner force to withstand, exemplifies the terrors of credible fictions. Likewise, images of Saint Anthony beset by hybrid demons figure the challenge of mimetic art and of apparitions simultaneously. The interest in producing such images, such illusions, was conditioned and offset by the ways in which such illusory presentations were mitigated by claims for the falsehood of images and tamed by judgment. To make an image of mixed-up creatures never before seen was a signal act of artistic creativity in the sixteenth century north and south of the Alps, and to do so required extraordinary imaginative resources. To see such an image as "merely painted," or as a Traumwerck or a counterfeit, required judgment and discretion, tinged by pleasure and horror alike.

Notes

1. Stephens, *Demon Lovers*, 112.

2. Some of the many examples of copies made of the Schongauer composition include Monogrammist IE, dated ca. 1480–1500 (Hollstein 54.d, after Schongauer); Monogrammist FVB, dated ca. 1475–1500 (Hollstein 54.e, after Schongauer); and Raphael de Mey, published by Johannes Bus, ca. 1590 (Hollstein 54.f).

3. See Höfler, "Zur frühen Schongauer-Rezeption in Florenz."

4. Vasari, *Lives of the Most Eminent Painters*, 2:646. Gaston du C. de Vere's edition, used here, is a translation of the 1568 "Giuntina" edition of Vasari's *Lives;* cf. the original-language edition, in both the Torrentiana (1550) and the Giuntina (1568) editions, "Giorgio Vasari, le Vite: Edizioni Giuntina e Torrentiana," available online at http://vasari.sns.it, 6:8. All further references to the original editions of Vasari's text indicate volume and page number as used on this site. I owe my fascination with this passage to Peter Parshall, who wrote of it, "Whatever the truth of Vasari's story, for him it stood to reason that fantasy required the glistening animation of nature in order to be made convincing." Parshall, "Introduction: Art and Curiosity," 328.

5. Condivi, *Life of Michelangelo*, 9–10.

6. Vasari's description of the early painting follows directly an account of Domenico Ghirlandaio's amazement at "the novel manner and the novel method of imitation that a mere boy of such tender age [roughly thirteen?] displayed by reason of the judgment bestowed upon him by Heaven." Vasari, *Lives of the Most Eminent Painters*, 2:645.

7. Cennini, *Craftsman's Handbook*, 1, 2.

8. Parshall, "Imago Contrafacta."

9. See Clark, "Demons, Natural Magic, and the Virtually Real" and *Vanities of the Eye*.

10. Leonardo, *Selections from the Notebooks*, 167.

11. The drawing is thought to be a sketch for a costume produced at the French court. Consider also Leonardo's *Studies of Dragons* (ca. 1478–80), stylus, black chalk, and pen and brown ink (recto), stylus, pen and ink (verso), 15.9 × 24.4 cm (6 1/4 × 9 5/8 in.), Royal Collection Trust, RL 12370.

12. Vasari, *Lives of the Most Eminent Painters*, 1:629.

13. The Italian original reads, "da la moltitudine de' quali variamente adattata insieme cavò uno animalaccio molto orribile e spaventoso" (see http://vasari.sns.it, 4:21).

14. Vasari, *Lives of the Most Eminent Painters*, 1:638.

15. Payne, "*Mescolare, Composti*, and Monsters," 273; and Panofsky, *Life and Art of Albrecht Dürer*, 278. The literature on the topos is mighty; see Payne, "*Mescolare, Composti*, and Monsters," 275n5. Although Payne's focus is on architectural theory and practice, her analysis of the principles and processes of assemblage and *mescolanza*, as a function of Horatian combinatory artistry, and of *licenzia* and its governing role in products of these processes, is more widely relevant, and has been crucial in formulating portions of my argument.

16. Dürer, *Schriftlicher Nachlass*, 3:292, §18. It is noteworthy that Joachim Camerarius translated the passage concerning the "Traumwerck" using the terms "monster" and "dream": "nisi forte propositum sit monstra et quasi somniorum gentem effingere, ubi omnium rerum naturae confundantur." Ibid., 3:301, §18. All translations are my own unless otherwise noted.

17. Ibid., 3:291, §11. Cf. Panofksy, *Life and Art of Albrecht Dürer*, 280.

18. Clark, *Vanities of the Eye*, 43, in a discussion of the shift in valuation of the imagination in the early modern period and the increasing significance of its combinatory function. See, more generally, his chap. 2, "Fantasies: Seeing Without What Was Within," 39–77.

19. Bacon, "A Description of the Intellectual Globe," *Works of Francis Bacon*, 5:503–4. See, recently, Corneanu and Vermeir, "Idols of the Imagination."

20. See Summers, *Michelangelo and the Language of Art*, esp. "L'alta fantasia," 103–43; Kemp, "From '*mimesis*' to '*fantasia*'"; Kemp, *Leonardo da Vinci*, esp. chap. 3, "The Exercise of Fantasy," 152–212; Couliano, *Eros and Magic in the Renaissance*; and Fermor, *Piero di Cosimo*, chap. 1, "Vasari's Life of Piero di Cosimo," esp. 29–37. See also Swan, "Eyes Wide Shut."

21. Kemp, "From '*mimesis*' to '*fantasia*.'"

22. See Bundy, *Theory of Imagination*, esp. chap. 2, "Plato," 19–59; Bundy, "'Invention' and 'Imagination' in the Renaissance"; Wolfson, "Internal Senses"; Pagel, "Medieval and Renaissance Contributions to Knowledge"; Harvey, *Inward Wits*; Fattori and Bianchi, *Phantasia-Imaginatio*; Carruthers, *Book of Memory*; and Kemp, *Cognitive Psychology in the Middle Ages*. Bundy's plea, in his 1927 book, for a history of the imagination has yet to be answered in full.

23. On the predominance of Aristotelian over Platonic notions of imagination and *phantasia*, see Bundy, *Theory of Imagination*, 19.

24. Aristotle *De anima* 3.7.l.431. See Couliano, *Eros and Magic in the Renaissance*, chap. 1, "History of Phantasy," 3–27.

25. For example, "Intelligere sine conversione ad phantasmata est (animae) praeter naturam." *Summa Theologica*, I, Q. 89, Art. l.

26. Ibid., I, Q. 78, Art. 4.

27. Plutarch, "Marcus Brutus," 5:342. Also available at http://classics.mit.edu/Plutarch/m_brutus.html.

28. The three treatises on optics are *Tractatus Opticus I* (ca. 1640), *Tractatus Opticus II* (1644), and *A Minute or First Draught of the Optiques* (1646). See Prins, "Hobbes on Light and Vision"; and Giudice, *Luce e visione*. For an exception to the art-historical rule, which remedies the neglect of "das gesamte Gebiet des Optischen" in Hobbes's work, see Bredekamp, *Thomas Hobbes visuelle Strategien*. The section titled "Hermetische Phantasie," 68–72, treats Hobbes's conception of memory and, indirectly, imagination.

29. Hobbes, *Leviathan*, chaps. 2, 3, and 45 in particular. All quotations are from Sir William Molesworth's edition of Hobbes's *Collected English Works*. Hobbes also discusses imagination in his *Human Nature, or the Fundamental Elements of Policy* (London, 1650), esp. chap. 3.

30. Hobbes, *Leviathan*, vol. 3, parts 1–2, 6.

31. Ibid., 9.

32. See Clark, "Demons, Natural Magic, and the Virtually Real."

33. Krämer and Sprenger, *Malleus Maleficarum*, 50.

34. The first edition of Weyer's *De praestigiis daemonum* (1563) was followed rapidly by augmented editions and translations. This quotation is from George Mora's edition, translated by John Shea, *Witches, Devils, and Doctors in the Renaissance*, bk. 3, chap. 11, p. 195. See also the edition by Kohl and Midelfort, also translated by Shea, *On Witchcraft*, 105.

35. Quoted in Weyer, *On Witchcraft*, 106.

36. Cole, "Demonic Arts," 621.

37. Ibid., 623, 621. See also Hults, *Witch as Muse*; Campbell, "'*Fare una Cosa Morta Parer Viva*,'" on the proximity of claims for the relationship between human and divine creation/animation; Swan, *Art, Science, and Witchcraft*, esp. part II; and Swan, "Eyes Wide Shut."

38. Wood, "Countermagical Combinations," 156.

39. Clark, *Vanities of the Eye*, 228; see generally chap. 6, "Apparitions," and also 142–43.

40. Parshall, "Imago Contrafacta," 564.

41. Vasari, *Lives of the Most Eminent Painters*, 2:646. This passage occurs only in the Giuntina edition of the *Lives* (see http://vasari.sns.it, 6:9).

42. Giuntina edition, http://vasari.sns.it, 6:9–10.

43. Bright, *Treatise of Melancholy*, 101–2.

44. Quoted in Clark, *Vanities of the Eye*, 128.

45. Casaubon, *A treatise proving witches, witches and supernatural operations by pregnant instances* (London, 1672), 157–58, quoted in ibid., 130.

46. Parshall, "Imago Contrafacta," 555.

47. Bright, *Treatise of Melancholy*, 100–101.

48. Vasari, *Lives of the Most Eminent Painters*, 1:629.

49. Ibid., 1:630.

50. Julius Caesar Scaliger, *Exotericarum exercitationum liber quintus decimus* (Paris, 1557), fols. 395v–396r, quoted in Grafton, "Renaissance Histories of Art and Nature," 194. Another, compatible reading of Bosch's fantastic imagery and of its status as a radical form of mimesis is explored in Koerner, "Impossible Objects."

51. See Clark, *Vanities of the Eye*, esp. chap. 2; Wood, "Countermagical Combinations"; and Moxey, "Hieronymus Bosch and the 'World Upside Down.'"

Bibliography

Bacon, Francis. *The Works of Francis Bacon*. Edited by James Spedding, Robert Leslie Ellis, and Douglas Denon Heath. 14 vols. London: Longman, 1857–74.

Bredekamp, Horst. *Thomas Hobbes visuelle Strategien: Der Leviathan, Urbild des modernen Staates, Werkillustrationen und Portraits*. Berlin: Akademie Verlag, 1999.

Bright, Timothy. *A Treatise of Melancholy*. New York: For the Facsimile Text Society by Columbia University Press, 1940.

Bundy, Murray Wright. "'Invention' and 'Imagination' in the Renaissance." *Journal of English and Germanic Philology* 29 (1930): 535–45.

———. *The Theory of Imagination in Classical and Medieval Thought*. Urbana: University of Illinois Press, 1927.

Campbell, Stephen J. "'*Fare una Cosa Morta Parer Viva*': Michelangelo, Rosso, and the (Un)Divinity of Art." *Art Bulletin* 84 (2002): 596–620.

Carruthers, Mary. *The Book of Memory: A Study of Memory in Medieval Culture*. Cambridge: Cambridge University Press, 1990.

Cennini, Cennino d'Andrea. *The Craftsman's Handbook, "Il libro dell'arte."* Translated by Daniel V. Thompson. New Haven: Yale University Press, 1960.

Clark, Stuart. "Demons, Natural Magic, and the Virtually Real: Visual Paradox in Early Modern Europe." In *Paracelsian Moments: Science, Medicine, and Astrology in Early Modern Europe*, ed. Gerhild Scholz-Williams and Charles D. Gunnoe Jr., 223–45. Kirksville, Mo.: Truman State University Press, 2002.

———. *Vanities of the Eye: Vision in Early Modern European Culture*. Oxford: Oxford University Press, 2007.

Cole, Michael. "The Demonic Arts and the Origin of the Medium." *Art Bulletin* 84 (2002): 621–40.

Condivi, Ascanio. *The Life of Michelangelo*. Edited by Helmut Wohl, translated by Alice Sedgwick Wohl. Baton Rouge: Louisiana State University Press, 1976.

Corneanu, Sorana, and Koen Vermeir. "Idols of the Imagination: Francis Bacon on the Imagination and the Medicine of the Mind." *Perspectives on Science* 20 (2012): 183–206.

Couliano, Ioan P. *Eros and Magic in the Renaissance*. Translated by Margaret Cook. Chicago: University of Chicago Press, 1987.

Dürer, Albrecht. *Dürer: Schriftlicher Nachlass*. Edited by Hans Rupprich. 3 vols. Berlin: Deutscher Verein für Kunstwissenschaft, 1956–69.

Fattori, M., and M. Bianchi, eds. *Phantasia-Imaginatio: Vo Colloquio internazionale del lessico intelletuale europeo*. Rome: Ateneo, 1988.

Fermor, Sharon. *Piero di Cosimo: Fiction, Invention, and Fantasia*. London: Reaktion Books, 1993.

Giudice, Franco. *Luce e visione: Thomas Hobbes e la scienza dell'ottica*. Florence: Leo S. Olschki, 1999.

Grafton, Anthony. "Renaissance Histories of Art and Nature." In *The Artificial and the Natural: An Evolving Polarity*, ed. Bernadette Bensaude-Vincent and William R. Newman, 185–210. Cambridge: MIT Press, 2007.

Harvey, E. Ruth. *The Inward Wits: Psychological Theory in the Middle Ages and Renaissance*. London: Warburg Institute, 1975.

Hobbes, Thomas. *The Collected English Works of Thomas Hobbes*. Edited by Sir William Molesworth. 12 vols. London: Routledge, 1997.

———. *Leviathan, or the Matter, Forme and Power of a Commonwealth Ecclesiasticall and Civill*. London: Andrew Crooke, 1651.

Höfler, Janez. "Zur frühen Schongauer-Rezeption in Florenz: Ein Verrocchio-Nachtrag." *Mitteilungen des Kunsthistorischen Institutes in Florenz* 49 (2005): 227–32.

Hults, Linda. *The Witch as Muse: Art, Gender, and Power in Early Modern Europe*. Philadelphia: University of Pennsylvania Press, 2005.

Kemp, Martin. "From '*mimesis*' to '*fantasia*': The Quattrocento Vocabulary of Creation, Inspiration, and Genius in the Visual Arts." *Viator* 8 (1977): 347–98.

———. *Leonardo da Vinci: The Marvellous Works of Nature and Man.* Oxford: Oxford University Press, 2007.

Kemp, Simon. *Cognitive Psychology in the Middle Ages.* Westport, Conn.: Greenwood Press, 1996.

Koerner, Joseph. "Impossible Objects: Bosch's Realism." *RES* 46 (Autumn 2004): 73–97.

Krämer, Heinrich, and Jakob Sprenger. *Malleus Maleficarum.* 1486. Translated by M. Summers. New York: Dover, 1971.

Leonardo da Vinci. *Selections from the Notebooks of Leonardo da Vinci.* Edited by Irma A. Richter. London: Oxford University Press, 1952.

Moxey, Keith. "Hieronymus Bosch and the 'World Upside Down': The Case of *The Garden of Earthly Delights.*" In *Visual Culture: Images and Interpretations*, ed. Norman Bryson, Michael Ann Holly, and Keith Moxey, 104–40. Hanover: University Press of New England, 1994.

Pagel, Walter. "Medieval and Renaissance Contributions to Knowledge of the Brain and Its Functions." In *The History and Philosophy of the Brain and Its Functions: An Anglo-American Symposium.* Edited by F. N. L. Poynter, 95–114. Oxford: Oxford University Press, 1958.

Panofsky, Erwin. *The Life and Art of Albrecht Dürer.* Princeton: Princeton University Press, 2005.

Parshall, Peter. "Imago Contrafacta: Images and Facts in the Northern Renaissance." *Art History* 16 (1993): 554–79.

———. "Introduction: Art and Curiosity." *Word and Image* 11 (1995): 327–31.

Payne, Alina. "*Mescolare, Composti*, and Monsters in Italian Architectural Theory of the Renaissance." In *Disarmonia bruttezza e bizzarria nel Rinascimento: Atti del VII Convegno internazionale*, ed. Luisa Secchi Tarugi, 273–94. Florence: Franco Cesati, 1998.

Plutarch. "Marcus Brutus." In *Plutarch's Lives: The Translation Called Dryden's*, trans. John Dryden, ed. A. H. Clough. 5 vols. Boston: Little Brown, 1906. http://oll.libertyfund.org/title/1775/93979/2123424.

Prins, Jan. "Hobbes on Light and Vision." In *The Cambridge Companion to Hobbes*, ed. Tom Sorell, 129–56. Cambridge Cambridge University Press, 1996.

Stephens, Walter. *Demon Lovers: Witchcraft, Sex, and the Crisis of Belief.* Chicago: University of Chicago Press, 2003.

Summers, David. *Michelangelo and the Language of Art.* Princeton: Princeton University Press, 1981.

Swan, Claudia. *Art, Science, and Witchcraft in Early Modern Holland: Jacques de Gheyn II (1565–1629).* New York: Cambridge University Press, 2005.

———. "Eyes Wide Shut: Early Modern Imagination, Demonology, and the Visual Arts." *Zeitsprünge: Forschungen zur Frühen Neuzeit* 7 (2003): 156–81.

Vasari, Giorgio. *Lives of the Most Eminent Painters, Sculptors, and Architects.* Translated by Gaston du C. de Vere. 2 vols. New York: Knopf, 1996.

Weyer, Johannes. *De praestigiis daemonum, & incantationibus ac veneficiis.* Basel: Joannes Oporinus, 1563.

———. *On Witchcraft: An Abridged Translation of Johann Weyer's De praestigiis daemonum.* Edited by Benjamin G. Kohl and H. C. Erik Midelfort, translated by John Shea. Asheville: Pegasus Press, 1998.

———. *Witches, Devils, and Doctors in the Renaissance: Johann Weyer, De praestigiis daemonum.* Edited by George Mora, translated by John Shea. Binghamton, N.Y.: Center for Medieval and Early Renaissance Studies, 1991.

Wolfson, Harry A. "The Internal Senses." *Harvard Theological Review* 28 (1935): 69–133.

Wood, Christopher. "Countermagical Combinations by Dosso Dossi." *RES* 49–50 (2006): 151–70.

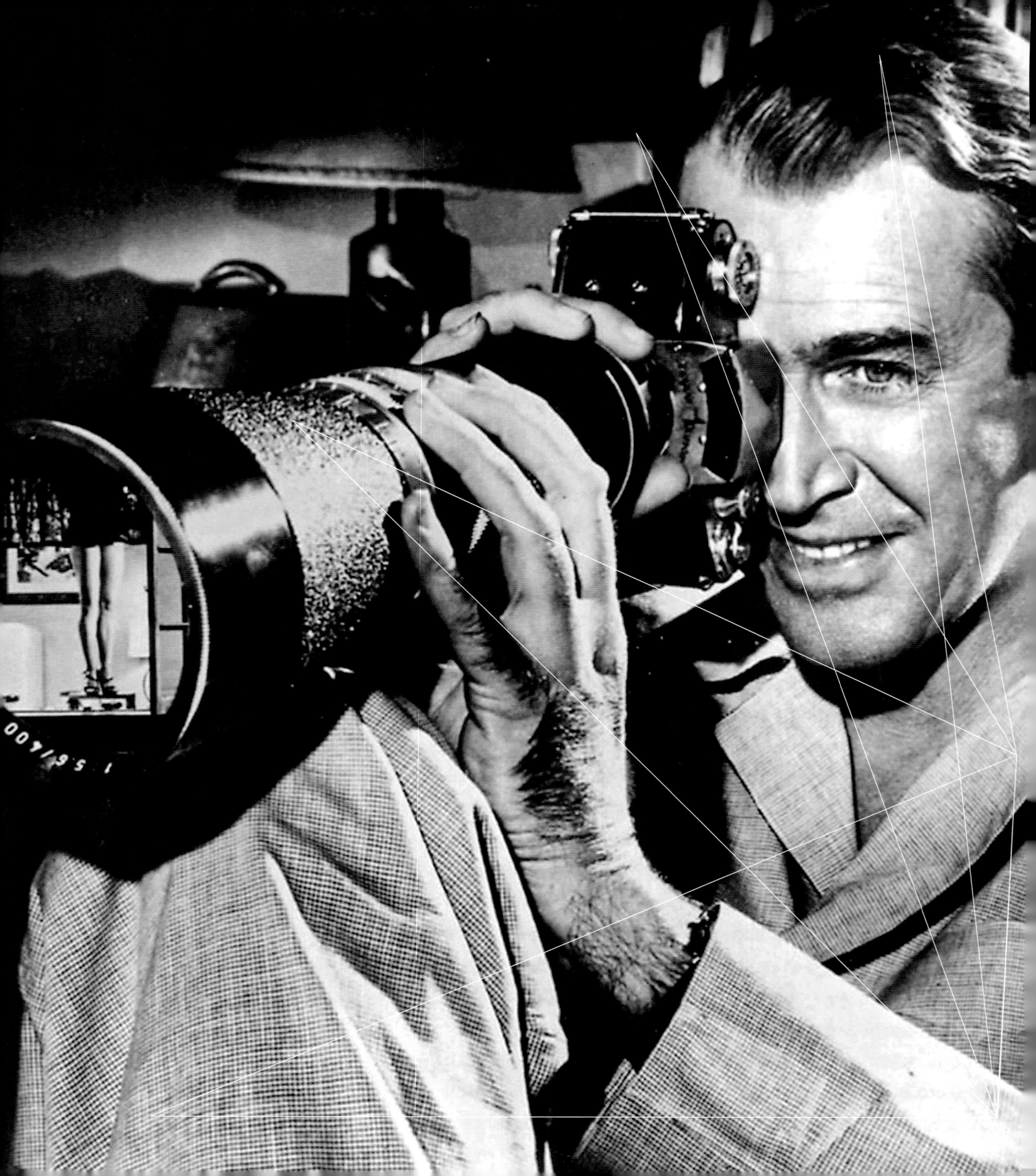

IV.

LOOKING BACK: FROM PHOTOGRAPHY AND FILM TO ALBERTI

11.

Sculpture Before Photography

MICHAEL COLE

At the turn of the twentieth century, Mary Ann Doane has argued, "time became palpable in a quite different way—one specific to modernity and intimately allied with its new technologies of representation." Doane's book, *The Emergence of Cinematic Time*, concerned the relationship between film and such contemporary fields as physiology, semiology, thermodynamics, and psychoanalysis; its arguments, though, could well be extended to other areas. The years first measured by pocket watches, train schedules, and workers' punch cards were also those that saw the formation of art history as a modern discipline, a discipline that shared many of the impulses

This chapter began with material presented at a 2004 Clark/Getty workshop called "Art History and the Moving Image." I owe thanks to Michael Ann Holly for the invitation to speak and especially to Tom Gunning and Davide Stimilli for helpful discussion at the two meetings. I developed the essay in its present form in the context of the workshops Alina Payne hosted in Florence and Cambridge in 2007 and early 2008. Karen Beckman kindly read a draft of the essay at that time and offered much useful criticism; Noam Elcott provided important suggestions and references as well. A number of publications from the past six years bear on the topics developed here; were I to start again now, my approach would have to be different. Much in my discussion of Michelangelo's *Moses*, in particular, will seem

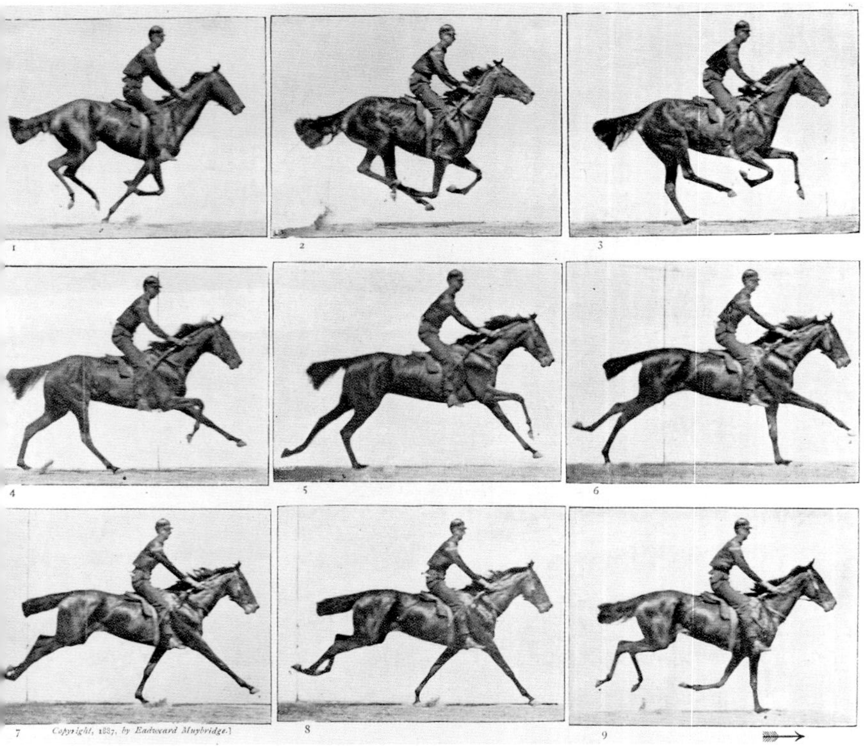

11.1 Eadweard Muybridge, *Transverse Gallop.* From Muybridge, *Animals in Motion*, 1887.

Doane saw elsewhere: a "new obsession with temporal exactitude," "a concerted attempt to fix a moment, to grasp it as precisely, fugitive," a "drive to condense time," an "aspiration for instantaneity," an "ability to arrest the ephemeral, to represent the contingent."[1]

Nothing encouraged this more than art history's dependence on the photograph, which played a double role as a stand-in for its objects of study and as the supporting evidence in its published texts.[2] Photography made it possible for historians writing in different and distant places to study what could not easily be consulted in person, to put side by side what could not be seen together, and to scrutinize objects under conditions that even firsthand observation did not always allow. Photography facilitated the earliest art historians' comparative techniques, whether these were connoisseurial (as in the case of Morelli), formal (as with Wölfflin) or iconological (as with Warburg). But it also helped make apprehensible the temporalities that art history internalized.

The photography that art history inherited was one that had itself been subject to a dynamic transformation.[3] The development of cameras with lower exposure rates made it possible to shoot things that did not remain stationary for minutes on end, such that by the 1870s, photographers could for the first time "snatch [postures] from a flow of movement."[4] Muybridge's grids (fig. 11.1) and Marey's chronophotography would turn these postures into a sequence, encouraging viewers to see any given figural arrangement as one with a before and an after.[5] Louis Lumière's invention of an improved dry plate process and George Eastman's mass-market devices with preloaded roll film and fixed shutter speed and focal length made

redundant in light of Mary Bergstein's excellent *Mirrors of Memory: Freud, Photography, and the History of Art* (Ithaca: Cornell University Press, 2010); my own recent book, *Ambitious Form: Giambologna, Ammanati, and Danti in Florence* (Princeton: Princeton University Press, 2011) likewise expanded on themes in this chapter, providing a different reading of the Castello birds and elaborating on the importance of Pomponius Gauricus's text in relation to sixteenth-century sculpture. I have added references to a few recent studies but refrained from extensively rewriting my original essay, which would distance it from the 2007–8 events of which this volume is a record. All translations are mine unless otherwise noted.

it possible, beginning in the 1880s, for even nonprofessionals to capture moments they witnessed around them in snapshots.[6] Doane remarks that the significance of the cinema lay in its "apparent capacity to perfectly *represent* the contingent, to provide the pure record of time." Yet as Tom Gunning has argued, the technology of the still camera was already, to early cinema, both a condition and a mirror: it was the possibility of collapsing time within the single frame that allowed those frames to be strung together into a virtual version of the elapsed moments from which they were taken.

The camera's new speed enabled its operators to consider potential material that extended beyond the landscapes and posed arrangements that had featured in the medium's earliest years, making time "palpable" in the way Doane describes. Yet the period during which Muybridge was photographing Leland Stanford's horses also saw an enormous expansion in a category of imagery that was much older than photography itself: the reproductive image of the still art object.[7] Other technical innovations, meanwhile, including the carbon printing process, allowed such images to be put to lucrative new ends, as Alinari, Braun, and other commercial firms began to sell photographs depicting historical paintings and sculptures in quantity.[8] Usually produced by taking very long exposures in controlled lighting conditions (fig. 11.2), these aspired no less than the new images of figures in motion to the sharpness that advanced cameras could achieve, even though they recorded nothing momentary, or sequential, at all.[9] The Alinari or Braun picture should remind us that photographs do not necessarily arrest the ephemeral any more than art history's other objects do. Yet as Thierry de Duve has observed, the experience of the picture and the experience of the event "coexist in our perception of any photograph, whether snapshot or time exposure."[10] And the familiarity of the snapshot helped lend a sense of the instantaneous even to earlier, prephotographic objects.

Nowhere is this more evident than in the history of early modern sculpture. The black-and-white photograph—until very recently the primary kind of picture on which art historians relied—does different things to sculptures than it does to works in other media. By graying the world around the (often already monochromatic) figure, by flattening and integrating the sculpture into its environment, photography compensates for the limits of sculpture's own illusion, suppressing what seems most artificial in marble or bronze. Theorists from Kracauer to Barthes have meditated on the photograph's aura of death, but photographs of sculpture can have the opposite effect, hinting that the petrified man in the image might once have been animate.[11]

At the same time, no premodern medium seems more distant than sculpture from photograph's capacity to fix the moving image. The cost of materials, the formal limits they imposed, and the rather narrow conventions constraining subject matter long ensured that sculpture could never quite indulge in photography's experimentalism. Since most of the public sculptures that photographs captured stood at two removes from the things they represented—they reproduced models that were themselves depictions of something else—rarely did they, like the photograph, actually document or index anything present before the artist.[12] Most significantly, perhaps, sculpture was slow. Even with a large studio involved, the cutting and polishing of a stone or the casting and chasing of a bronze was laborious work, with forms produced over months, even years. There is nothing in

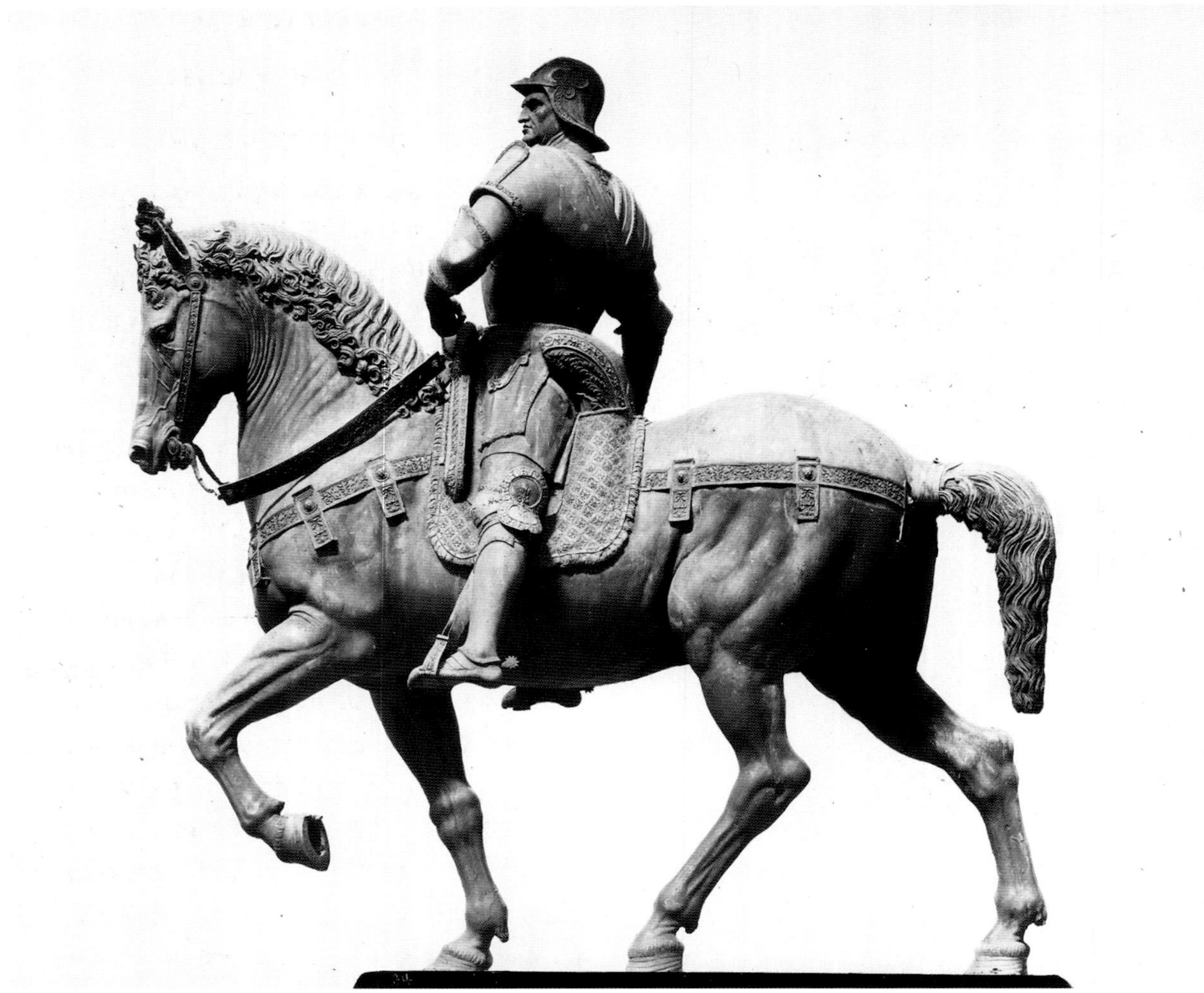

11.2 Andrea del Verrocchio, *Monumento a Bartolomeo Colleoni,* 1480–88. Bronze, 395 cm in height without base.

the *technology* of sculpture to make the viewer conscious of the simultaneous or the momentary.

In many respects, then, the sculpture and the photograph articulate against each other. Yet photographs of sculptures make the differences easy to forget. Photographs like figure 11.1 announce that exposure time has been collapsed to a mere instant. Photographs like figure 11.2, by contrast, make the mechanism behind the image and the speed with which the image was registered much harder to see. Photographs produced in the latter mode but with content that resembles that of the former project instantaneity onto something with no definable instant at all; they transfer an imagined instant, the time of the shutter, from the photograph to the thing that it shows.

We see this transference, notably, in the literature on the work that early twentieth-century writers regarded as the greatest of all Renaissance sculptures, Michelangelo's *Moses* (fig. 11.3). Here is Marcel Brion, writing in 1939: "The prophet was seated, weary and discouraged. He had just come down from Sinai. . . . He had been face to face with God and God had spoken to him. Now he had returned, bearing the stone tablets with God's laws written thereon. But his people did not listen to him, for they were engrossed in vain and wicked pleasures. It was this moment of mounting anger that Michelangelo chose to portray."[13] Brion presents the tablets Moses holds and the angry expression he wears as evidence that Michelangelo showed the patriarch at a single transitory juncture, an *event* in Exodus 32:19: "And when he came nigh to the camp, he saw the calf, and the dances: and being very angry, he threw the tablets out of his hand and broke them at the foot of the mount." In this, Brion was following a standard line of interpretation. As early as the 1850s, Burckhardt had remarked that "Moses seems to be shown in that moment in which he catches sight of the worship of the Golden Calf," and when Herman Grimm, in 1864, suggested that "[Moses] sits there as if on the point of starting up," he underscored what he regarded as the transient quality of the scene.[14] Writers in the decades that followed, however, especially those studying the work from a distance through photographs, uniformly tried to narrow the depicted instant, seeking precision not only about the verse from Exodus to which the sculpture corresponded but also about how the action unfolded. Carl Justi, in 1900, suggested that "[Moses's] turn can only indicate his sudden distraction by an event that riles him, and the tablets of the law allude to his descent from Sinai."[15] *For an instant*, Justi went on, "it seems to Moses that his work is destroyed and he despairs for his people. . . . He lets the tablets slip from his right hand onto the stone seat; they have been brought up sideways there, pressed by his forearm against the side of his body. His hand, however, comes in contact with his breast and beard and thus, at the turning of the head to the spectator's right, it draws the beard to the left."[16] Fritz Knapp, writing in 1912, saw the statue differently, yet his mode of description was much the same: "The cry of song and dance wakes [Moses] from his dream. His eye and his head turn to look. In a single moment, fear, wrath, a whole fury of wild passions penetrate his giant form. He has seen the people worshipping the golden calf. The tablets of the law begin to slip downward—they will fall and break on the ground when he rises to hurl his thundering words of anger into the mass of apostates. Already he rebels wildly inside; his hair stands up like horns."[17] Knapp ultimately concluded that Michelangelo opted to portray "the moment of highest tension" in the Moses story. And in this Knapp anticipated the most dramatic example of such an interpretative sensibility, the essay on the *Moses* that Sigmund Freud published in 1914.

Freud had seen the statue for the first time at the beginning of the century, and he had read widely on it since then.[18] He considered his essay amateurish enough that he decided to publish it anonymously, but he would not have arrived at the conclusions he did had he not thought carefully about the most recent art-historical literature. He agreed, significantly, with the Burckhardtian view that the statue depicted a single moment in the Moses narrative, as well as with the intuition of Justi and Knapp that

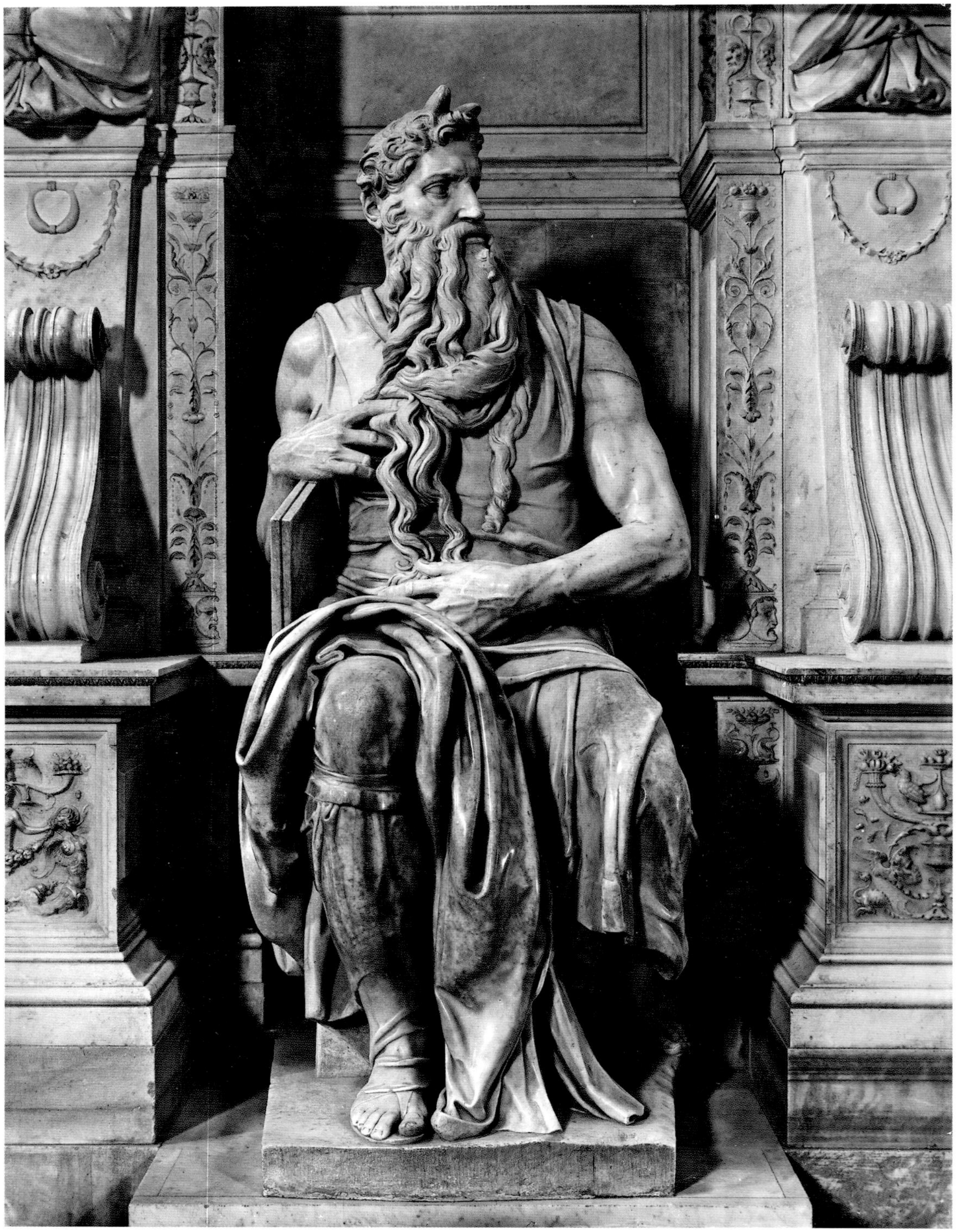

11.3 Michelangelo, *Moses,* from the tomb of Pope Julius II, ca. 1513–15. Marble, 235 cm in height. San Pietro in Vincoli, Rome.

the statue embodied one in a sequence of clearly implied, definable instants. Whereas most earlier critics had emphasized the statue's prospective dimension, however—Moses is about to rise, the tablets are about to fall—Freud looked almost exclusively to the past. He observed, for example, that the figure's right index finger pulls at strands from the left side of its beard "and that this oblique hold prevents the beard from accompanying the turn of the head and eyes to the left." Such an arrangement was difficult to rationalize, Freud thought, unless we imagined that Moses's "hand had seized his beard . . . had reached across to its left edge, and, in returning to that position in which the statue shows it, had been followed by a part of this beard, which now testifies to the movement that has just taken place."[19] Moses's pose, that is, only makes sense if we imagine it following from something that transpired earlier. And returning to the assumption, gleaned from art history, that the episode Michelangelo portrays was that of the worship of the golden calf, Freud proposed the following reconstructed plot:

> [Moses] was sitting there calmly . . . his head with its flowing beard facing forward, and his hand in all probability not near it at all. Suddenly the clamour strikes his ear; he turns his head and eyes in the direction from which the disturbance comes, sees the scene and takes it in. Now wrath and indignation lay hold of him; and he would fain leap up and punish the wrongdoers, annihilate them. His rage, distant as yet from its object, is meanwhile directed in a gesture against his own body. His impatient hand, ready to act, clutches at his beard which has moved with the turn of his head, and presses it between his thumb and palm in [an] iron grasp. . . . But now an alteration takes place, as yet we do not know how or why. The hand that had been put forward and had sunk into his beard is hastily withdrawn and unclasped, and the fingers let go their hold; but so deeply have they been plunged in that, in their withdrawal, they drag a great piece of the left side of the beard across to the right, and this piece remains lodged over the hair of the right under the weight of one finger, the longest and uppermost one of the hand. And this new position, which can only be understood with reference to the former one, is now retained.[20]

This is not the place to rehearse the rest of Freud's elaborate and imaginative interpretation—his explanation, for example, of how, during this motion of Moses's upper body, the tablets of the law fell into their current position, or his account of how the statue corresponds to the biblical text, which he quotes at length. Notice, though, that Freud's description of the statue is not simply the eccentric product of a psychoanalyst who came across an artwork and decided to treat it as a patient. The fundamental question Freud asks—"Did Michelangelo intend to create a 'timeless study of character and mood' in his Moses, or did he portray him at a particular and, if so, at a highly significant moment of his life?"—is one that he lifted almost verbatim from Henry Thode's great 1908 study *Michelangelo: Kritische Untersuchungen über seine Werke*.[21] And interesting though it is that Freud rejected Thode's own answer to the question, what is more important here is how close those particular alternatives look to what de Duve characterized as the two opposite ways in which every photograph is perceived.[22] By 1914, Michelangelo's *Moses* was strung between time exposure and snapshot, picture and event.

Michelangelo scholars had been using photographs for several decades, but in the years when Freud was writing, the inclusion of a photographic survey had become a major selling point of new books. Titles called upon readers to linger over the illustrations: 1912, the year Freud first commented to Ernest Jones about his interest in the statue, was the year of Max Sauerlandt's *Michelangelo . . . mit hundert Abbildungen* as well as Knapp's *Michelangelo, des meisters Werke in 169 Abbildungen*. Freud no doubt knew both, and he quotes Knapp in his essay. To be sure, Freud had also seen the statue in person on visits to Italy. It was through photographs, however, that he studied it. (Jones reports that he sent Freud photos from Rome, and even had new ones commissioned for him.)[23] That Freud relied heavily on photographs in his writing is hardly surprising, given his remarks on how much Morelli's methods inspired him. The issue here, though, is that Freud's inferences about the sculpture are inferences that photographic technology, more than Morelli's example as such, had encouraged.

Among the clichés of the Renaissance *paragone* debate is the assertion that painting is an additive, sculpture a subtractive process.[24] Yet to differentiate Renaissance sculpture from its photographic double, it is helpful to notice the degree to which Renaissance and later sculptures were always conceptually additive as well. Most statues from the period consist principally of a figure, perhaps a pair of figures, placed into a pose and *outfitted* with attributes that establish the figure's identity. The artist gives his figures gesture and expression—and "gives" these to them in the most literal sense, whenever he makes models—and often enough the poses count among the attributes, characterizing their subject as fierce, or graceful, or modest, or associating it with a defining action. Neptune calms the seas, and the river god spills forth streams: it is the difference in what they *do* that allows us to tell one from the other.

Early modern statues frequently consist of little more than the figure and its attributes, and the attributes are the first features of the work to move us away from any kind of documentary paradigm. Michelangelo's *Moses*, for example, is seated, or rather enthroned: this, along with his possession of the laws, establishes him as a typological forerunner of Saint Peter and thus of the pope whose tomb the statue was to ornament. The same attributes interfere with the perception of a single extension of time, extracted from a before and after—the inference, for example, that the lawgiver has just descended from Mount Sinai, come "nigh to the camp," and discovered that his people have fallen into idolatry. The pairing of tablets and enthronement take the figure outside historical time altogether. Here and elsewhere, we need to differentiate the question of what is happening in a painting or sculpture from the question of what episode it illustrates.

Mary Bergstein observes that Erwin Panofsky, writing in 1939, had already rejected the idea that Michelangelo's *Moses* showed a particular narrative moment.[25] But Panofsky could hardly end the conflation of the time of the shutter and the time of the sculpture, the transposition of attribute into event. Rather, the effect of his essay was to make Renaissance art look somewhat more temporally layered than its chronological neighbors, relocating the fantasy of the sculptural "instant" to a later historical period. There it has thrived ever since. The most important recent writer on the bronze birds that Giambologna made for the grotto in the Medici garden at Castello (fig. 11.4), for example, is Detlef Heikamp:

> [Giambologna] captures the animal while it is in movement, or better, captures a single phase of a transitory act that takes place in a fraction of a second. The operation is facilitated in modern times by the technology of the photograph or the cinema, with which the various reflexes of motion can be fixed with ease: when, for example, a pigeon is preparing to fly and extends its wings, even just barely, the tail by reflex opens like a fan and points downward. Giambologna, who knew well the dynamics of concatenated movements thanks to his experience representing horses and other large animals on the run, was now given occasion to extend his abilities to birds as well.[26]

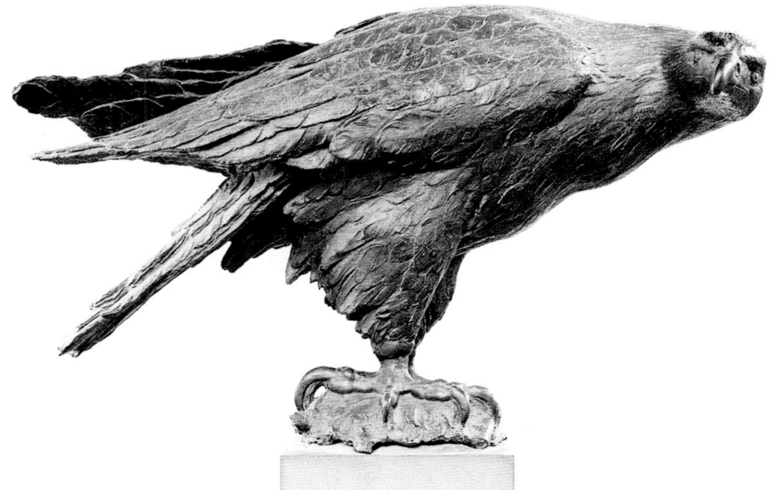

11.4 Giambologna, *Eagle,* ca. 1567. Bronze, 83.12 x 50.4 cm. Museo nazionale del Bargello, Florence.

In Giambologna, as Heikamp sees it, we marvel at the fact that the eye of a premodern artist could record what we might assume to have been made visible only by a later instrument. "If we knew nothing about the history of the Grotto and its avian bronzes," he remarks, "we might think that they had been created in the positivistic nineteenth century."

Heikamp himself is perfectly conscious of the anachronism in his comparison, and he uses it to work against the weighty historiographical tradition identifying Giambologna with "mannerism." We cannot say the same of all the similar remarks in the literature on Gianlorenzo Bernini, the premodern artist who has become virtually synonymous with the protophotographic, with baroque instantaneity. Alessandro Angelini's 2005 *Baroque Sculpture in Rome*, to take a recent example, suggests that in the *Pluto and Proserpina* (fig. 11.5) "Bernini experiments with the kinetic action of the statue, summing up various moments of the *istoria* as in a snapshot. This representation of the abduction in the midst of the act actually suggests the preceding instant in which the maiden was quietly picking flowers before being seen and desired by Ditis."[27] The simile in this case brings a string of contradictions in tow. But in evaluating such a reading of the work, we should first note what it stands against. To say that the statue is an experiment in "kinetic action" is to take something like the opposite of the materialist view, which would instead underscore the virtuosity involved in stabilizing a monumental composition in stone.[28] To see in the *Pluto and Proserpina* a reference to, or even a depiction of, a preceding instant is to play down the extent to which the composition rather reworked a preceding *statue*, returning to the exercise in equilibrium staged by Giambologna's famous marble *Sabine* and by Bernini's own earlier *Aeneas*, *Anchises*,

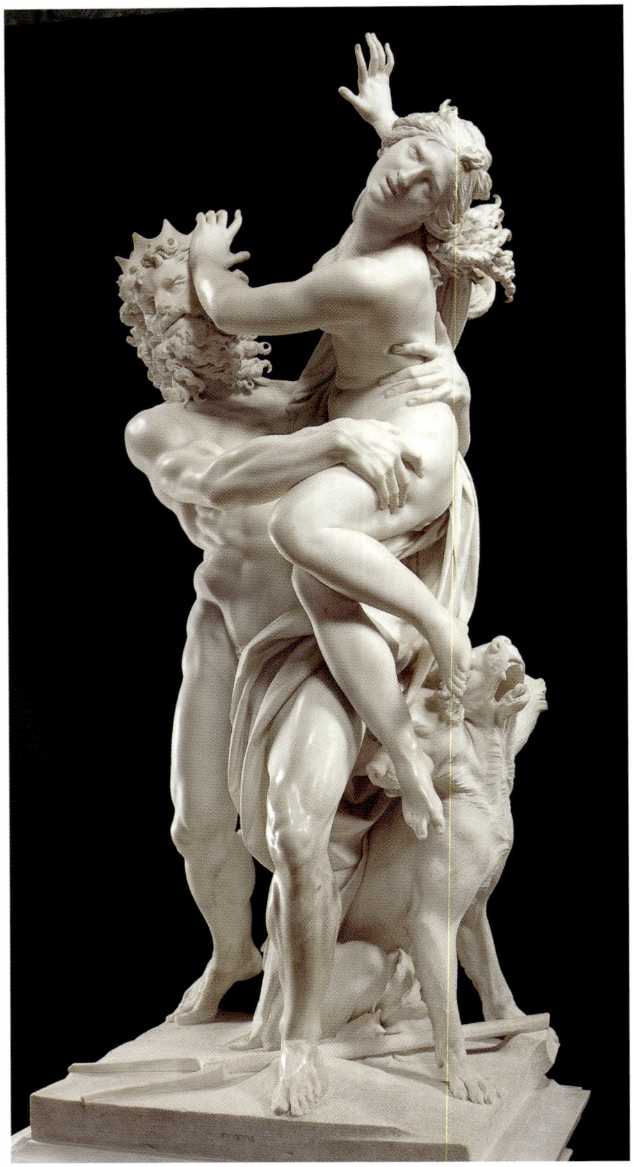

11.5 Gianlorenzo Bernini, *Pluto and Proserpina*, 1621–22. Marble, 225 cm in height. Galleria Borghese, Rome.

and Ascanius. And to characterize the encounter Bernini depicts as a snapshot is to crop out a not insignificant detail like the three-headed god Cerberus at Pluto's feet. As a marble brace, this device literally prevents the stone from moving—from falling over. As a creature that faces multiple directions, moreover, it draws the viewer around the object, encouraging that *viewer* to move and study an object that itself remains still for inspection. What it does not do is reinforce the impression of a moment, since Cerberus does not even appear in the primary narrative reference behind Bernini's work, Ovid's account of Proserpina's abduction in book 5 of the *Metamorphoses*.

Angelini's take on the *Pluto and Proserpina* illustrates what was, by 2005, a widespread trope in the literature on Bernini. Four decades earlier, Howard Hibbard's classic monograph on the artist had already proposed that Bernini's aim in his *David* (fig. 11.6) was to convey "a single, heightened moment in time."[29] No one could deny that *this* statue associates its subject with a defining action—it is because the figure throws a rock that we know it to be David. Yet if we take the action as attributive, we also separate it from any performance Bernini would have witnessed or even imagined. In fact, the statue records no moving figure at all, but rather instantiates a generic type, descending from Leonardo's *Treatise on Painting*, that Annibale Carracci's *Polyphemus* in the Farnese Gallery had newly made famous. And it is a testament to what Doane calls our "obsession with temporal exactitude" that even those who noticed this could still look for an unperformed movement, a moment that never was.[30] Thus Rudolf Wittkower wrote of Bernini's response to Annibale that "almost alone among his contemporaries, he found in [the *Polyphemus*] the representation of unambiguous and energetic action; he found physical concentration rendered at the split-second before release."[31]

The concept of the "split second" was almost certainly foreign to Bernini. Though contemporaries were equipped to express something like a concept of instantaneity, that concept looked quite different from what instantaneity has become in the time of photography. One Roman who certainly knew Bernini, for example, was the physician and art critic Giulio Mancini: he cited Pliny as an authority on the principle that "the painter imitates things in that indivisible instant in which there is neither time nor motion, but only stillness."[32] The formulation reminds us that the Italian *istante*, like the English "instant," derives from the Latin *stare*, "to stand." Stasis was built into the very word.

And can we even follow Wittkower's suggestion that Bernini's *David* represents an "unambiguous" action? Once again, the statue's attributes evidence the artist's willingness to work against an impression of momentary singularity. Needing, as he had a year earlier with the *Pluto*, to prop up the figure from below, Bernini once again added a brace. In this case, the reference is to 1 Samuel 17:38–40: "Saul . . . armed [David] with a coat of mail. And David, having girded his sword upon his armor, began to try if he could walk in armor: for he was not accustomed to it. And David said to Saul: I cannot go thus, for I am not used to it. And he laid them off, And he took his staff . . . and chose five smooth stones out of the brook, and put them into the shepherd's scrip . . . and he took a sling in his hand, and went forth against the Philistine." The armor, structurally necessary to keep the marble leg from breaking, shows what David "laid off" at another time in another place. The same is true of the harp, also at David's feet: this alludes to the boy's performances at Saul's court and

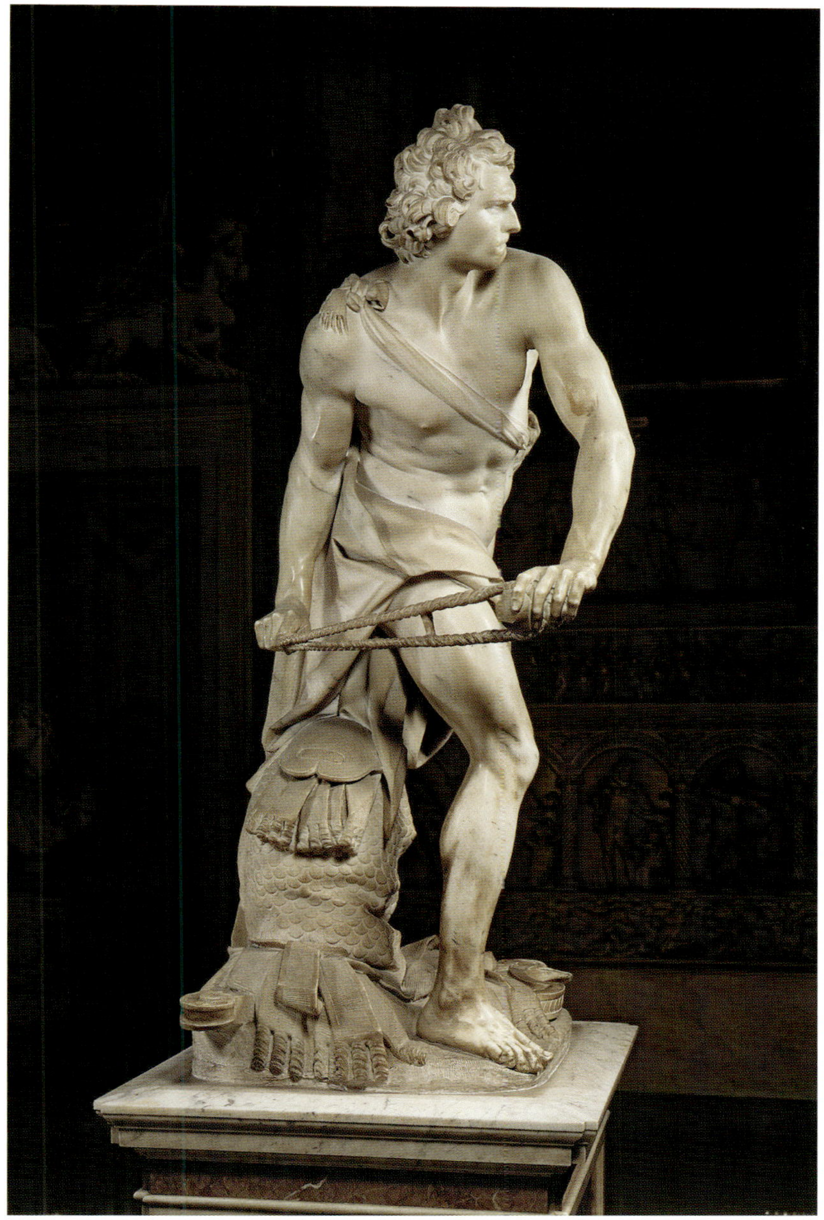

Gianlorenzo Bernini, *David*, 1623–24. Marble, 170 cm in height. Galleria Borghese, Rome. **11.6**

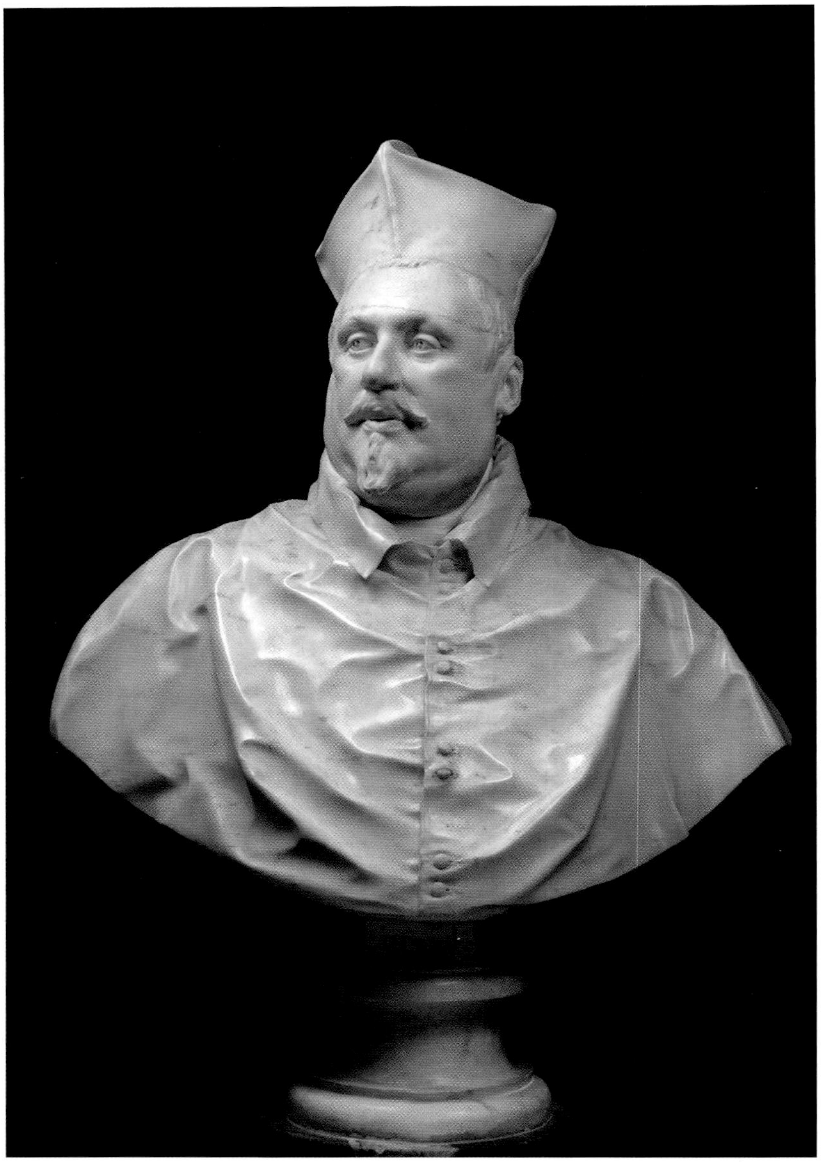

11.7 Gianlorenzo Bernini, *Bust of Scipione Borghese,* 1632. Marble, 78 cm in height. Galleria Borghese, Rome.

to his future as a psalmist, but it has little to do either with the armor that encumbered David or with the battle that followed its removal. Taking the episode of the slung rock to be the center of the statue, we might say that the harp and armor are anachronistic, except that the very concept of anachronism depends on an expectation or standard of compositional simultaneity for which we have yet to see clear evidence in any example of Italian statuary. It is the idea of anachronism itself that is anachronistic.

Much the same can be said of Bernini's portraiture. Here, too, scholars have frequently drawn attention to spontaneous expressions, animated draperies, parted lips that suggest speech. Wittkower went so far as to compare the artist's marble bust of Scipione Borghese (fig. 11.7) to what he called the "snapshot sketch," now in the Morgan Library.[33] Robert Torsten Petersson, writing in 2002 of the same work, and no doubt with Wittkower's discussion in mind, loosened the specific reference to the Morgan drawing while preserving the photographic comparison: "For details Bernini would make dozens of sketches, each one from life, with none serving as a model. He did not work from his sketches and bozzetto but from the conceptual image in his head. It may sound easy, like a snapshot, but the image must be held intact, unaffected by changes both internal and external."[34] The comment in this case depends on period accounts of the artist's working method. In describing Bernini's procedures, however, the seventeenth-century writers do not frame an effect of instantaneity but rather emphasize the *duration* of Bernini's contact with his subject: the repeated sittings, the courtly banter, the serial preparatory works on paper and in clay. And when Bernini's son Domenico wrote that after these sessions the artist would set his studies aside, executing the marble from memory, the point was not to suggest a registration of the visible but just the opposite—to reject the notion that Bernini's portraits were mere *copies* of things seen.[35]

None of this is to deny that seventeenth-century sculptors did, on occasion, try to represent a specific action, or that their viewers understood this. Consider Giovanni Battista Passeri's well-known report on the *Veronica* (fig. 11.8) that Francesco Mochi unveiled in the crossing of St. Peter's in 1639:

> [Pope] Urban ordered that Mochi be assigned one of the large statues, each twenty-three palms high, which are located in the four large niches in the big pilasters of St. Peter's, those that hold up the dome. He was given the Veronica, which is a figure of immense spirit and mastery. He represented her in a pose of movement, and in a pose of violent movement, not only walking, but running, and in this (and let this be said in peace), the statue lost its proper essence, for the noun form of the word "statue" derives from the Latin *sto stas*, which means to remain firm, stably standing. [Mochi's statue] is not a permanent and immobile statue, as it ought to be, if it is to form a simulacrum that is there to be enjoyed and admired by those who look at it. It is a character [personaggio] that passes by, and that does not remain.[36]

Passeri, who was ambivalent about Mochi's achievement, makes paradoxical references to the *Veronica*'s "pose of movement" (atto di moto). What he does not try to do, though, is align the statue with any particular narrative. And how could he have? There was no passage in the hagiographic literature that had Veronica, Christ's face already registered on her cloth, run like this. What

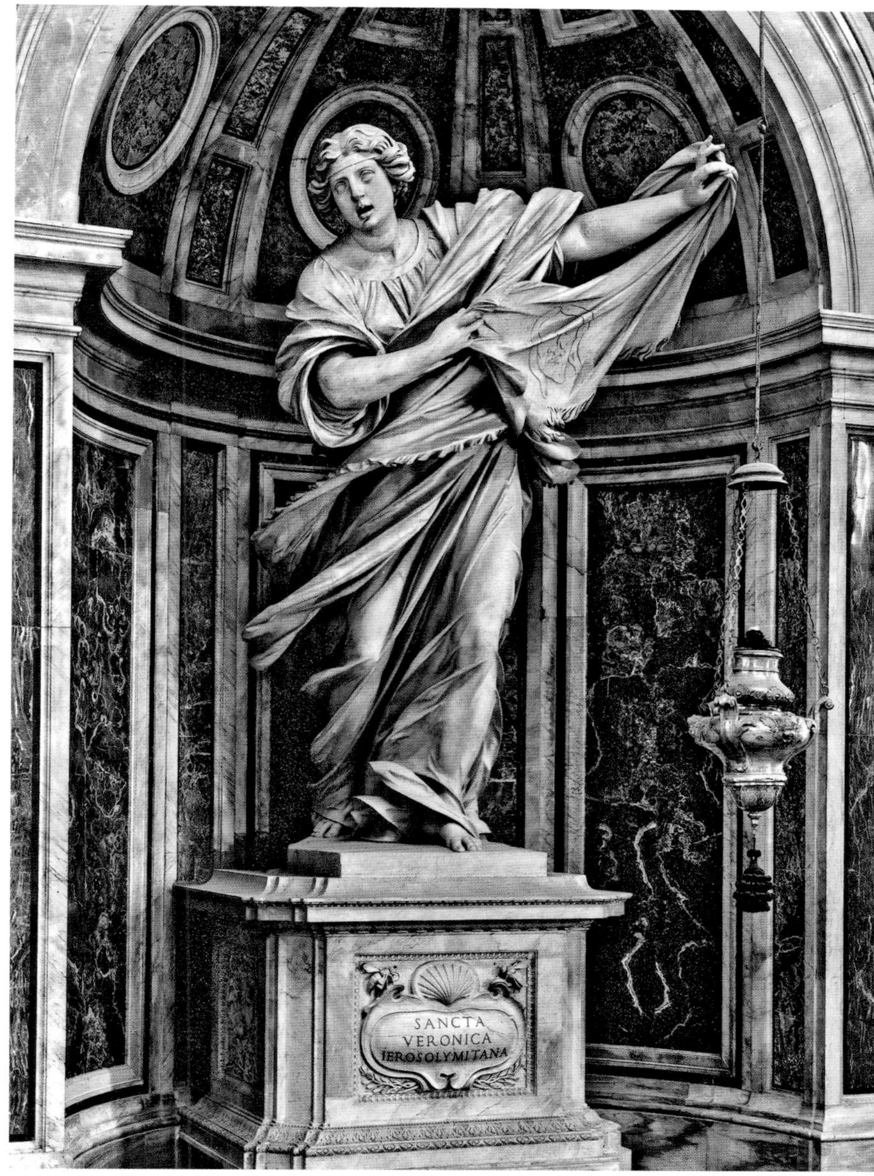

Francesco Mochi, *Veronica*, 1629–32. Marble, 500 cm in height. St. Peter's Basilica, Vatican. **11.8**

substitutes for narrative in Passeri's account is the notion that Mochi endowed the figure with "immense spirit," the signature quality that Mochi's admirers suggested he brought to all his statues.[37]

One might take Mochi to exemplify an early modern aesthetic of *Lebendigkeit*.[38] Nonetheless, it is relevant to emphasize that to the extent that Mochi aimed to bespirit his work, this took him outside chronology, even outside the assumption that the movement he represented should be controlled by any pictorial frame. Bernini himself, mocking the *Veronica*, allegedly asked after the source of the wind that blew her drapery.[39] The implication is that the figure shows evidence of the ambient conditions within St. Peter's itself, that where the statue represents an event, that event belongs to something other than the historical time of the character Veronica herself.

Still more important is the fact that it was Mochi's depiction of movement that Passeri liked least about the statue. As he saw it, the very attempt to show what "passes by" was a vice. This was hardly an idiosyncratic view. Gotthold Ephraim Lessing is now best known as a theorist of the sculptural moment—Mary Bergstein has even suggested that Freud's reading of Michelangelo's *Moses* adhered to Lessing's precept—yet the paragraph following his famous statement that "the artist can never make use of more than a single moment in ever-changing nature" adds a crucial qualification: "Furthermore, this single moment, if it is to receive immutable permanence from art, must express nothing transitory."[40] Sculpture and painting, Lessing believed, had to devote themselves to the perpetuation of the things they depicted. Viewers, he thought, tend to regard what they see in a sculpture not as an aspect of an instant but rather as a permanent feature of the body it ornaments. This is why Laocoön—though bitten by snakes—cannot scream: "When a man of firmness and endurance cries out he does not do so unceasingly, and it is only the seeming perpetuity of such cries when represented in art that turns them into effeminate helplessness or childish petulance. This, at least, the artist of the Laocoön had to avoid." The doctrine of the pregnant moment is not the rule that the artist should arrange his figure so as to telegraph what comes immediately before or after, depicting a condition between two others; rather, it is the idea that the artist should compose a cohesive sculpture that "gives the freest reign to the imagination" of the spectator.

One familiar reading of Lessing would have him anticipate a modern photographic (or even psychoanalytic) paradigm. But Lessing was also an early modern writer, and his notion of sculptural permanence has a substantial prehistory. An important exposition of much the same principle, notably, appears in the writings of the seventeenth-century Roman sculptor Orfeo Boselli. In a chapter titled "What you need to think about before making a statue," from the second book of his *Observations on Ancient Sculpture*, Boselli distinguishes between what he terms the *atto*, the pose or attitude that an individual character strikes, and the *attione*, or action, that motivates it. He continues, "I hold that action is nothing other than the operation that the statue you are making carries out, an operation that the sculptor must first consider and establish in his own mind. [Such] a figure can do one of two things: it can either show its expressive signs [i.e., its attributes] to the spectator, or it can contemplate them itself."[41] Action, that is, does not entail narrative, or even *motion*. Either the figure can make a gesture of demonstration—Boselli's example is a Mercury who shows a pouch to the beholder—or it can affect some variety of meditation. "Under these two headings [i.e., demonstration and contemplation], *attione* is finished."[42]

To modern readers, this may sound like a rather reduced set of options, yet as far as sixteenth- and seventeenth-century sculpture is concerned, they cover a lot of territory. The action of Mochi's *Veronica*, after all, consists of little more than the dramatic display of an attribute: the sculpture essentially labels the pier in which it stands, which contains the relic—Veronica's veil—that the statue holds before it. And might the same not be said of Michelangelo's *Moses*—that its hands *display* the beard that identifies the man? The choice to make Moses the centerpiece of the tomb must, in the end, have depended on Pope Julius's identification with the Old Testament figure, and Julius, as Loren Partridge and Randolph Starn showed, was famous for his beard, alternatively growing and cutting it as a symbolic gesture as his political and military fortunes shifted.[43]

Boselli's and Passeri's ways of looking at sculptures, moreover, would not have surprised Michelangelo or his contemporaries. Anyone who knew the lines that Leonardo associated with what he calls "the man who wants to throw a rock," for example, would have been aware that the painter and theorist associated this kind of figure with *potential* force, not with motion per se. And in perhaps the most important Renaissance theorization of sculpture, Pomponius Gauricus anticipated Passeri in tracing the word "statue" to the Latin "status," literally, "made to stand still," writing that his contemporaries especially praised those sculptures showing "states that seem to precede or follow from a motion"—that is to say, statues that embody no motion at all.[44]

Early modern writers like these help alienate the obsession with the instant evident in much of the art-historical literature on early modern sculpture, not to mention other media. They leave us the task of writing about the early modern eye as an instrument of vision, a task that must involve the differentiation of that instrument from the technology behind the photograph. The goal of such a project cannot, of course, be the abandonment of photography as a tool—which would be both pointless and impossible—so much as an awareness of how the evidence for the critical faculties of early modern viewers might put pressure on photography itself. Rather than reinforce the comfortable truism that photography made art newly visible, early modern writers help us see what photography has distorted. To look from the Renaissance toward photography is to ask what, beyond the blink of the shutter, the camera renders.

Notes

1. Doane, *Emergence of Cinematic Time*, 4, 10, 14–15.
2. For a general discussion of the role of the photograph in shaping late nineteenth-century art history, see Dilly, *Kunstgeschichte als Institution*, 155–60.
3. See Tom Gunning's three essays, "'Animated Pictures'"; "New Thresholds of Vision"; and "Never Seen This Picture Before."
4. The phrase is Kracauer's. See his *Theory of Film*, 44.
5. See the discussion in Doane, *Emergence of Cinematic Time*, esp. 33–68, with further references.
6. Gunning writes on the importance of reduced exposure time for amateur photography in "New Thresholds of Vision"; see esp. 81–87, on amateurs' fascination with the instantaneous. On early snapshots, see Nickel, *Snapshots;* and Smith, "Roll Over."
7. For the importance of the still art object as a category, see Snyder, "Nineteenth-Century Photography of Sculpture"; Bergstein, "Mystification of Antiquity Under Pius IX"; and Billeter, *Skulptur im Licht der Fotografie*. Billeter notes that early photographers were interested in sculpture precisely because it did not move (17).

8. For the Alinari brothers' early photographs of artworks, see Conti, Ferretti, and Spalletti, "Documentzione dell'arte"; Messina, *Scultura e fotografia*; and Quintavalle, *Gli Alinari*, 283–399, and on sculpture esp. 325–31. For Braun, see Bergstein and O'Brien, *Image and Enterprise*.

9. Photographers working in interior spaces often kept the shutter open for hours when producing an image: Melissa Lemke, in a talk she gave at the National Gallery in conjunction with the study days on Desiderio da Settignano, observed that Clarence Kennedy left the shutter open for as long as twenty-four hours when making his own professional photos in the early twentieth century. The Alinari photograph would thus resist the otherwise very useful categorical distinction that Thierry de Duve makes in his classic essay "Time Exposure and Snapshot." De Duve writes that "the aesthetic ideal of instantaneous photographs is *sharpness*" (119), while "the aesthetic ideal of time exposure is . . . a slight *out-of-focus*" (121).

10. Ibid., 113.

11. Eugenia Parry Janis makes a similar point, considering the appeal of the photograph itself rather than the sculpture in it: "Through this new technology, dedicated to the here and now, sculpture offered the viewer a means to contemplate the romantic obsession with *then*. To photograph was not only to record, but in some strange fashion to raise the dead from complex and discontinuous time." Janis, "Fabled Bodies," 19.

12. On the documentary value of photography and its limits in the late nineteenth century, an essential reference is Daston and Galison, *Objectivity*, esp. 115–90.

13. "Le prophète est assis, las et découragé. Il descend du Sinaï, aveuglé encore per les éclairs, assourdi par la tempête divine qui faisait rage autour du sommet de la montagne. Il a vu Diue face à face et Dieu lui a parlé. Voilà qu'il revient parmi les hommes, maintenant, portant, écrites sur la pierre, les lois de Dieu. Et son people, occupé à des plaisirs futiles et pervers, ne l'écoute pas. Tel est le moment que Michel-Ange a choisi pour représenter Moïse." Brion, *Michel-Ange*, 238.

14. Cf. Burckhardt, *Cicerone*, 634: "Moses scheint in dem Moment dargestellt, da er die Verehrung des goldenen Kalbes erblickt und aufspringen will." Grimm, *Leben Michelangelos*, 2:95: "Er sitzt da, als wollte er eben aufspringen, das Haupt stolz aus den Schultern in die Höhe regreckt, mit der Hand, unter deren Arme die Gestzestafeln ruhen, in den Bart greifend, der in schweren Strömen auf die Brust sinkt, mit weit athmenden Nüstern und mit einem Munde, auf dessen Lippen die Worte zu zittern scheinen."

15."Die geschichtlich sachliche Deutung ist nicht schwer zu finden. . . . Diese Wendung kann nur die plötzliche Ablenkung durch ein aufregendes Ereignis bedeuten, und die Gesetzestafeln weisen auf die Herabkunft vom Sinai." Justi, *Michelangelo*, 326.

16. "Einen Augenblick scheint ihm sein Werk zerstört, er verzweifelt an diesem Volke. In solchen Augenblicken verräth sich der innere Aufruhr in unwillkürlichen kleinen Bewegungen. Er lässt die beiden Tafeln, die er in der Rechten hielt, auf den Steinsitz herabrutschen, sie sind über Eck zu stehen gekommen, vom Unterarm an die Seite der Brust gedrückt. Die Hand aber fährt an Brust und Bart, bei der Wendung des Halses nach rechts muss sie den Bart nach der linken Seite ziehen." Ibid.

17."Er hört Lärm, das Geschrei von gesungenen Tanzreigen weckt ihn aus dem Traum. Das Auge, der Kopf wenden sich hin. Schrecken, Zorn, die ganze Furie wilder Leidenschaften durchfahren im Moment die Riesengestalt. Er hat das Volk in Anbetung des goldenen Kalbes erblickt. . . . Die Gestezestafeln fangen an herabzugleiten, sie werden zur Erde fallen und zerbrechen, wenn die Gestalt auffährt, um die donnerneden Zornesworte in die Massen des abtrünnigen Volkes zu schleludern. Schon bäumt sich das Innere wild auf, die Haare steigen wie hörner empor." Knapp, *Michelangelo*, xxxvii.

18. See Jones, *Life and Work of Sigmund Freud*, 2:20. Freud went to see the *Moses* for the first time on Tuesday, 3 September 1902. Freud's essay cites much of what was then the basic literature, including Grimm, Thode, Springer, Justi, Müntz, and Burckhardt. See the discussion in Bergstein, "Freud's 'Moses of Michelangelo,'" and Bergstein, *Mirrors of Memory*, 46, 91, 98, and passim.

19. Freud, "Moses of Michelangelo," 532.

20. Ibid., 532–33.

21. "Handelt es sich um einen dramatischen Moment, der durch ein bestimmetes äusseres Eriegniss hervorgerufen wird oder um eine Chrakterschilderung?" Thode, *Michelangelo*, 1:205.

22. "Whereas the snapshot refers to the fluency of time without conveying it, the time exposure petrifies the time of the referent and denotes it as departed. Reciprocally, whereas the former freezes the superficial time of the image, the latter releases it. It liberates an autonomous and recurrent temporality, which is the time of remembrance. While the portrait as *Denkmal*, monument, points to a state in a life that is gone forever, it also offers itself as the possibility of staging that life again and again in memory." De Duve, "Time Exposure and Snapshot," 116.

23. "In the summer of 1912 he told me of his interest in the meaning of the statue, with his interpretation, and in September he reported from Rome to his wife that he was visiting Moses every day. On returning to Vienna he plunged into the extensive literature on the subject, and in October had managed to get the English book he particularly wanted whose interpretation had come nearest to his own. . . . I then sent him two photographs from Rome, and at his request had some also specially taken of the lower edge of the tablets." Jones, *Life and Work of Sigmund Freud*, 2:365.

24. Michelangelo put the idea succinctly in a letter to Benedetto Varchi: "Io intendo scultura quella che si fa per forza di levare; quella che si fa per via di porre è simile alla pittura." See Barocchi, *Scritti d'arte del Cinquecento*, 1:523.

25. Bergstein, *Mirrors of Memory*, 52. Bergstein's footnote refers to "Panofsky 1939"; no work by Panofsky with that date appears in her

bibliography, but the reference must be to the first edition of *Studies in Iconology*.

26. "Coglie l'animale mentre è in movimento, o per meglio dire coglie la singola fase di un atto transitorio che si svolge con la velocità di una frazione di secondo. Questa operazione è facilitata, nei tempi moderni, dalla tecnica, dalla fotografia e dalla cinematografia, così che alcuni riflessi del moto si possono fissare con agio: quando ad esempio il piccione sta preparandosi per il volo e distende le ali, anche solo minimamente, la coda per riflesso si apre a ventaglio e punta verso il basso. Giambologna, che conosceva bene le dinamiche del concatenarsi dei movimenti grazie alla sua consuetudine con la rappresentazione di cavalli e di altri grossi animali in corsa, ora è messo nelle condizioni di estendere le sue competenze anche agli uccelli." Heikamp, "Uccelli di bronzo," 251.

27. Angelini, *Baroque Sculpture in Rome*, 24.

28. See, for example, Preimesberger's discussion of equilibrium, in Coliva and Schütze, *Bernini scultore*, 110–23, as well as the same author's "Themes from Art Theory in the Early Work of Bernini."

29. Hibbard, *Bernini*, 55.

30. On the "unperformed movement," see de Duve, "Time Exposure and Snapshot," 114.

31. Wittkower, *Gian Lorenzo Bernini*, 5.

32."Dice [Plinio] che con la pittura viene dimostrato il moto delle cose corporee che va immitando . . . et il pittore immita le cose in quello istante indivisibile nel quale non vi è tempo nè moto, ma sol quiete." Mancini, *Considerazioni sulla pittura*, 1:159. See the discussion in Spagnolo, "Barn Owl Painters in St. Peter's," esp. 269.

33. Wittkower, *Gian Lorenzo Bernini*, 13.

34. Petersson, *Bernini and the Excesses of Art*, 103.

35. Bernini, *Vita del Cavalier Gio*, 134.

36. "[Urbano] ordinò che se gli desse da fare una delle statue grandi d'altezza di 23 palmi l'una, che sono collocate nelle quattro nicchie maggiori ne' pilastroni di S. Pietro che sostengono la cupola. Fu data a lui quella della Veronica la quale è una figura di tutto spirito, e maestria. La rappresentò in atto di moto, e d'un moto violento non solo di camminare, ma di correre, e qui mancò (e sia detto con sua pace) della sua propria essenza, perche se la parola denominativa di statua deriva dal moto latino *sto stas*, che significa esser fermo, stabile in piedi, quella figura non è più statua permanente, ed immobile come esser deve per formare un simulacro da esser goduto, ed ammirato da' riguardanti, ma un personaggio che passa, e non rimane." Passeri, *Künstlerbiographien*, 133–34.

37. See, for example, the rare booklet *La Veronica Vaticana del Signor Francesco Mochi: Componimenti poetici*. On p. 6, Bartolomeo Tortoletti addresses the *Veronica* as the "statua spiritosissima del Signor Francesco Mochi."

38. Compare the motifs Frank Fehrenbach has studied in "Bernini's Light."

39. See Pascoli, *Vite de' pittori, scultori ed architetti moderni*, 2:416, as well as Montagu, *Roman Baroque Sculpture*, 33–35.

40. Lessing, *Laocoön*, 19–20.

41. "Ma tornando al primo proposto della Attione dico, che ella altro non e, che l'operatione la quale deue fare a la statua da farsi; la quale lo scultore deue prima nella mente pensare et stabilire. Una delle doi cose puo fare una figura; o mostrare li proprij segni espressiui, a lo spettatore: o per se stessa contemplarli." Boselli, *Osservazioni della scoltura antica*, fols. 36v–37r.

42. "Sotto questi doi capi si termina l'attione." Ibid., fol. 37r.

43. See Partridge and Starn, *Renaissance Likeness*, esp. 43–46.

44. Gauricus, *De sculptura*, 192–93.

Bibliography

Angelini, Alessandro. *Baroque Sculpture in Rome*. Milan: 5 Continents, 2005.

Barocchi, Paola, ed. *Scritti d'arte del Cinquecento*. 3 vols. Milan: Riccardo Ricciardi, 1971.

Bergstein, Mary. "Freud's 'Moses of Michelangelo': Vasari, Photography, and Art Historical Practice. *Art Bulletin* 88 (2006): 158–76.

———. *Mirrors of Memory: Freud, Photography, and the History of Art*. Ithaca: Cornell University Press, 2010.

———. "The Mystification of Antiquity Under Pius IX: The Photography of Sculpture in Rome, 1846–1878." In *Sculpture and Photography: Envisioning the Third Dimension*, ed. Geraldine Johnson, 35–50. Cambridge: Cambridge University Press, 1998.

Bergstein, Mary, and Maureen C. O'Brien, eds. *Image and Enterprise: The Photographs of Adolphe Braun*. London: Thames & Hudson, 2000.

Bernini, Domenico. *Vita del Cavalier Gio: Lorenzo Bernino*. Perugia: Editart, 1999.

Billeter, Erika, ed. *Skulptur im Licht der Fotografie: Von Bayard bis Mapplethorpe*. Bern: Benteli, 1997.

Boselli, Orfeo. *Osservazioni della scoltura antica*. Edited by Phoebe Dent Weil. Florence: SPES, 1978.

Brion, Marcel. *Michel-Ange*. Paris: Albin Michel, 1939.

Burckhardt, Jakob. *Der Cicerone: Eine Anleitung zum Genuss der Kunstwerke Italiens*. Stuttgart: Kröner, 1986.

Coliva, Anna, and Sebastian Schütze, eds. *Bernini scultore: La nascita del barocco in Casa Borghese*. Exh. cat. Rome: Edizioni de Luca, 1998.

Conti, Alessandro, Massimo Ferretti, and Ettore Spalletti. "La documentazione dell'arte." In *Gli Alinari fotografi a Firenze: 1852–1920*, ed. Wladimiro Settimelli and Filippo Zevi, 101–74. Florence: Alinari, 1977.

Daston, Lorraine, and Peter Galison. *Objectivity*. New York: Zone Books, 2007.

de Duve, Thierry. "Time Exposure and Snapshot: The Photograph as Paradox." *October* 5 (1978): 113–25.

Dilly, Heinrich. *Kunstgeschichte als Institution: Studien zur Geschichte einer Disziplin*. Frankfurt am Main: Suhrkamp, 1979.

Doane, Mary Ann. *The Emergence of Cinematic Time: Modernity, Contingency, the Archive*. Cambridge: Harvard University Press, 2002.

Fehrenbach, Frank. "Bernini's Light." *Art History* 28 (2005): 1–42.

Freud, Sigmund. "The Moses of Michelangelo." In *The Freud Reader*, ed. Peter Gay, 522–39. New York: W. W. Norton, 1995.

Gauricus, Pomponius. *De sculptura (1504)*. Edited by André Chastel and Robert Klein. Geneva: Droz, 1969.

Grimm, Herman. *Das Leben Michelangelos*. 3rd ed. 3 vols. Hannover: Rümpler, 1868.

Gunning, Tom. "'Animated Pictures': Tales of Cinema's Forgotten Future, After 100 Years of Films." In *Reinventing Film Studies*, ed. Christine Gledhill and Linda Williams, 316–31. Oxford: Oxford University Press, 2000.

———. "Never Seen This Picture Before: Muybridge in Multiplicity." In *Time Stands Still: Muybridge and the Instantaneous Photography Movement*, ed. Phillip Prodger, 222–72. Oxford: Oxford University Press, 2003.

———. "New Thresholds of Vision: Instantaneous Photography and the Early Cinema of Lumière." In *Impossible Presence: Surface and Screen in the Photogenic Era*, ed. Terry Smith, 71–99. Chicago: University of Chicago Press, 2001.

Heikamp, Detlef. "Uccelli di bronzo." In *Giambologna: Gli dei, gli eroi*, ed. Beatrice Paolozzi Strozzi, 249–52. Florence: Giunti, 2006.

Hibbard, Howard. *Bernini*. New York: Penguin, 1965.

Janis, Eugenia Parry. "Fabled Bodies: Some Observations on the Photography of Sculpture." In *The Kiss of Apollo: Photography and Sculpture, 1845 to the Present*, exh. cat., 9–23. San Francisco: Fraenkel Gallery in association with Bedford Arts Publishers, 1991.

Jones, Ernest. *The Life and Work of Sigmund Freud*. 3 vols. New York: Basic Books, 1953–57.

Justi, Carl. *Michelangelo: Beiträge zur Erklärung der Werke und des Menschen*. Leipzig: Breitkopf & Härtel, 1900.

Knapp, Fritz, ed. *Michelangelo, des meisters Werke in 169 Abbildungen*. Stuttgart: Deutsche Verlags-Anstalt, 1912.

Kracauer, Siegfried. *Theory of Film: The Redemption of Physical Reality*. Edited by Miriam Bratu Hansen. Princeton: Princeton University Press, 1997.

Lessing, Gotthold Ephraim. *Laocoön: An Essay on the Limits of Painting and Poetry*. Translated by Edward Allen McCormick. Baltimore: Johns Hopkins University Press, 1962.

Mancini, Giulio. *Considerazioni sulla pittura*. Edited by Adriana Marucchi. 2 vols. Rome: Accademia Nazionale dei Lincei, 1956–57.

Messina, Maria Grazia, ed. *Scultura e fotografia: Questioni di luce*. Florence: Alinari, 2001.

Montagu, Jennifer. *Roman Baroque Sculpture: The Industry of Art*. New Haven: Yale University Press, 1989.

Nickel, Douglas R. *Snapshots: The Photography of Everyday Life, 1888 to the Present*. San Francisco: San Francisco Museum of Modern Art, 1998.

Partridge, Loren, and Randolph Starn. *A Renaissance Likeness: Art and Culture in Raphael's Julius II*. Berkeley: University of California Press, 1980.

Pascoli, Lione. *Vite de' pittori, scultori ed architetti moderni*. 2 vols. Rome: Soc. Multigrafica Ed. SOMU, 1965.

Passeri, Giovanni Battista. *Die Künstlerbiographien*. Edited by Jacob Hess. Worms am Rhein: Wernersche Verlagsgesellschaft, 1995.

Petersson, Robert T. *Bernini and the Excesses of Art*. Florence: M&M Maschietto & Ditore, 2002.

Preimesberger, Rudolf. "Themes from Art Theory in the Early Work of Bernini." In *Gian Lorenzo Bernini: New Aspects of His Art and Thought*, ed. Irving Lavin, 1–24. University Park: Pennsylvania State University Press, 1985.

Quintavalle, Arturo Carlo. *Gli Alinari*. Florence: Alinari, 2003.

Smith, Joel. "Roll Over: The Snapshot's Museum Afterlife." *Afterimage* 29 (2001): 8–11.

Snyder, Joel. "Nineteenth-Century Photography of Sculpture and the Rhetoric of Substitution." In *Sculpture and Photography: Envisioning the Third Dimension*, ed. Geraldine Johnson, 21–34. Cambridge: Cambridge University Press, 1998.

Spagnolo, Maddalena. "Barn Owl Painters in St. Peter's in the Vatican, 1604: Three Mocking Poems for Roncalli, Vanni, and Passignano (and a Note on the Breeches-Maker)." *Journal of the Warburg and Courtauld Institutes* 73 (2010): 257–96.

Thode, Henry. *Michelangelo: Kritische Untersuchungen über seine Werke*. 6 vols. Berlin: G. Grote, 1902–13.

La Veronica Vaticana del Signor Francesco Mochi: Componimenti poetici. Rome: Lodovico Grignani, 1641.

Wittkower, Rudolf. *Gian Lorenzo Bernini: The Sculptor of the Roman Baroque*. London: Phaidon Press, 1955.

12.

From Alberti's *Finestra Aperta* to Hitchcock's *Rear Window*

Avatars of the Scopic Drive in Painting and Film

VICTOR I. STOICHITA
Translation by Aaron Wile

The main character in *Rear Window* (1954), L. B. "Jeff" Jefferies (James Stewart), is a daredevil photographer who has broken a leg while attempting to capture an action shot at a motor race. Confined to a wheelchair for seven weeks, he spends his time spying on his neighbors, oblivious to the charms of his fiancée (Lisa Caro Freemont, played by Grace Kelly). As Jeff peers out his window, the entire microcosm of a Greenwich Village apartment building comes before his gaze.

First drawn to Miss Torso, a beautiful young dancer, then to Miss Lonelyheart, a spinster who holds romantic dinners for an imaginary guest, he turns his interest next to the arrival of new neighbors—a young married couple—and the boisterous parties organized by a young composer in his loft. Ultimately, it is the marital conflicts of his neighbors Lars and Anna Thorwald that come to occupy all of his attention.

When his eyes alone prove insufficient, he avails himself of a pair of binoculars and then a powerful telephoto lens (fig. 12.1). His optical prostheses, his close scrutiny, and finally his imagination conspire together to persuade Jeff that things have turned sour at the Thorwald apartment and that a crime might even have taken place. In search of evidence, he enlists the aid of his

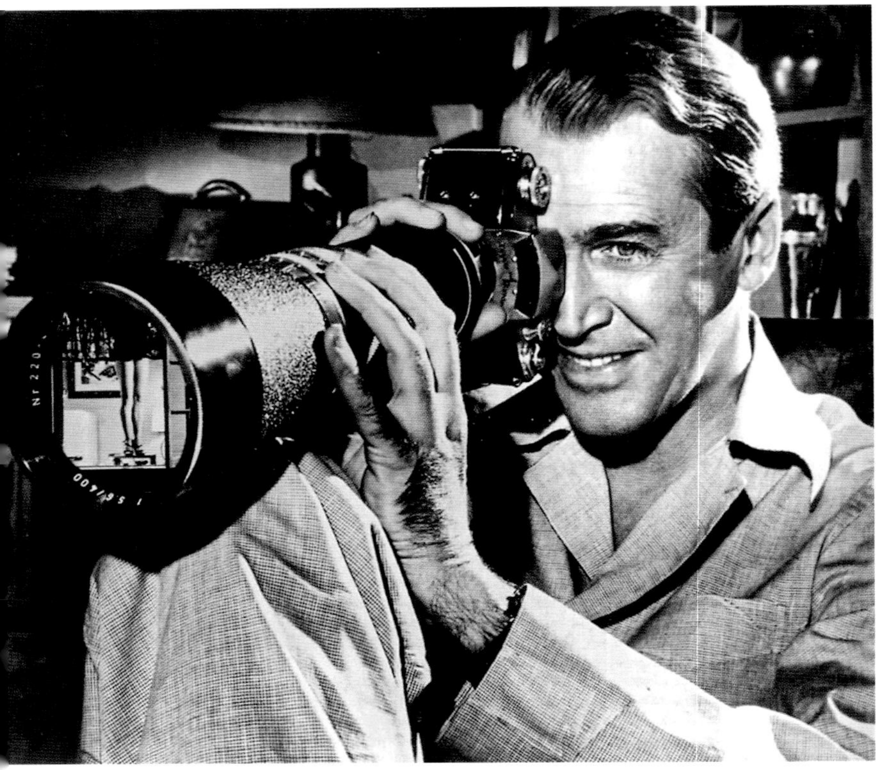

12.1 Advertising photography for *Rear Window* (1955).

fiancée, Lisa, his nurse, Stella, and, finally, his longtime friend, Tom Doyle.

Rear Window, as many have noted, is conceived in its totality as a "critique of the gaze," as a film about scopophilia and voyeurism taken to excess, about the pleasures and perils of looking. In his famous interview with François Truffaut, Hitchcock himself defined it as "a purely cinematic film" and offered the following summary: "You have an immobilized man looking out. That's one part of the film. The second part shows how he reacts. This is actually the purest expression of a cinematic idea."[1]

I would like to take this disclosure as a departure point for my reflections on Hitchcock's cult film. Well known by the public and commented on countless times by scholars—to the point where one might wonder whether a new reading is even possible—I propose nonetheless to examine the film anew by considering it not only as a "film about film" but also as a film that lays bare the persistence of an entire tradition of ocular spectacle to which cinema is heir.[2] In this chapter, I follow several tracks, the first of which is to consider the optical devices at work in *Rear Window* and the dialogue they establish with tradition. Next, I explore the relationship between voyeurism, eroticism, and crime, both in the film itself and in the larger tradition of which it is a part. Hitchcock's particular gloss on this relationship is brought face to face with the director's profound knowledge of the art of "static images," of both his time and the past. Finally, I examine the film's ironic wink at the mixed structures of word and image found in popular illustrated thrillers.

Let us begin with optical devices. Pride of place belongs, of course, to the window, featured in the title of the film itself. But in fact there is not only one window in the film (Jeff's), but several. Every hole in the wall, every transparent surface, every opening in the opaque expanse of the apartment building—all become windows through which the eye can infiltrate the private lives of the people within.

Hitchcock's originality here is commensurate with his debt to a tradition (fig. 12.2) that had long rendered the spectator a kind of voyeur, a tradition that presented, with the most convincing naturalness possible, the spectacle of a man praying, undressing, or sleeping—or of a woman receiving or reading a love letter.[3] Hitchcock's "window," in short, is the heir to the foundational apparatus of all Western art—the painting—and a metaphor for the new apparatus of the twentieth century, the cinematic screen.[4] In a celebrated passage from his famous 1435–36 treatise *On Painting*, Leon Battista Alberti defined paintings as "an open window through which I see what I want to paint" (una finestra aperta per donde io miri quello che quivi sarà dipinto).[5]

Leonardo da Vinci and Albrecht Dürer, among others, develop Alberti's idea and offer significant figurations. Let us look more closely at Dürer's engraving representing an artist at work (fig. 12.3).[6] The painter is seated, immobile, his gaze fixed. The lower half of his body is cut off on one side by the drawing table and on the other by the frame of the engraving itself: he is, so to speak, "all gaze." The eye that perceives and the hand that transforms the visual impression into a drawing are, as many commentators have noted, part of an apparatus founded on voyeurism.[7] The object on which this apparatus is centered is a woman's body and, in Dürer's engraving in particular, a passive model, discreetly covered up, apparently asleep—an object of pure contemplation. Modern painting would fundamentally challenge the passivity of the object of contemplation, but it would be chronophotography and, finally, film that would signal a true rupture.

The device employed by Jeff (fig. 12.1) is an element of transition between old and new media, and it has two functions. As a photographic telephoto lens, it is supposed to capture and "congeal" images, but the way Jeff uses it is clearly heretical: instead of capturing fixed images, the telephoto lens here captures life in motion and thus, in a sense, creates "cinema."[8] The injured photographer watching at a window is an emblematic "captor of images" and incorporates features of all image makers, old and new (figs. 12.2, 12.3). Both tradition and its culmination are marked here by drives at the limits of the

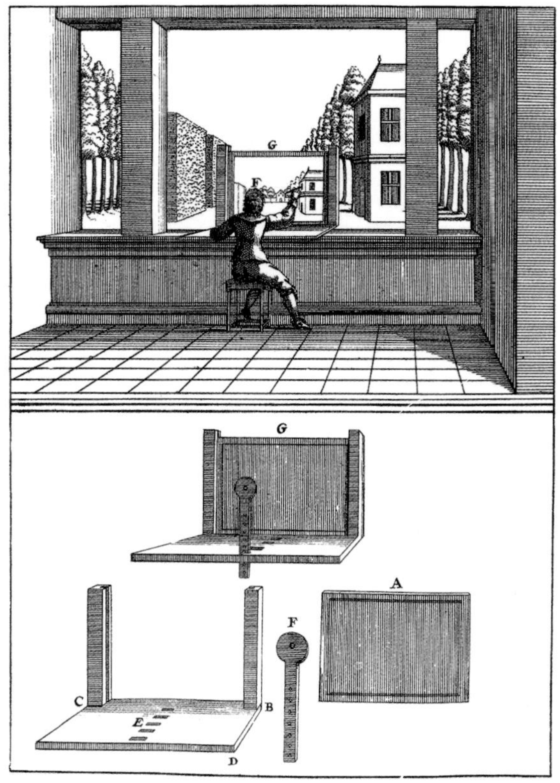

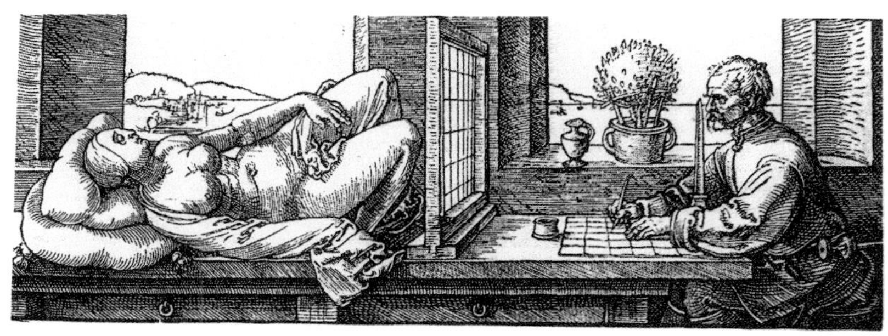

12.2 Jean Dubreuil, *An Artist Using a Drawing Frame*, 1651. Etch ng from *La perspective practique* (Paris: M. Tavernier / F. Langlois, 1651).

12.3 Albrecht Dürer, *Artist Drawing a Nude*, 1525. Woodcut for *Underweysung der Messung, mit dem Zirckel und Richtscheyt, in Linien, Ebenen unnd gantzen corporen* (Nuremberg, 1538).

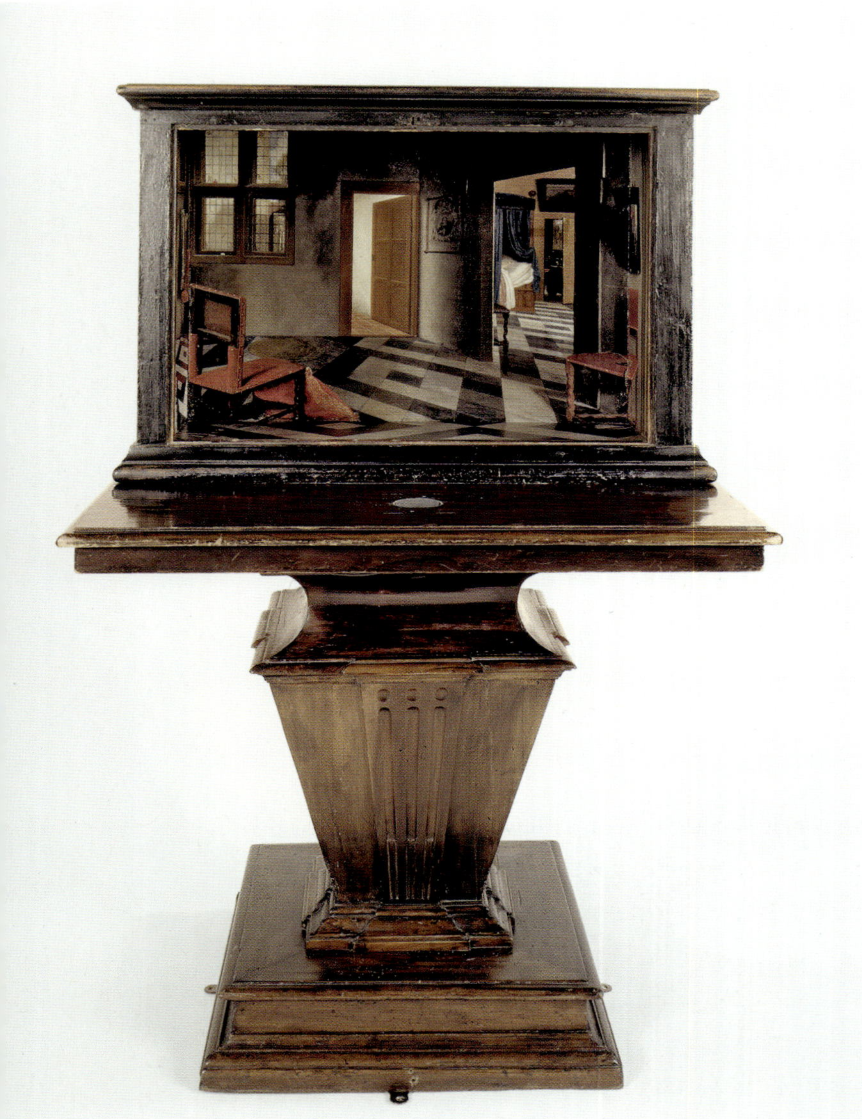

12.4 Samuel van Hoogstraten, *Perspective Box with Views of a Dutch Interior,* ca. 1655–60. Wooden box on pedestal from left front, oil on panel, 55 x 88 x 63.5 cm. The National Gallery, London.

acceptable, and it was the pioneering work of Jacques Lacan that allowed us to see the libidinal element latent in the visual enterprise itself.[9] We need to consider only briefly the points where the history of painting and the prehistory of film intersect to realize how art and the visual media have thematized and integrated the theme of the voyeur.

In this "prehistory" of film, the perspective boxes inherited from seventeenth-century Dutch art play a special role (fig. 12.4). These devices expand the dimensions of the Dutch interior painting by translating them into three dimensions.[10] But we must also note the fundamental fact that this transformation involves neither the suggestion of movement nor the creation of a "real" space. One cannot "enter" a perspective box, or "stroll around" in one, or "breathe" in one—one cannot, in short, "live" in a perspective box, because the reality it offers is of a visual order, of a virtuality that offers itself essentially to the gaze or, more precisely, to the eye. Just like the stereoscope later,[11] the perspective box is fitted with tiny openings that designate where the viewer should look. Nothing about this activity appears unwholesome or unsavory; the interiors that open up before the viewer are as immaculate as a Calvinist sermon. But if we pay attention to the outer decoration of the box, we realize that the whole apparatus is accompanied by emblematic images of the god of love.

In *Rear Window*, the cinematic spectacle initiates a tacit dialogue with the tradition of older stereoscopic instruments by alternating in a similar way between scopophilia, eroticized looking, and emblems of love. At the level of plot, *Rear Window* is a film centered, as many have pointed out, on the issue of the couple.[12] Jeff is indecisive about his relationship with his beautiful and elegant fiancée: in a sign of his irresolution, he looks for points of reference for his relationship by spying on the lives of others. The film is nonetheless imbued with a Victorian spirit (as are all of Hitchcock's films) in that the couple's private life is displayed with the utmost modesty. In *Rear Window*, what is not shown is often more important and more significant than what is—and ultimately it is this element of reticence that contributes to the buildup of suspense. The curiosity of the protagonist and of all who get caught up in his game (the spectator included) is brought to a brilliantly calculated crescendo.

The protagonist's voyeurism is treated from the beginning of the film with a mixture of irony and indulgence.[13] We understand, however, that this impotent photographer has, so to speak, an "erect" gaze (as all voyeurs do) (figs. 12.1, 12.5), that his telephoto lens and binoculars are charged with thinly veiled phallic connotations. The multiple and permanent narrative caesuras that result from Hitchcock's skillful manipulation of the fixed frames of the film's many "windows," the alternation between day and night, and the existential dynamic of this rent-controlled apartment building on 9th Street in Greenwich Village are matched by an equal number of optical caesuras. Jeff's telephoto lens is an instrument of intrusion, but it is also an instrument of power that pierces walls and sees through windows to reveal the mysteries and secrets hidden on the other side.

It is this shifting between accessibility and occlusion that structures the entire film, which would be nothing without the game of spatial syncopations, which Hitchcock creates with the same skill with which he creates suspense. The opaque brick walls that separate the windows are elements of a mechanism of obstruction that obfuscate the essential unities of the story. Others before Hitchcock have glossed what could be called the paradox of "zero representation." Among them is the painter René Magritte, whose reflections on the "glories and miseries" of the "painting apparatus" constitute one of his principal warhorses (figs. 12.6 and 12.7).[14] Like certain frames in Hitchcock's *Rear Window*, certain paintings show "nothing" but their own negation. Other paintings (fig. 12.7) play with the frame/window/painting relationship and appear to us today as the true progenitors of Hitchcock's conception of suspense, a conception based not only on temporal expansion and acceleration but also—and above all, in *Rear Window*—on the dialectic between opacity and transparency.

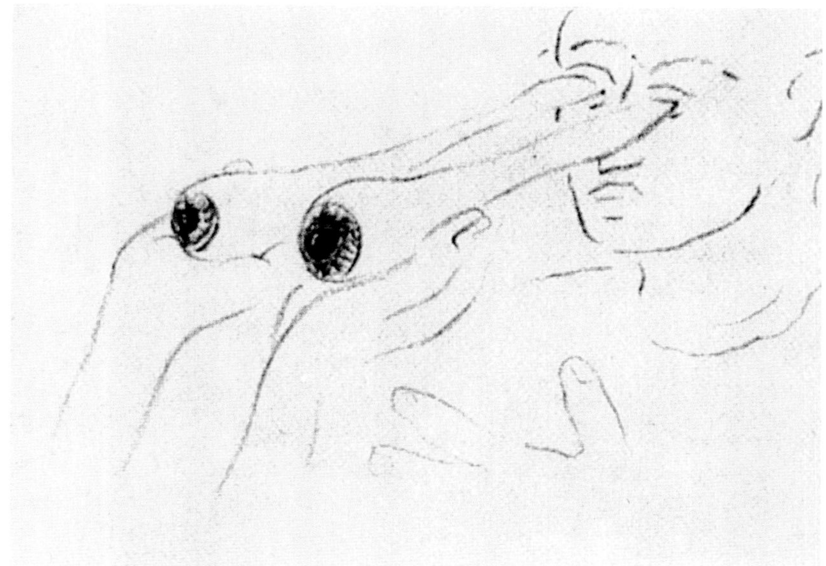

Markus Raetz, *o. T.* (*Without Title*), 1985. Pencil on paper. 24.7.1985/15. © 2013, ProLitteris, Zurich. **12.5**

Yet there is another work that is engaged in an even more profound dialogue with Hitchcock's film: Marcel

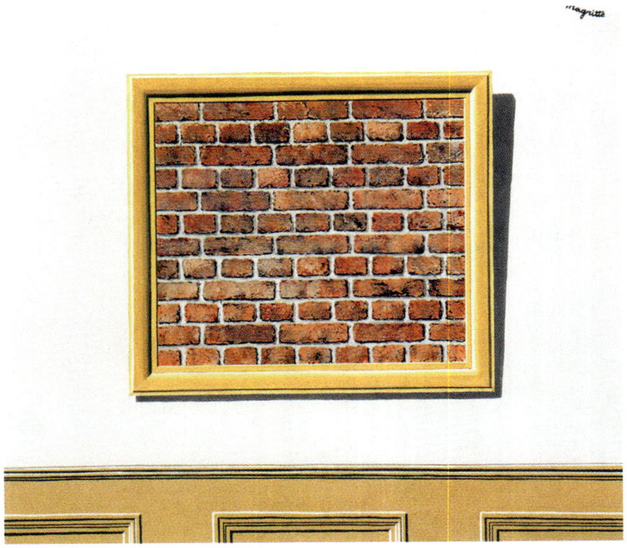

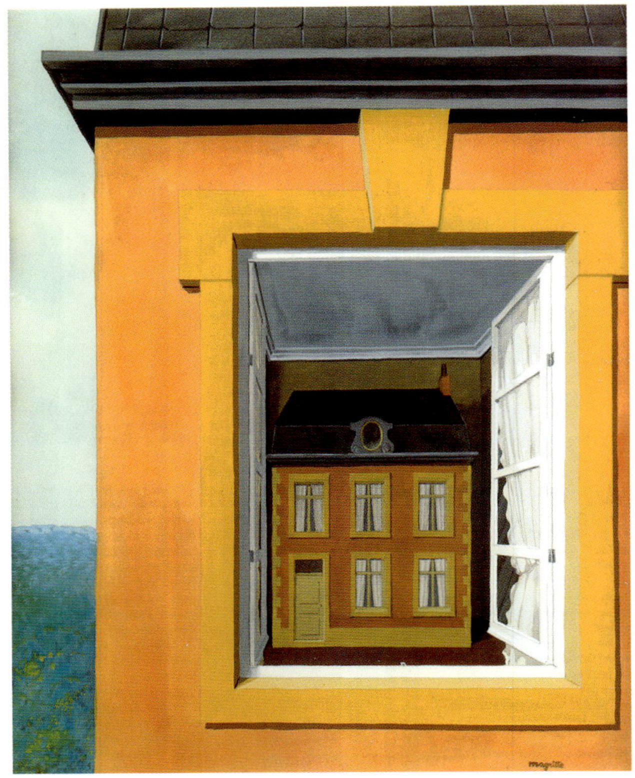

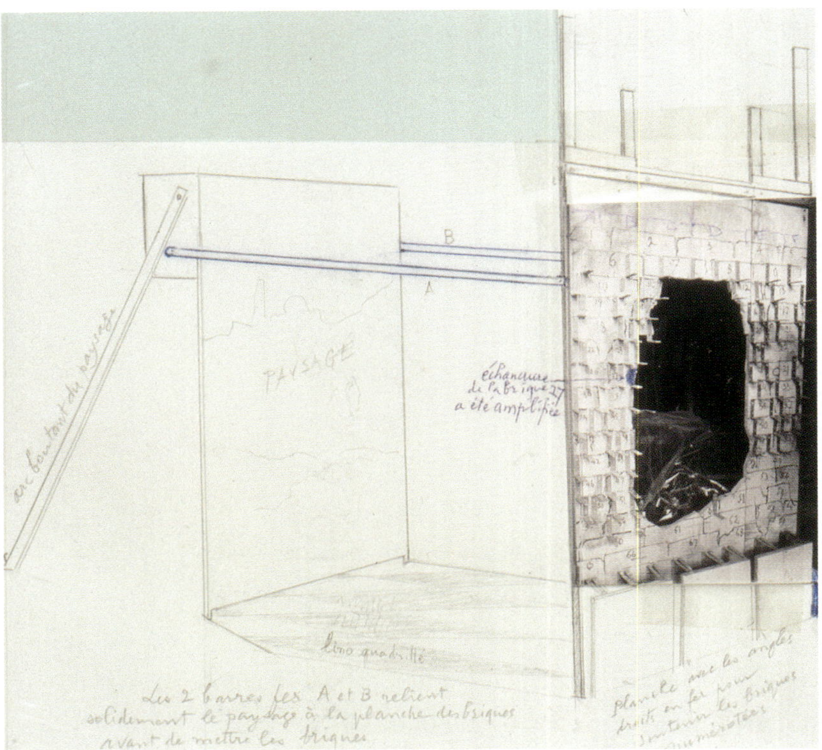

12.6 (*top left*) René Magritte, *La Saignée*, 1938–39. Gouache on paper, 37 x 43 cm. Museum Boijmans Van Beuningen, Rotterdam. © 2013, ProLitteris, Zurich.

12.7 (*top right*) René Magritte, *Eloge de la dialectique,* 1936. Gouache painting, 38 x 32 cm. Musée d'Ixelles, Brussels. © 2013, ProLitteris, Zurich.

12.8 (*left*) Marcel Duchamp, sketch for *Étant donnés,* detail, ca. 1950–55. Pencil on paper. © Succession Marcel Duchamp / 2013, ProLitteris, Zurich.

Duchamp's *Étant donnés* (fig. 12.8). Although the director could not have seen the installation, a jealously guarded secret still being constructed in Duchamp's New York studio at 210 West 14th Street when Hitchcock was filming, it is one of the seminal works of twentieth-century art. It also encompasses the entire range of allusion in the scopic tradition of the West. One of the best descriptions of the work is given by Juan Antonio Ramirez:

> Surrounded by a heavy brick frame is a large door made of wood, eaten away by time, and crude, hand-crafted nails. The tiny holes in the centre, at eye-level, are noticeable only when we move in closer to examine the door's physical characteristics carefully. And if we look through them?
>
> This spectacle, then, is as unexpected in a museum as it would be expected in some deep root of the collective unconscious. On the first level is a brick wall with an irregular opening which looks as if it had been produced by a violent explosion or battering ram. Through it one can see almost the entire naked body of a woman, legs spread wide apart and pointing towards the viewer in an ostentatious display of genitals and shaved pubic hair; her left hand (the only visible one) holds a lamp which gives off a greenish light. The whitish body lies on an autumnal bed of dry branches. In the background, against a brilliant country landscape, we see a waterfall whose imaginary waters fall with genuine perpetual motion.[15]

Ramirez concludes, "Duchamp, who spent almost his entire life reviling 'retinal' painting, here exceeded all extremes of visual illusion." Duchamp did so, one might add, by creating a voyeuristic device, the heir to the classic perspective box—to which he brings, however, an element of excessive *suspense*, thanks to an intriguing visual plot and the virtual reconstitution of a crime.[16] Had Hitchcock been able to visit Duchamp's secret studio, he would surely have been astonished.

But let us leave suppositions aside and return to the film. In the first part of *Rear Window*, which takes place during the day, Jeff's interest is focused above all on Miss Torso's seductive gyrations. The woman herself remains nonetheless inaccessible and, so to speak, fragmentary: despite the fact that she is the character whose body is exhibited most frequently and prominently, neither Jeff nor the spectator (who functions as both his innocent and his perverse double) ever sees her entirely or even partially naked. The distance here from Magritte and Duchamp is clear; the only space that would have allowed the kind of view offered by their works, the bathroom, features a window so small that any furtive observation is impossible.

Hitchcock's focus here addresses cinematically one of the central dilemmas of Western painting, itself also long confronted with the dialectic between the scopic drive and propriety. A return once again to tradition might prove instructive. Let us look at a painting from the fifteenth century, *Bathsheba at Her Bath* (fig. 12.9), which depicts the story of the wife of Uriah: while walking on the terrace of his palace, King David catches sight of Bathsheba while she is bathing and instantly falls in love (2 Samuel 11:2–5). It is one of the most impressive scenes of a nude in northern fifteenth-century art and at the same time one of the first voyeur scenarios conveyed through tradition. The painting is not the same as it was in its initial state. Today, the spectator, the voyeur outside the painting, is in a far more advantageous position than

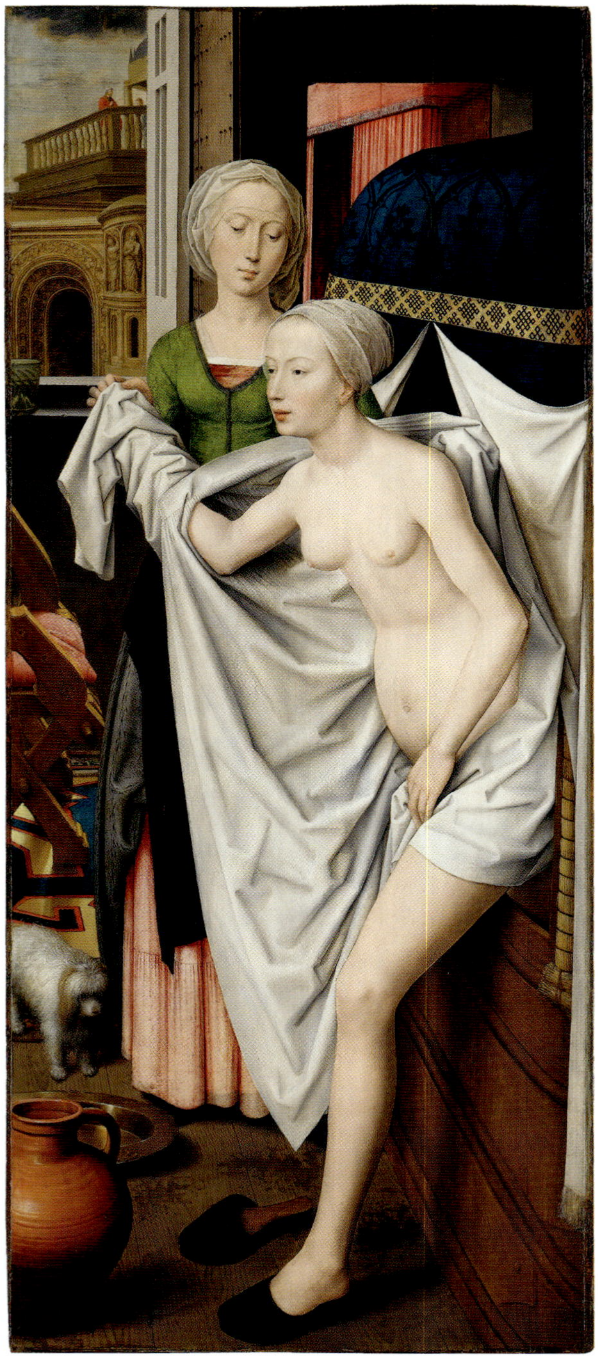

12.9 Hans Memling, *Bathsheba at Her Bath*, ca. 1480. Oil on wood, 191 x 84.5 cm. Staatsgalerie, Stuttgart.

King David, the voyeur inside the painting, who appears as a miniscule silhouette, barely discernable in the background. This situation, however, is the result of a later intervention: the "minimization" of the adulterous king is the result of "censorship"—not directly because of the nude but because of her scopic situation. In other words, at a certain moment in its history, the painting was mutilated by the elimination of the interior voyeur (King David, accompanied by his servant), to be replaced today with two miniscule silhouettes on the terrace. The small painting of King David (fig. 12.10), held for many years at the Art Institute of Chicago and returned recently to Stuttgart, was originally an integral part of the work. We will never know for sure whether the initial removal of King David was done for economic reasons (for example, so the small painting could be sold as a work in itself) or aesthetic reasons (to enhance the painting's proportions and perspective), but we can say with certainty that the next operation—the addition of a new, miniature King David—irrevocably pushed the spectator away by diminishing the painting's internal scopic tension almost to the point of annihilation.[17] In the film, by contrast, the visual plot is confined to the relationship between shot and reverse shot, and the spectator has accesses to it only through a subjective focusing, which highlights, with much aplomb, the pleasure (*jouissance*) of the viewer.

The second episode of the film to thematize Jeff's scopic quest is the arrival of the young married couple (fig. 12.11). Their arrival is significant in that it is the only episode—though it unfolds only briefly—that provokes in Jeff a complex reaction in which curiosity, frustration, and guilt intermingle. We can only appreciate Jeff's discomfort and the film's restraint with the intimacy of the couple, all the more so since the erotic imaginary (as much Western as Eastern) would have allowed for exploits of an entirely different tenor. A few examples from the past could prove enlightening here. They leave us with the impression of Hitchcock's continual dialogue with tradition, all the while allowing us to discern the means by which he breaks free of it. Let us take the example of a mysterious painting of the Renaissance, dated to around 1500, whose authorship remains an object of dispute among scholars (figs. 12.12, 12.13).[18] It offers the distinction of being painted on two sides. On one side is a portrait of a gentleman, while on the other a scene of erotic intimacy, seen through a window, is represented. The relationship between the two sides of the painting is not clear, and it will probably never be explained completely. There is no direct thematization of a voyeuristic relationship between the gentleman and the couple, and the two-sided arrangement rather undermines such an intention. Nevertheless, this double-sided portrait exposes several elements of a scopic scenario whose arrangement can only trouble the viewer. The framework of the window is the most explicit. The red curtain behind which a severe gentleman looks at the viewer is the most obscure part of this framework, and for this reason the most troubling. This curtain is not peculiar in itself (curtains of this sort were a convention of Renaissance portraits), but it becomes so at the moment when the viewer begins to question the relationship between "unveiling" and "dissimulation," and especially when a second question arises regarding the relationship between the two red curtains, the one showing a face, and the other, on the reverse of the painting, concealing an alcove. Even if they remain without answers (or perhaps, more precisely, *because they are without answers*), these questions leave us disconcerted, disturbed.

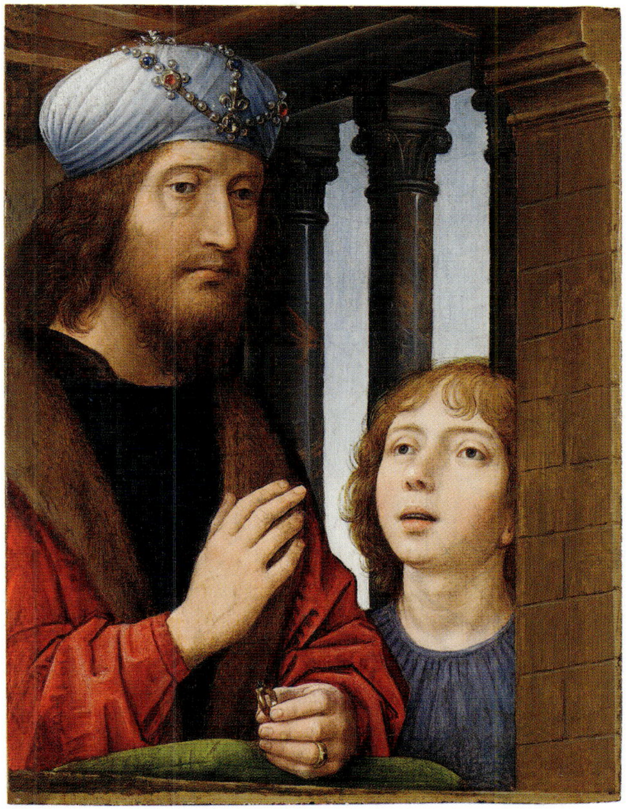

Hans Memling, *King David and a Servant*, ca. 1480. Oil on wood, 25.4 x 19.7 cm (fragment of the former painting). Staatsgalerie, Stuttgart. **12.10**

Two other examples, from the art of the Far East, bring us from the soft to the hard register (fig. 12.14). I will refrain from commenting on them[19] and observe only that everything leads us to believe that Hitchcock, whose rich visual culture is no longer in doubt, was heading, at the moment when he confronted the scopic devices of the past (windows, blinds, screens), toward the exploitation and accentuation of their taboo, forbidden aspect. In spite of the wonderful 2001 exhibition *Fatal Coincidences*, devoted to Hitchcock's artistic culture,[20] we do not know

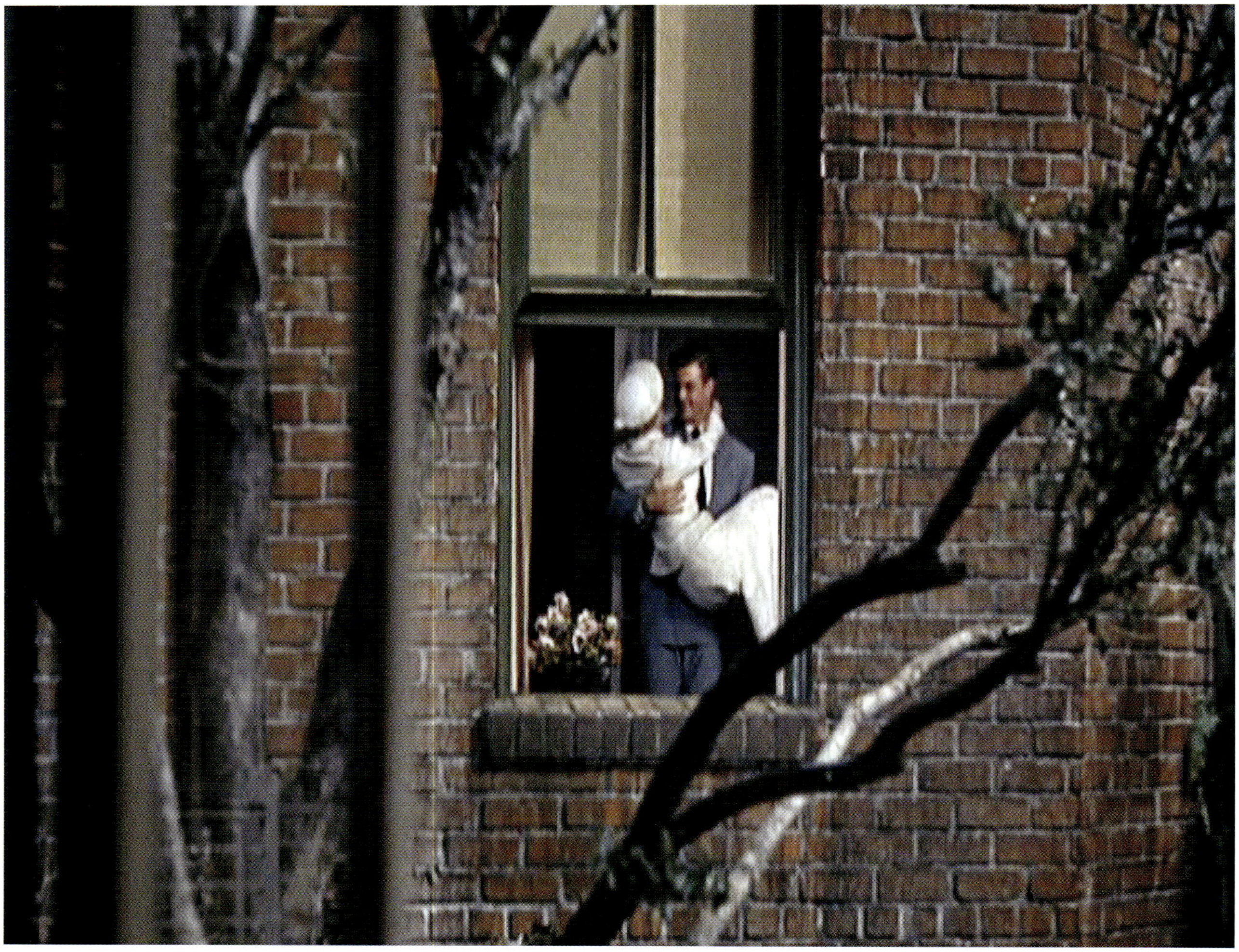

12.11 Alfred Hitchcock / Robert Burks, *The Bride and Groom*. Still photograph from Alfred Hitchcock's *Rear Window*, 1954–55.

exactly what his sources of inspiration and dialogue were, and this fact in itself is secondary. It seems to me more important to note that, for the most part in *Rear Window*, the window of the young married couple remains closed and the shades drawn. This situation, with the exception of the brief moments when Larry opens the window to get some fresh air or smoke a cigarette before being called back by his wife, stands in sharp contrast to the scenarios of traditional erotic art. The blinds that open and close play a far from negligible role in this story, and we cannot help but admire the fine web of symmetries and correspondences they weave. They are present in the most

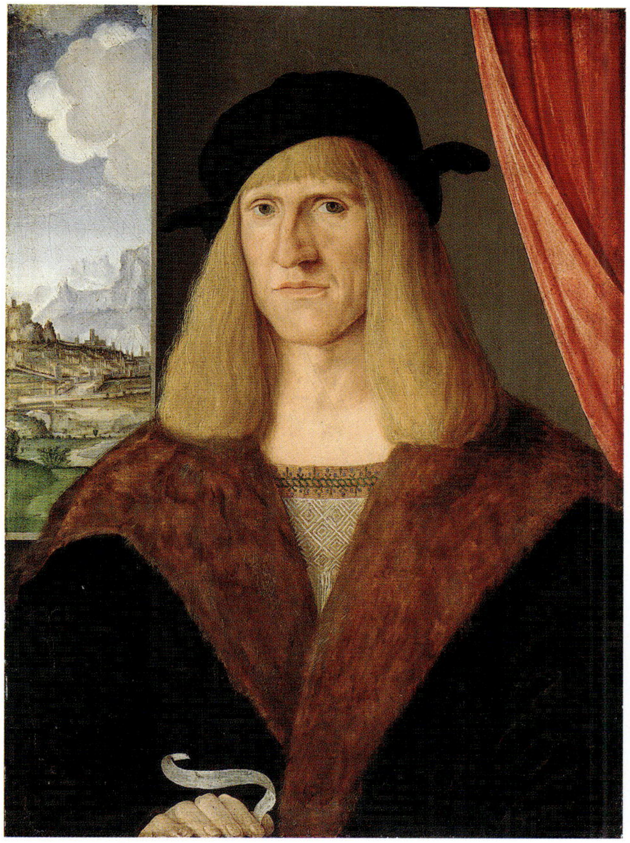

12.12 Unknown (northern Italian?), *Portrait of a Man,* ca. 1500. Oil on wood, 55 x 45 cm. Gemäldegalerie, Berlin. © bpk, Berlin / Bildagentur für Kunst, Kultur und Geschichte.

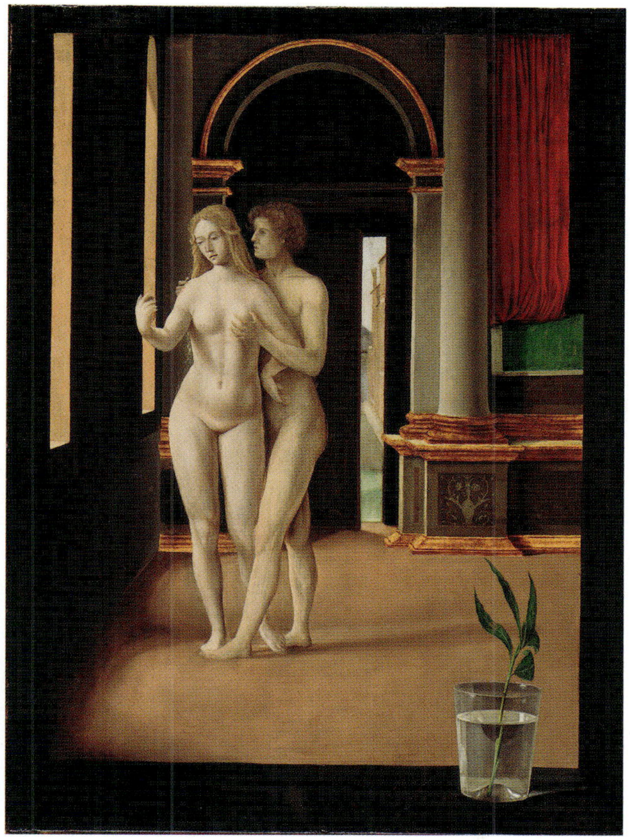

12.13 Unknown (northern Italian?), *Naked Couple,* ca. 1500. Oil on wood, 55 x 45 cm. Gemäldegalerie, Berlin. © bpk, Berlin / Bildagentur für Kunst, Kultur und Geschichte.

beautiful secondary narrative of the film, that of Miss Lonelyheart, on which I would like to pause for a moment (fig. 12.15).

Apart from offering some of the most memorable sequences in Hitchcock's entire filmography, owing in large part to the poetry of gestural language and the indisputable mastery with which the director and the cameraman orchestrate the game of optical caesuras, this story is significant for the way in which Hitchcock incorporates suggestions from the great American painting of his time. The first sequence shot, where Miss Lonelyheart imagines that she has received a guest (evidently a man), is formed by an extraordinary series of *tableaux vivants.* In the second sequence, more complex still, Miss Lonelyheart confronts reality. She gets ready, leaves her apartment, and goes into the café across the street in order to return, finally, with a man. The lowered blinds do not have the capacity to obscure the event entirely, but they are part of a

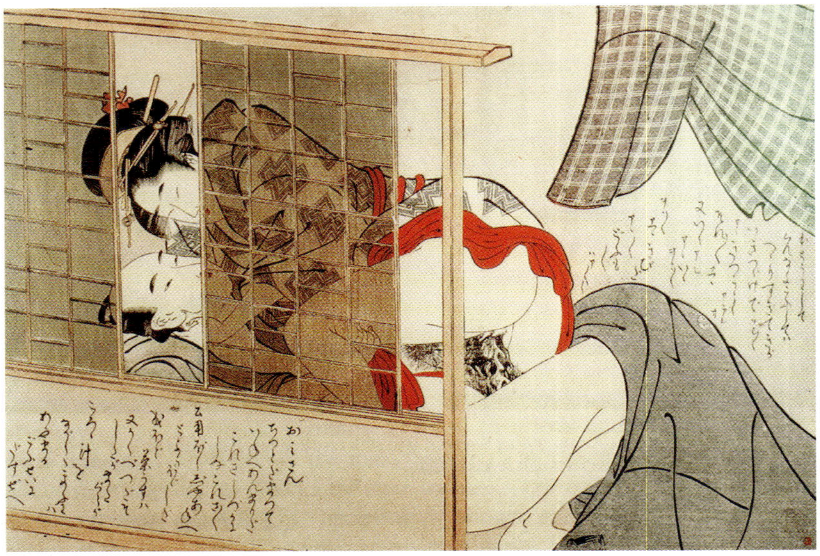

12.14 (*left*) Rekisen-tei-Eiri, *Lovers Behind a Sliding Door,* ca. 1785. Multicolored woodblock, page from an untitled *shunga* album.

12.15 (*below*) Alfred Hitchcock / Robert Burks, *Miss Lonelyheart.* Still photograph from Alfred Hitchcock's *Rear Window,* 1954–55.

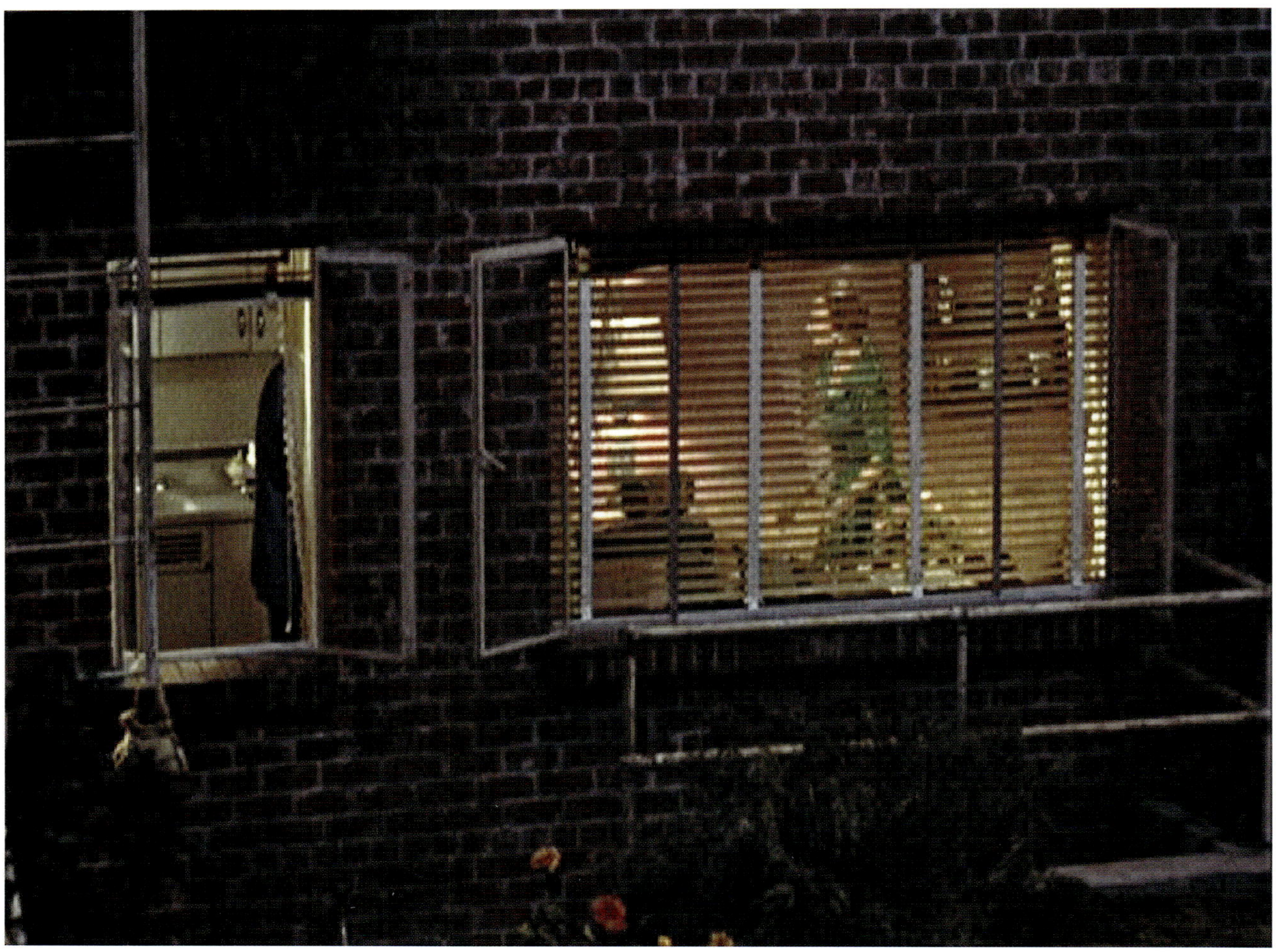

system of filters through which the affair is transmitted. This system, as has often been remarked, owes a debt to Edward Hopper (fig. 12.16).[21] One of Hopper's great innovations was to provide in his painting an insight into daily life, using views from a window, nocturnal lighting, and visual syncopations. This last element, obtained through the integration of intermediary spaces (frames, window frames, doorways, partition walls), leaves the spectator the task (and pleasure) of completing and reconstituting a story, an operation rendered all the more frightening and dangerous by the fact that the spectator never knows precisely whether he is confronting a story (*une histoire*) or rather an absence of one (*un vide d'histoire*). In *Rear Window*, Hitchcock's dialogue with painting is direct and works on several levels. The first level proves the director's awareness of the great, poignant silence that dominates the closed spaces of great cities. Another shows him attentive to the spatial gaps and accidental frames operating in the common spaces, which function in turn as signs of shared solitude. But undoubtedly the most interesting allusion comes when the director's cinematic eye exploits the way in which Hopper's technique of cutting manipulates the interplay between the *simultaneity* of the windows/screens and the *principle of succession* implied in filmic action. This is one of the reasons why "the Hopper model" is not only exploited in the Miss Lonelyheart subplot (undoubtedly the episode most indebted to Hopper) but proves fundamental in the principal subplot of *Rear Window*, that of the Thorwalds.

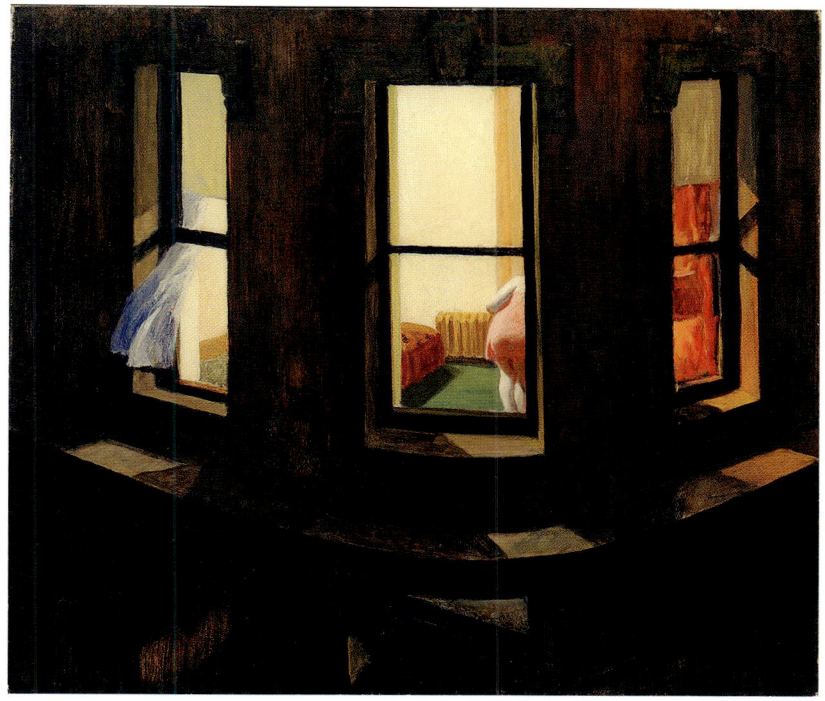

12.16 Edward Hopper, *Night Windows*, 1928. Oil on canvas, 73.7 x 86.4 cm. Museum of Modern Art, New York. © 2013. Digital image © The Museum of Modern Art, New York / Scala, Florence.

Finally—and here we are dealing with a stroke of genius on Hitchcock's part—we arrive at the sequence of maximum tension: when Lisa goes "to the other side of the courtyard" and everything seems to be headed toward a final disaster. The suspense is manifested in the extraordinary entanglement of two plots, that of Lisa's confrontation with the presumed assassin and that of Miss Lonelyheart's suicide attempt, two stories that unfold simultaneously not through three windows—which is the maximum cutting offered in Hopper's paintings—but through no less than six narrative screens at the same time (fig. 12.17).

In order to understand the stakes of this narrative and optical escalation, it is necessary to consider the entirety of the film and notably one of the anterior sequences that emphasize the ambiguous relationship between exhibitionism, voyeurism, and privacy. The scene in question

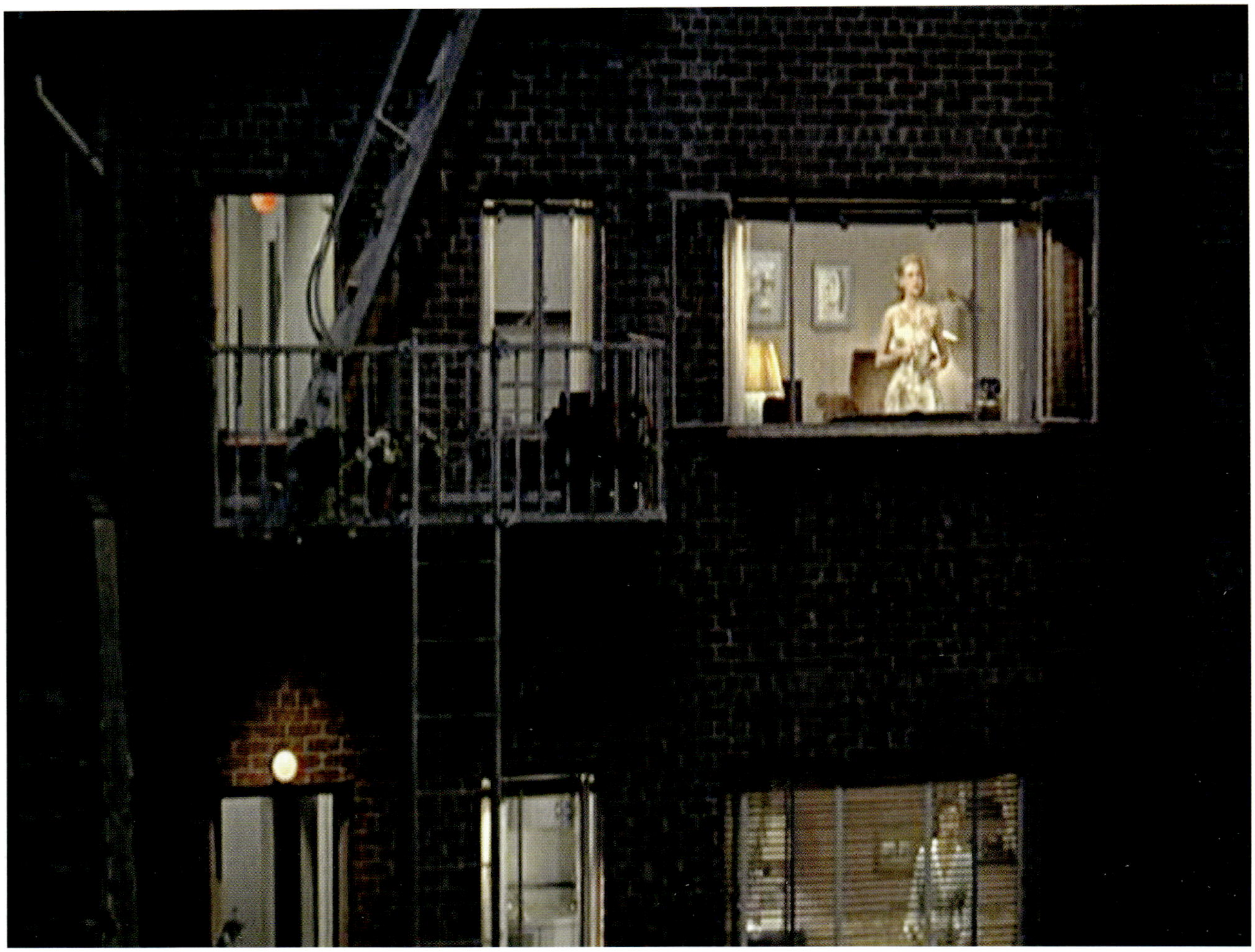

12.17 Alfred Hitchcock / Robert Burks, *The Flats of the Thorwalds and of Miss Lonelyheart.* Still photograph from Alfred Hitchcock's *Rear Window,* 1954–55.

unfolds when the beautiful Lisa ventures to transform the lookout post—Jeff's loft—into an optically protected space in order to show off her splendid silk negligee in complete privacy. But, because of the irony of fate, Hitchcock's cruelty, or, perhaps more exactly still, the fatality of the plot, Jeff does not deign to throw even one glance—all Lisa has ever wanted—toward his fiancée. Rather than look at her in the magnificent negligee, he prefers—stupidly and irreparably—to look "further," "through Lisa" and through the blinds behind her (fig. 12.18). The moment when he finally decides to look at *her* takes place when Lisa herself makes the decision to "go to the other side"—in other words, to leave the loft, traverse the courtyard, and climb across the thresholds and barriers to penetrate the world of visual phantasms held dear by her fiancé. It is then and only then that Jeff *finally* looks at her.

We have thus arrived at the core of the film, the part that concerns the relationship between Jeff's scopic passion and the process of detection. It is a relationship that the film takes up from the earlier genre of detective novels, whose lesson Hitchcock assimilates in an exemplary way. In doing so, he initiates what seems to us a very personal dialogue with one of the founders of the genre, Arthur Conan Doyle, and one of its most famous characters, Sherlock Holmes. In the case of *Rear Window*, the duplicity of the director, the "captor of images" and creator of visual plots, converges with the plurality of functions assumed by Jeff, protagonist of the film, photographer, voyeur, and detective all at the same time. With him, as with Holmes, the process of detection corroborates the interpretation of visible signs with secret logical operations. The window as a semiotic space is one of Doyle's favorite motifs, and his first illustrator, Sidney Paget, understood this well, as he demonstrated in his illustrations for the serials published in *Strand Magazine*. If Hitchcock drew his inspiration from these illustrations, as I believe he did, his references are at once deferential, ironic, and transgressive. In the first place, he pays homage to his predecessor, while at the same time establishing a substantive dialogue between the medium that (at least partially) inspired him—the illustrated serial story—and the new art to which he devoted himself—film. If the substance and attraction of the illustrated serial resided in its *mise-en-page*, the substance and attraction of film, as Hitchcock well knew, resides in "montage." He appreciated nonetheless the adroit way in which the Doyle-Paget duo used the layout of the page as a source of suspense. The integration of engravings into the text is never linear—it is syncopated.

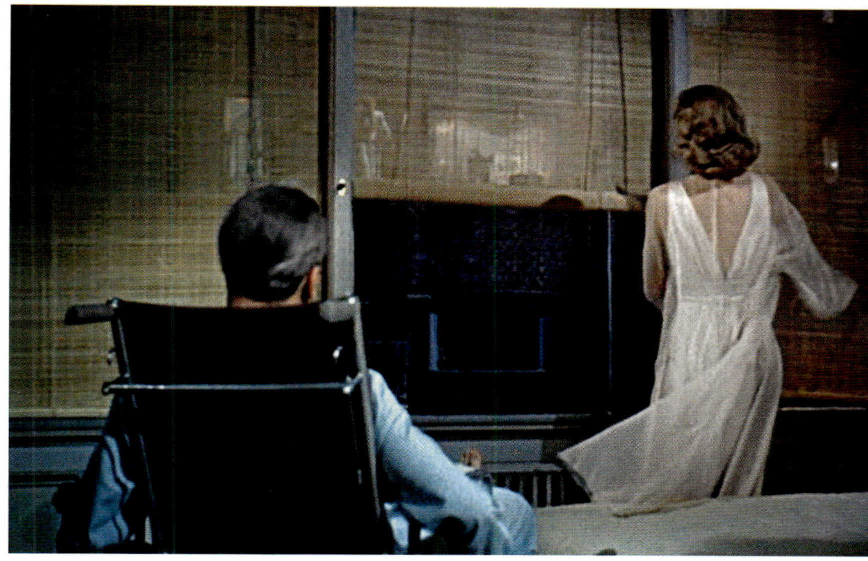

12.18 Alfred Hitchcock / Robert Burks, *Lisa Closing the Blinds.* Still photograph from Alfred Hitchcock's *Rear Window,* 1954–55.

The image is never completely integrated in the text—and never could be—but instead cuts the page, divides it, fragments it. Even more cunningly, the insertion of the illustration, always accompanied by a caption taken from the text, almost always comes in advance. In other words, the illustration does not repeat or visualize the event but precedes it, anticipating its textualization. In *The Return of Sherlock Holmes*, serialized in *Strand Magazine* in October 1903, for example, the text that recounts the window scene (fig. 12.19) is not found on the same page as the engraving, and not even on the one that faces it, but on the following page. The ruse is even more effective in that the story is about doubles, which the

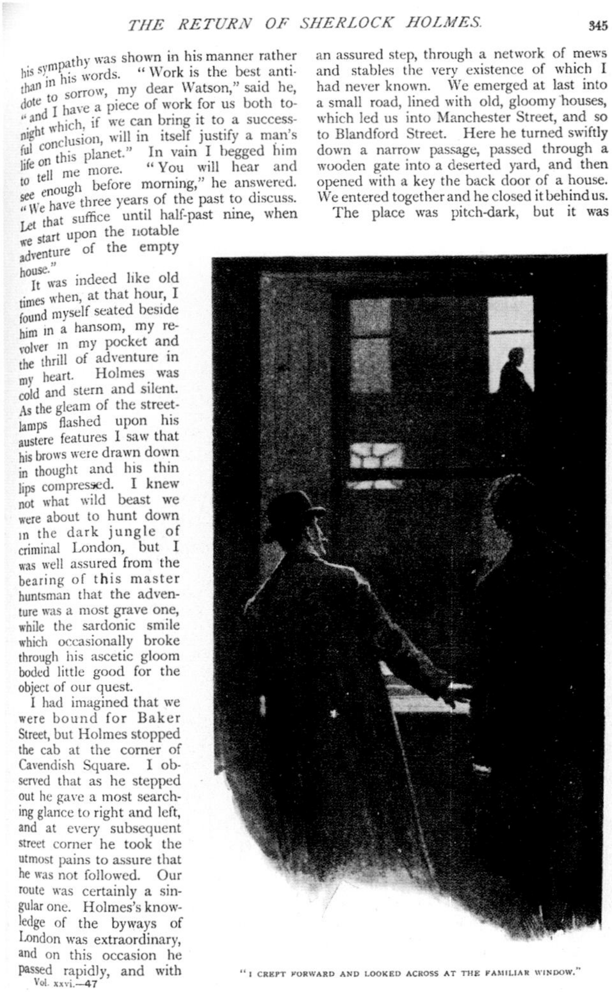

THE RETURN OF SHERLOCK HOLMES. 345

his sympathy was shown in his manner rather than in his words. "Work is the best antidote to sorrow, my dear Watson," said he, "and I have a piece of work for us both to-night which, if we can bring it to a successful conclusion, will in itself justify a man's life on this planet." In vain I begged him to tell me more. "You will hear and see enough before morning," he answered. "We have three years of the past to discuss. Let that suffice until half-past nine, when we start upon the notable adventure of the empty house."

It was indeed like old times when, at that hour, I found myself seated beside him in a hansom, my revolver in my pocket and the thrill of adventure in my heart. Holmes was cold and stern and silent. As the gleam of the street-lamps flashed upon his austere features I saw that his brows were drawn down in thought and his thin lips compressed. I knew not what wild beast we were about to hunt down in the dark jungle of criminal London, but I was well assured from the bearing of this master huntsman that the adventure was a most grave one, while the sardonic smile which occasionally broke through his ascetic gloom boded little good for the object of our quest.

I had imagined that we were bound for Baker Street, but Holmes stopped the cab at the corner of Cavendish Square. I observed that as he stepped out he gave a most searching glance to right and left, and at every subsequent street corner he took the utmost pains to assure that he was not followed. Our route was certainly a singular one. Holmes's knowledge of the byways of London was extraordinary, and on this occasion he passed rapidly, and with an assured step, through a network of mews and stables the very existence of which I had never known. We emerged at last into a small road, lined with old, gloomy houses, which led us into Manchester Street, and so to Blandford Street. Here he turned swiftly down a narrow passage, passed through a wooden gate into a deserted yard, and then opened with a key the back door of a house. We entered together and he closed it behind us.

The place was pitch-dark, but it was

"I CREPT FORWARD AND LOOKED ACROSS AT THE FAMILIAR WINDOW."

Vol. xxvi.—47

12.19 Sidney Paget, "I crept forward and looked across at the familiar window." Illustration from Arthur Conan Doyle, *The Return of Sherlock Holmes,* serialized in *Strand Magazine,* October 1903, reproduced in *The Illustrated Sherlock Holmes Treasury* (New York: Chatham River Press, 1986), 345.

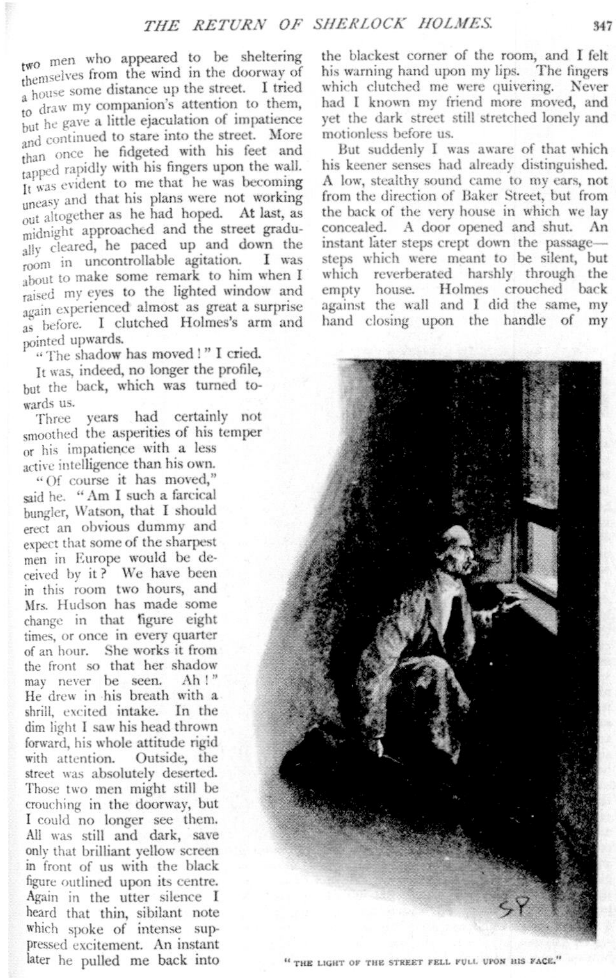

THE RETURN OF SHERLOCK HOLMES. 347

two men who appeared to be sheltering themselves from the wind in the doorway of a house some distance up the street. I tried to draw my companion's attention to them, but he gave a little ejaculation of impatience and continued to stare into the street. More than once he fidgeted with his feet and tapped rapidly with his fingers upon the wall. It was evident to me that he was becoming uneasy and that his plans were not working out altogether as he had hoped. At last, as midnight approached and the street gradually cleared, he paced up and down the room in uncontrollable agitation. I was about to make some remark to him when I raised my eyes to the lighted window and again experienced almost as great a surprise as before. I clutched Holmes's arm and pointed upwards.

"The shadow has moved!" I cried.

It was, indeed, no longer the profile, but the back, which was turned towards us.

Three years had certainly not smoothed the asperities of his temper or his impatience with a less active intelligence than his own.

"Of course it has moved," said he. "Am I such a farcical bungler, Watson, that I should erect an obvious dummy and expect that some of the sharpest men in Europe would be deceived by it? We have been in this room two hours, and Mrs. Hudson has made some change in that figure eight times, or once in every quarter of an hour. She works it from the front so that her shadow may never be seen. Ah!" He drew in his breath with a shrill, excited intake. In the dim light I saw his head thrown forward, his whole attitude rigid with attention. Outside, the street was absolutely deserted. Those two men might still be crouching in the doorway, but I could no longer see them. All was still and dark, save only that brilliant yellow screen in front of us with the black figure outlined upon its centre. Again in the utter silence I heard that thin, sibilant note which spoke of intense suppressed excitement. An instant later he pulled me back into the blackest corner of the room, and I felt his warning hand upon my lips. The fingers which clutched me were quivering. Never had I known my friend more moved, and yet the dark street still stretched lonely and motionless before us.

But suddenly I was aware of that which his keener senses had already distinguished. A low, stealthy sound came to my ears, not from the direction of Baker Street, but from the back of the very house in which we lay concealed. A door opened and shut. An instant later steps crept down the passage—steps which were meant to be silent, but which reverberated harshly through the empty house. Holmes crouched back against the wall and I did the same, my hand closing upon the handle of my

"THE LIGHT OF THE STREET FELL FULL UPON HIS FACE."

12.20 Sidney Paget, "The light of the street fell full upon his face." Illustration from Arthur Conan Doyle, *The Hound of the Baskervilles,* serialized in *Strand Magazine,* January 1902, reproduced in *The Illustrated Sherlock Holmes Treasury* (New York: Chatham River Press, 1986), 347.

text unveils, but only after the reader turns the page. The silhouette in the window, we learn, is only the simulacrum of Holmes, his double. In order to realize this fact, the reader *must turn the page*, which creates an effect of delayed action, of a kind of slow motion. This type of syncopation of image and text recurs throughout the serial novel, and we have every reason to believe that Hitchcock took his greatest pleasure (even if half-secretly) not just

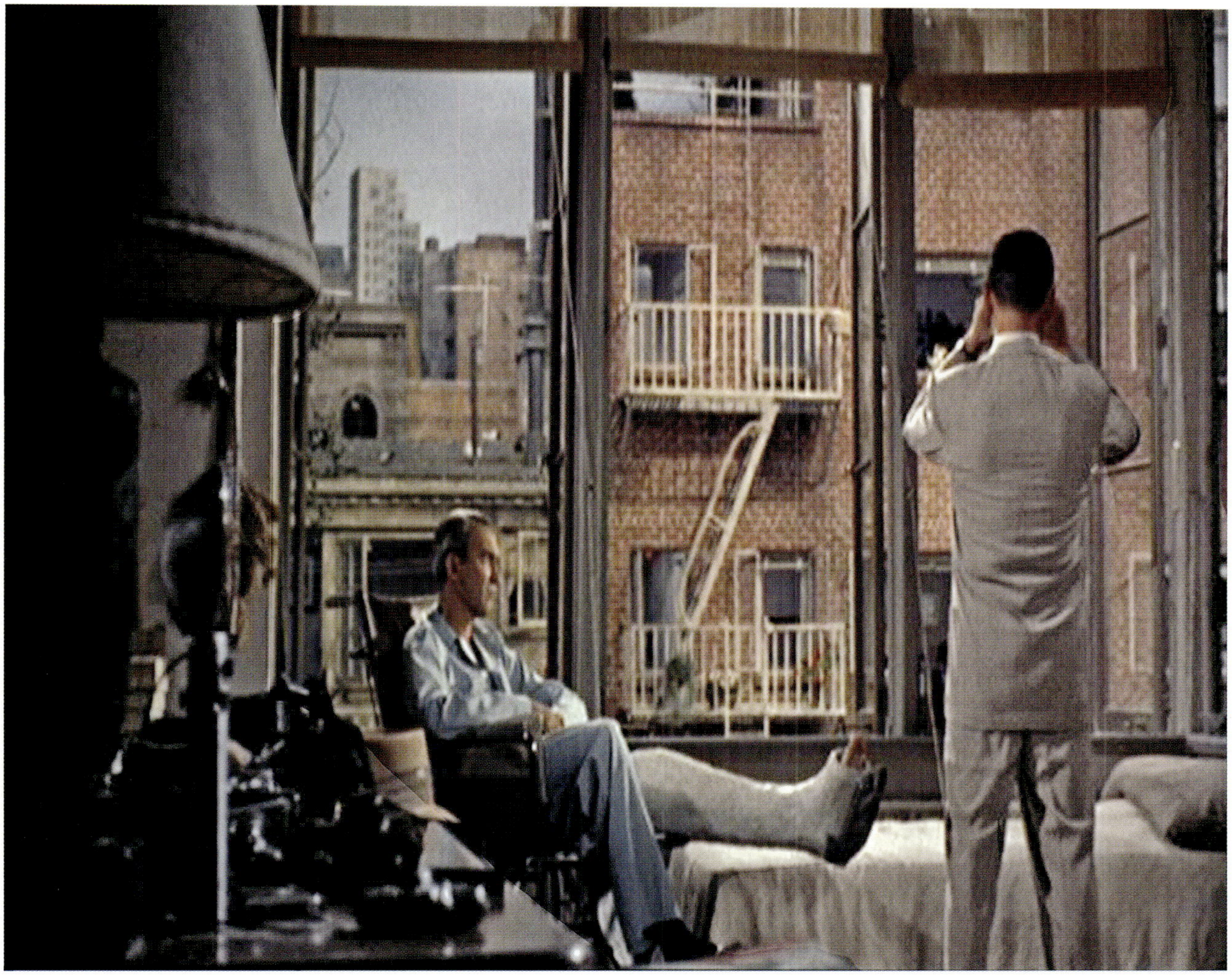

Alfred Hitchcock / Robert Burks, *Jeff and His Friend Thomas J. Doyle.* Still photograph from Alfred Hitchcock's *Rear Window,* 1954–55. **12.21**

from his many winks at the old iconography of Sherlock Holmes (figs. 12.19, 12.20) but from reworking it cinematically. He nonetheless takes care to unveil, though only "in passing" and only figuratively, his sources when he introduces into his narrative the character of "Jeff's longtime friend"—first name Tom, detective by profession, last name *Doyle* (fig. 12.21).

In spite of the fact that the photographer-voyeur succeeds in enlisting his friend, his nurse, and even his fiancée in his quest, the process of *detection* that leads to

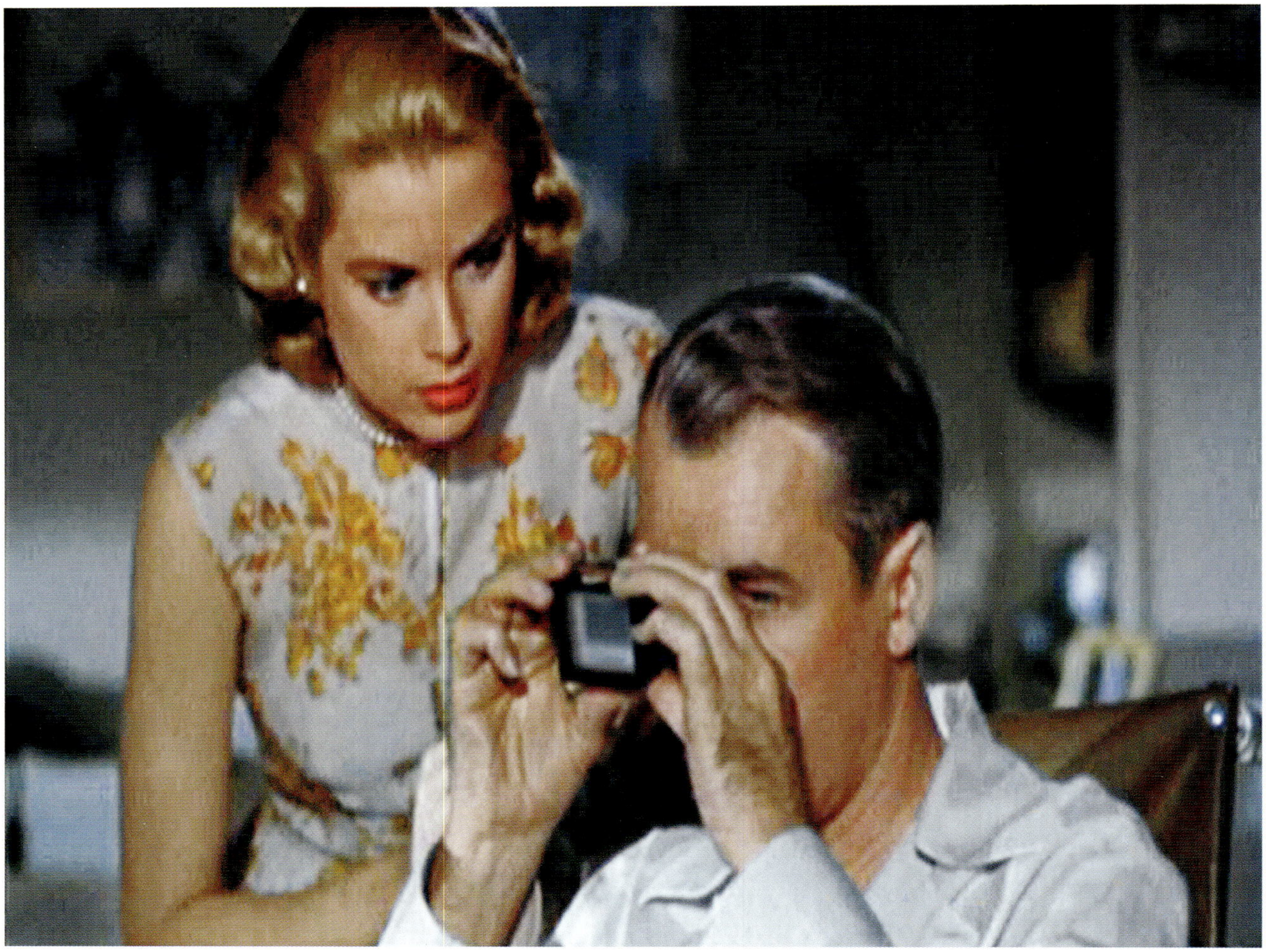

12.22 Alfred Hitchcock / Robert Burks, *Lisa, Jeff, and the Little Yellow Box.* Still photograph from Alfred Hitchcock's *Rear Window,* 1954–55.

the discovery of Lars Thorwald's crime is a solitary one, which, although carried out in steps, culminates in a single critical point. This critical point must not be confused with the moments of maximum suspense in the film, since, being purely mental and purely optical, there is no possibility of actual danger. Instead, it is the emblematic moment when Jeff finds the truth and "solves the crime" (fig. 12.22). The sequence joins the motif of concentrated observation with that of deduction, and avails itself of the great topoi of the detective story, with the difference that the relationship between observation and *detection* here emphasizes once again the optical instruments to which the film attributes almost preternatural qualities. In truth, Jeff's recourse in extremis to "the little yellow box" that gives him "the key to the truth" fits awkwardly within the strict logic of the narrative, so much so that the object appears absurd. Where do the photo slides showing the appearance of the garden "a few weeks ago" come from? How did Jeff (immobilized for more than a month) make them? And to what end? All of these questions expose the fact that this object—the box of photo slides—hardly relates to the coherence of the *story* and that its true place concerns the strict logic of the *narrative*, or, to be more exact, *the strict logic of the cinematic narrative*. The photo slides allow Jeff (that is, Hitchcock) to stage a confrontation that joins *succession* and *simultaneity*. If Jeff (or Hitchcock) had needed simple documentation of the state of the garden before and after the presumed crime, a photograph would have sufficed and would, moreover, have been more in line with Jeff's habits. But it is the transparency of film that Jeff (and Hitchcock) needs to uncover the minimum distance necessary for unveiling the crime, which is revealed in the interstice between the two images: that of the film and that of the "window." I do not believe I am mistaken in seeing in the sequence a tender homage of the film director to the master of the fixed image who most inspired him in the creation of optical enigmas: René Magritte (fig. 12.23). But if in Magritte's work we can hardly tell where the painting begins and where it ends (because ultimately *everything* is a painting), Hitchcock's central point in *Rear Window* is that we see neither the crime nor the window.

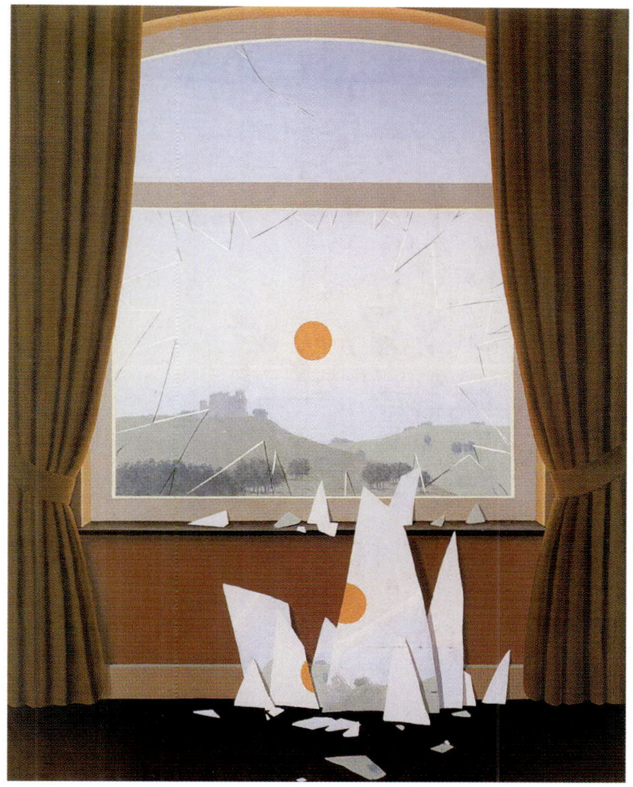

René Magritte, *Le soir qui tombe*, 1964. Oil on canvas, 162 x 130 cm. The Menil Collection, Houston, Texas. © 2013, ProLitteris, Zurich. **12.23**

What we see is a film.

Notes

1. Truffaut, *Hitchcock*, 214.
2. The best and most complete analysis remains Sharrf, *Art of Looking*. A good selection of articles is also available in Belton, *Alfred Hitchcock's Rear Window*.
3. For more on this theme, see Stoichita, *Self-Aware Image*, 44–52.
4. See Bonitzer, *Peinture et cinéma*; and Charbonnier, *Cadre et regard*.
5. Alberti, *On Painting*, 56. On the Albertian visual pyramid and its manifestations in film, see the fine observations in Bordwell, *On the History of Film Style*. On the Albertian window and cinematic screen, see Wajcman, *Fenêtre*, 27–50 (on *Rear Window*, 43–45).

6. See Kemp, *Science of Art*, 171.

7. On voyeurism in Western culture, see Stadler and Wagner, *Schaulust*; and Sprinter, *Voyeurismus in der Kunst*.

8. On the relationship between photography and film in *Rear Window*, see the excellent study by Michael Diers, *Fotografie, Film, Video*, 140–73.

9. Miller, *Séminaire de Jacques Lacan*, 65–108. For a historico-artistic perspective, see Summers, *Vision, Reflection, and Desire*.

10. See Brusati's fine *Artifice and Illusion*, 169–217.

11. For more on the stereoscope, see Crary, *Techniques of the Observer*, 133–36; and Mannoni, *Grand art de la lumière*, 197–245.

12. Interesting observations on this subject can be found in Saba, *Alfred Hitchcock*, 89–105.

13. On the general problem of voyeurism in cinematic spectacle, see Mulvey, "Visual Pleasure and Narrative Cinema"; and Koch, "*Was ich erbeute sind Bilder*." On voyeurism and Hitchcock in particular, see Bordwell, *Narration in the Fiction Film*, 40–47; and Gunning, "Desire and Pursuit of the Hole."

14. On Hitchcock's artistic culture, see Païni and Gogeval, *Hitchcock and Art*.

15. Ramirez, *Duchamp: Love and Death*, 200.

16. See Wallis, "Case Open and/or Unsolved"; and Belting, *Looking Through Duchamp's Door*, 18–75.

17. On the motivations and dating of the mutilations of Memling's painting, see Martens, "Mise au goût du jour des œuvres médiévales," esp. 48–49.

18. For more on this two-sided painting, see Dülberg, *Privatporträts*, 161–63.

19. But see Wu Hung, *Double Screen*; and Screech, *Sex and the Floating World*.

20. See the exhibition catalogue edited by Païni and Gogeval, *Hitchcock and Art*.

21. The most recent work on this subject is Matt and Stief, *Western Motel*. See also Païni, "Wandering Gaze." On the tradition, see Stoichita, "Bedingtes Sehen, gehindertes Sehen."

Bibliography

Alberti, Leon Battista. *On Painting*. Translated by John R. Spencer. London: Routledge & Kegan Paul, 1956.

Belting, Hans. *Looking Through Duchamp's Door*. Cologne: Walther König, 2009.

Belton, John, ed. *Alfred Hitchcock's Rear Window*. Cambridge: Cambridge University Press, 2000.

Bonitzer, Pascal. *Peinture et cinéma: Décadrages*. Paris: Editions de l'Etoile, 1987.

Bordwell, David. *Narration in the Fiction Film*. Madison: University of Wisconsin Press, 1985.

———. *On the History of Film Style*. Cambridge: Harvard University Press, 1999.

Brusati, Celeste. *Artifice and Illusion: The Art and Writing of Samuel van Hoogstraten*. Chicago: University of Chicago Press, 1995.

Charbonnier, Louise. *Cadre et regard: Généalogie d'un dispositif*. Paris: L'Harmattan, 2008.

Crary, Jonathan. *Techniques of the Observer: On Vision and Modernity in the Nineteenth Century*. Cambridge: MIT Press, 1991.

Diers, Michael. *Fotografie, Film, Video: Beiträge zu einer kritischen Theorie des Bildes*. Hamburg: Philo & Philo Fine Arts, 2006.

Dülberg, Angelica. *Privatporträts: Geschichte und Ikonologie einer Gattung im 15. und 16. Jahrhundert*. Berlin: Gebr. Mann, 1990.

Gunning, Tom. "The Desire and Pursuit of the Hole: Cinema's Obscure Object of Desire." In *Erotikon: Essays on Eros, Ancient and Modern*, ed. Shadi Bartsch and Thomas Bartscherer, 261–77. Chicago: University of Chicago Press, 2005.

Hung, Wu. *The Double Screen: Medium and Representation in Chinese Painting*. London: Reaktion Books, 1996.

Kemp, Martin. *The Science of Art: Optical Themes in Western Art from Brunelleschi to Seurat*. New Haven: Yale University Press, 1992.

Koch, Gertrud. "*Was ich erbeute sind Bilder*": *Zum Diskurs der Geschlechter im Film*. Basel: Stroemfeld / Roter Stern, 1988.

Mannoni, Laurent. *Le grand art de la lumière et de l'ombre: Archéologie du cinéma*. Paris: Editions Nathan, 1994.

Martens, Didier. "La mise au goût du jour des œuvres médiévales flamandes aux XVIIe et XVIIIe siècles." *Recherches Poïétiques* 3 (1995–96): 46–55.

Matt, Gerald, and Angela Stief, eds. *Western Motel: Edward Hopper und die Zeitgenossische Kunst*. Exh. cat. Vienna: Verlag für Moderne Kunst, 2008.

Miller, Jacques-Alain. *Le séminaire de Jacques Lacan, livre XI*. Paris: Seuil, 1973.

Mulvey, Laura. "Visual Pleasure and Narrative Cinema." *Screen* 16, no. 3 (1975): 6–18.

Païni, Dominique. "The Wandering Gaze: Hitchcock's Use of Transparencies." In *Hitchcock and Art: Fatal Coincidences*, exh. cat., ed. Dominique Païni and Guy Gogeval, 51–78. Paris: Centre Pompidou, 2001.

Païni, Dominique, and Guy Gogeval, eds. *Hitchcock and Art: Fatal Coincidences*. Exh. cat. Paris: Centre Pompidou, 2001.

Ramirez, Juan Antonio. *Duchamp: Love and Death, Even*. London: Reaktion Books, 1993.

Saba, Cosetta G. *Alfred Hitchcock: La Finestra sul Cortile*. Turin: Lindau s. r. l., 2001.

Screech, Timon. *Sex and the Floating World: Erotic Images in Japan, 1700–1820*. London: Reaktion Books, 1999.

Sharrf, Stefan. *The Art of Looking in Hitchcock's Rear Window*. New York: Limelight Editions, 1997.
Sprinter, Peter. *Voyeurismus in der Kunst*. Berlin: Reimer, 2008.
Stadler, Ulrich, and Karl Wagner, eds. *Schaulust: Heimliche und verpönte Blicke in der Literatur und Kunst*. Munich: Fink, 2005.
Stoichita, Victor I. "Bedingtes Sehen, gehindertes Sehen: Zur Geschichte der impressionistischen Bildauffassung." In *Modernisierung des Sehens: Sehweisen zwischen Künsten und Medien*, ed. Matthias Bruhn and Kai-Uwe Hemken, 189–206. Bielefeld: Transcript, 2008.
———. *The Self-Aware Image: An Insight into Early Modern Meta-Painting*. Cambridge: Cambridge University Press, 1997.
Summers, David. *Vision, Reflection, and Desire in Western Painting*. Chapel Hill: University of North Carolina Press, 2007.
Truffaut, François. *Hitchcock*. Rev. ed. New York: Touchstone, 1983.
Wajcman, Gérard. *Fenêtre: Chroniques du regard et de l'intime*. Lagrasse: Editions Verdier, 2004.
Wallis, Jonathan. "Case Open and/or Unsolved: *Étant donnés*, the Black Dahlia Murder, and Marcel Duchamp's Life of Crime." *Toutfait.com: The Marcel Duchamp Studies Online Journal* (2005). http://www.toutfait.com/articals.php?id=4310.

Contributors

MICHAEL COLE is Professor of Art History and archaeology at Columbia University. His recent books include *Ambitious Form: Giambologna, Ammanati, and Danti in Florence* (2011) and *Leonardo, Michelangelo, and the Art of the Figure* (2014). With Stephen Campbell, he has also published the recent textbook *Italian Renaissance Art* (2011). Currently, he is co-curating an exhibition of Italian Renaissance and baroque sculptors' drawings for the Isabella Stewart Gardner Museum, to open in the fall of 2014.

LORRAINE DASTON is Director of the Max Planck Institute for the History of Science in Berlin and Visiting Professor on the Committee on Social Thought at the University of Chicago. Her recent publications include (co-edited with Elizabeth Lunbeck) *Histories of Scientific Observation* (2011) and (with Paul Erikson et al.) *How Reason Almost Lost Its Mind: The Strange Career of Cold War Rationality* (2013), as well as essays on the history of scientific facts, objectivity, curiosity, probability, attention, and the moral authority of nature, which have appeared in various journals and collections.

FRANK FEHRENBACH is an art historian whose work focuses on the relationship between art, natural philosophy, and science in early modern Europe. He has published widely on Leonardo da Vinci; other publications include books on Roman baroque fountains and medieval Madonna statues, articles on Renaissance anatomy, monochrome sculpture, the medium of light in Bernini's work, baroque poetry, Goethe, Hans von Marées, and neorealist film. He is currently working on a monograph about the concept of "enlivenment" in early modern art. Fehrenbach was a professor at Harvard University until 2013, when he was awarded the Alexander von Humboldt Prize at Hamburg University.

SACHIKO KUSUKAWA is a fellow in history and philosophy of science at Trinity College, Cambridge. She is interested in the visual and graphic practices of naturalists and natural philosophers in early modern Europe, and is the author of *Picturing the Book of Nature: Image, Text, and Argument in Sixteenth-Century Human Anatomy and Medical Botany* (2012).

KARIN LEONHARD is Professor of Art History at the University of Bonn, with a focus on Dutch art history and theory. Her most recent publications, on the concept of the "Bildfeld" or "campo" in seventeenth-century nature pieces and still life paintings, are *Bild/Felder: Stilleben und Naturstücke des 17. Jahrhunderts* (2013), "Painted Poison: Venomous Beasts, Herbs, Gems, and Baroque Colour Theory" (2012), and "Pictura's Fertile Field: Otto Marseus van Schrieck and the Genre of Sottobosco Painting" (2009–10).

CARLA MAZZIO, Associate Professor of English at the State University of New York at Buffalo, is the author of *The Inarticulate Renaissance: Language Trouble in an Age of Eloquence* (awarded the 2010 Roland H. Bainton Book Prize for Literature) and co-author of *Book Use, Book Theory:*

1500–1700 (2005). She is also the editor of *Shakespeare and Science* (2009) and the co-editor (with Douglas Trevor) of *Historicism, Psychoanalysis, and Early Modern Culture* (2000) and (with David Hillman) *The Body in Parts: Fantasies of Corporeality in Early Modern Europe* (awarded the 1999 Beatrice White Prize).

ALINA PAYNE is Alexander P. Misheff Professor of History of Art and Architecture at Harvard University. Most recently, she published *From Ornament to Object: Genealogies of Architectural Modernism* (2012) and *The Telescope and the Compass: Teofilo Gallaccini and the Dialogue Between Architecture and Science in the Age of Galileo* (2012) and edited *Dalmatia and the Mediterranean: Portable Archaeology and the Poetics of Influence* (2013). In 2006 she received the Max Planck and Alexander von Humboldt Prize in the Humanities.

VICTOR I. STOICHITA is Professor of Art History at the University of Fribourg in Switzerland. Among his work are *Figures de la transgression* (2013), *The Pygmalion Effect: From Ovid to Hitchcock* (2008), *Goya: The Last Carnival* (with Anna Maria Coderch; 1999), *A Short History of the Shadow* (1997), *The Self-Aware Image* (1997), and *Visionary Experience in the Golden Age of Spanish Art* (1995).

NICOLA SUTHOR teaches art history at the Freie University of Berlin and recently accepted a professorship at Yale. Her field of research is early modern art and art theory, with a special focus on the articulation of pictorial intelligence in theory and practice. Her publications include *Augenlust bei Tizian: Zur Konzeption sensueller Malerei in der frühen Neuzeit* (2004), *Bravura: Virtuosität und Mutwilligkeit in der Malerei der Frühen Neuzeit* (2010), and *Rembrandts Rauheit: Eine phänomenologische Untersuchung* (2014).

CLAUDIA SWAN is Associate Professor of Art History at Northwestern University in Evanston, Illinois. Her publications include *Art, Science, and Witchcraft in Early Modern Holland: Jacques de Gheyn II (1565–1629)*, *Rariteyten van dese Landen: Encounters with the Exotic in Early Modern Holland* (forthcoming); and the edited volumes (with Londa Schiebinger) *Colonial Botany: Science, Commerce, and Politics in the Early Modern World* (2004) and (with Paul Bakker, Christoph Lüthy, and Claus Zittel) *Image, Imagination, and Cognition: Medieval and Early Modern Theory and Practice*.

CATHERINE WILSON is Anniversary Professor of Philosophy at the University of York (UK) with a concurrent appointment at the CUNY Graduate Center in New York. She is the author of books and articles on aesthetic experience, visuality, and early modern science and philosophy, including a study of early microscopy, *The Invisible World* (1995), and, most recently, a book on the recovery of ancient atomism, *Epicureanism at the Origins of Modernity* (2008).

GERHARD WOLF is Director of the Kunsthistorisches Institut / Max-Planck-Institut, Florence, and honorary professor at Humboldt University in Berlin. His research topics and projects include Mediterranean art histories; theories of the image and object in religion, art, and art history; sacred and profane topographies in an intercultural perspective (including museums); art, space, and mobility in the early ages of globalization (400–1650) (with Hannah Baader and Avinoam Shalem); artistic exchange between Mexico and Europe in the sixteenth and seventeenth centuries; and interrelations between artistic and scientific worldviews.

Index